Handbook of RNA Biochemistry

Handbook of RNA Biochemistry

Edited by Boston Carter

SYRAWOOD
PUBLISHING HOUSE
New York

Published by Syrawood Publishing House,
750 Third Avenue, 9th Floor,
New York, NY 10017, USA
www.syrawoodpublishinghouse.com

Handbook of RNA Biochemistry
Edited by Boston Carter

International Standard Book Number: 978-1-64740-363-8 (Hardback)

Cataloging-in-publication Data

Handbook of RNA biochemistry / edited by Boston Carter.
p. cm.
Includes bibliographical references and index.
ISBN 978-1-64740-363-8
1. RNA. 2. Biochemistry. 3. Biotechnology. 4. Bioengineering. I. Carter, Boston.
QP623 .H36 2023
572.88--dc23

TABLE OF CONTENTS

Preface VII

Chapter 1 ***Arabidopsis* Non-Coding RNA Regulation in Abiotic Stress Responses** 1
Akihiro Matsui, Anh Hai Nguyen, Kentaro Nakaminami and Motoaki Seki

Chapter 2 **How Can Plant DNA Viruses Evade siRNA-Directed DNA Methylation and Silencing?** 14
Mikhail M. Pooggin

Chapter 3 **Up-Regulation of microRNA* Strands by their Target Transcripts** 41
Sung-Min Kang, Ji-Woong Choi, Su-Hyung Hong and Heon-Jin Lee

Chapter 4 **Non-Coding RNAs: The "Dark Matter" of Cardiovascular Pathophysiology** 51
Claudio Iaconetti, Clarice Gareri, Alberto Polimeni and Ciro Indolfi

Chapter 5 **Involvement of miR-20a in Promoting Gastric Cancer Progression by Targeting Early Growth Response 2 (EGR2)** 82
Xiangsheng Li, Zhichao Zhang, Ming Yu, Liqi Li, Guangsheng Du, Weidong Xiao and Hua Yang

Chapter 6 **MicroRNA Expression Changes during Interferon-Beta Treatment in the Peripheral Blood of Multiple Sclerosis Patients** 96
Michael Hecker, Madhan Thamilarasan, Dirk Koczan, Ina Schröder, Kristin Flechtner, Sherry Freiesleben, Georg Füllen, Hans-Jürgen Thiesen and Uwe Klaus Zettl

Chapter 7 **Long Non-Coding RNAs and Complex Human Diseases** 120
Jing Li, Zhenyu Xuan and Changning Liu

Chapter 8 **Non-Coding RNA in Oncogenesis: A New Era of Identifying Key Players** 139
Guorui Deng and Guangchao Sui

Chapter 9 **Dynamic Nature of Non-Coding RNA Regulation of Adaptive Immune Response** 170
Graziella Curtale and Franca Citarella

Chapter 10 **Advances in the Techniques for the Prediction of microRNA Targets** 200
Hao Zheng, Rongguo Fu, Jin-Tao Wang, Qinyou Liu, Haibin Chen and Shi-Wen Jiang

Chapter 11 **Intercellular Communication by Exosome-Derived microRNAs in Cancer** **209**
Bethany N. Hannafon and Wei-Qun Ding

Permissions

List of Contributors

Index

PREFACE

Every book is a source of knowledge and this one is no exception. The idea that led to the conceptualization of this book was the fact that the world is advancing rapidly; which makes it crucial to document the progress in every field. I am aware that a lot of data is already available, yet, there is a lot more to learn. Hence, I accepted the responsibility of editing this book and contributing my knowledge to the community.

Non-coding RNAs (ncRNAs) are RNA molecules which are not translated into proteins. They are classified into two types based on their length, including short non-coding RNAs comprising microRNAs (miRNAs) and long non-coding RNAs (lncRNAs). They play a role in gene regulation at various levels, ranging from epigenetic gene silencing to post-transcriptional modulation of mRNA stability. The ncRNAs contribute in the regulation of cell physiology and shape cellular functions. They play an important role in brain development and function, and are also associated with the pathogenesis of psychiatric disorders. Several ncRNAs including circRNAs, miRNAs and lncRNAs have been implicated in regulatory functions. Furthermore, they are believed to play a role in Alzheimer's disease and cancer. This book explores all the important aspects of the biochemistry of RNA. It presents researches and studies performed by experts across the globe. The extensive content of this book provides the readers with a thorough understanding of this topic.

While editing this book, I had multiple visions for it. Then I finally narrowed down to make every chapter a sole standing text explaining a particular topic, so that they can be used independently. However, the umbrella subject sinews them into a common theme. This makes the book a unique platform of knowledge.

I would like to give the major credit of this book to the experts from every corner of the world, who took the time to share their expertise with us. Also, I owe the completion of this book to the never-ending support of my family, who supported me throughout the project.

Editor

1

Arabidopsis Non-Coding RNA Regulation in Abiotic Stress Responses

Akihiro Matsui [1], Anh Hai Nguyen [1], Kentaro Nakaminami [1] and Motoaki Seki [1,2,3,*]

[1] Plant Genomic Network Research Team, RIKEN Center for Sustainable Resource Science, 1-7-22 Suehiro-cho, Tsurumi-ku, Yokohama, Kanagawa 230-0045, Japan; E-Mails: akihirom@psc.riken.jp (A.M.); nguyenhaianh@psc.riken.jp (A.H.N.); nm_ken@psc.riken.jp (K.N.)

[2] Kihara Institute for Biological Research, Yokohama City University, 641-12 Maioka-cho, Totsuka-ku, Yokohama, Kanagawa 244-0813, Japan

[3] CREST, JST, 4-1-8 Honcho, Kawaguchi, Saitama 332-0012, Japan

* Author to whom correspondence should be addressed; E-Mail: motoaki.seki@riken.jp

---

Abstract: Plant growth and productivity are largely affected by environmental stresses. Therefore, plants have evolved unique adaptation mechanisms to abiotic stresses through fine-tuned adjustment of gene expression and metabolism. Recent advanced technologies, such as genome-wide transcriptome analysis, have revealed that a vast amount of non-coding RNAs (ncRNAs) apart from the well-known housekeeping ncRNAs such as rRNAs, tRNAs, small nuclear RNAs (snRNAs) and small nucleolar RNAs (snoRNAs) are expressed under abiotic stress conditions. These various types of ncRNAs are involved in chromatin regulation, modulation of RNA stability and translational repression during abiotic stress response. In this review, we summarize recent progress that has been made on ncRNA research in plant abiotic stress response.

Keywords: non-coding RNA; small RNA; antisense RNA; abiotic stress response

---

1. Introduction

Environmental stresses, such as drought, heat, salinity and low temperature, are major limiting factors for plant growth and productivity. Under natural conditions, plants are exposed to a variety of environmental stresses. In order to adapt and survive under the stresses, plants have evolved various molecular mechanisms for a fine-tuned control of adaptive responses [1]. Post-transcriptional regulatory mechanisms, as well as epigenetic and post-translational modifications, like ubiquitination and sumoylation, have been implicated to play an important role in the regulation of gene expression during stress conditions.

Recent genome-wide transcriptome analysis, such as tiling arrays and next generation sequencing, has revealed a large number of stress-responsive ncRNAs. Emerging evidence has revealed that ncRNAs are major products of the plant transcriptome with significant regulatory importance [2,3]. ncRNAs are transcribed from intergenic regions, antisense strands of protein-coding genes and also pseudogenes. According to their size, ncRNAs are classified as small ncRNAs (sRNAs) (<40 nt) and long ncRNAs (lncRNAs) (>200 nt). These ncRNAs are involved in the transcriptional and posttranscriptional regulation of gene expression and the modulation of RNA stability and translation under stress conditions [1,4–7].

2. Small RNAs (sRNAs)

sRNAs are known to have major functional roles in eukaryotic gene regulation. In plants, knowledge regarding the biogenesis and mechanisms of action of sRNA classes including microRNAs (miRNAs), transcriptional gene silencing (TGS)-related heterochromatic small interfering RNAs (siRNAs), trans-acting siRNAs (ta-siRNAs) and natural antisense transcript siRNAs (nat-siRNA) has been primarily gained through *Arabidopsis* studies (Figure 1; Table 1). These sRNAs are loaded into RNA-induced silencing complexes (RISC) and negatively regulate the expression of their target genes by affecting mRNA levels, chromatin remodeling and DNA methylation.

Figure 1. Regulation of ncRNAs in abiotic stress responses. ncRNAs (lncRNAs, miRNAs and siRNAs, *etc.*) are generated in response to abiotic stress, such as drought, low-temperature, heat and high-salinity. The ncRNAs are involved in various types of regulation, such as chromatin regulation, transcriptional regulation, RNA degradation, protection of miRNA targets, translational repression.

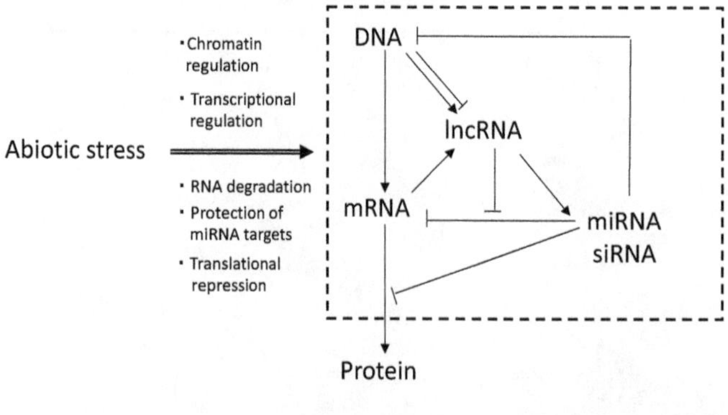

Table 1. Classification of non-coding RNAs (ncRNAs) [(1)] involved in abiotic stress responses.

Small ncRNAs (sRNAs)
• micro RNAs (miRNAs)
• small interfering RNAs (siRNAs)
TGS-related heterochromatic siRNAs
trans-acting siRNAs (ta-siRNAs)
natural antisense transcript siRNAs (nat-siRNA)
others
Long ncRNAs (lncRNAs)
• long antisense non-coding RNAs (lancRNAs)
lancRNAs generated during biogenesis of ta-siRNAs and S-PTGS-related siRNAs
lancRNAs involved in epigenetic silencing
lancRNAs co-expressed with sense-stranded protein-coding mRNAs
others
• long intergenic non-coding RNAs (lincRNAs)
RNAs generated from small open reading frame regions
RNA decoys mimicking miRNA targets
RNAs transcribed by Pol III
RNAs transcribed by Pol IV during biogenesis of TGS-related heterochromaticsiRNAs
others

(1) ncRNAs are classified except for well-known housekeeping ncRNAs, such as rRNAs, tRNAs, snRNAs and snoRNAs.

2.1. Micro RNAs (miRNAs)

Many stress-responsive miRNAs have been identified in various plants [6,8]. miRNAs are generated from pri-miRNAs that are transcribed by RNA polymerase II and the mature miRNAs guide to cleavage target mRNAs [9]. Therefore, the expression levels of pri-miRNAs are regulated by *cis*-elements in a similar manner as protein-coding genes. An abundance of positive cold stress-related *cis*-regulatory elements, such as the Dehydration Responsive Element/Low Temperature Responsive Element (DRE/LTRE) -core (A/GCCGAC) [10], Abscisic Acid-Responsive Element (ABRE) -core (ACGTGG/TC) [11] and W-box (TTGAC) [12] are found in the promoter region of several cold-inducible *MIRNA* genes in *Arabidopsis* [13]. Several miRNAs that are upregulated in response to various abiotic stresses, including cold [14], dehydration [14,15], salinity [14] and nutrient deficiency [16] have been reported.

An *Arabidopsis* miRNA involved in the detoxification of reactive oxygen species (ROS) has been reported [17]. Expression of miR398 was found to be decreased by copper (Cu(II)), which is an essential nutrient in photosynthesis and response to oxidative stress and abiotic stresses [18]. miR398 was also shown to suppress superoxide dismutase (SOD) and Cu/Zinc (Zn) superoxide dismutases (CSDs) under low Cu. Interestingly, it was reported as a rare instance in plants that miR398 not only digests *CSD1* and *CSD2* mRNAs but also negatively regulated the translation of their protein products [19]. This miRNA-dependent translational repression is effected in part by the ARGONAUTE proteins AGO1 and AGO10 [19]. It also requires the activity of the microtubule-severing enzyme katanin and the de-capping component VARICOSE (VCS)/Ge-1, as recently suggested from animal studies. *katanin1* and *vcs1* mutants that did not affect *CSD2* mRNA accumulation, exhibited an overaccumulation of CSD2 proteins under low Cu (II) conditions in comparison to WT plants. However, under high Cu (II) conditions, this trend was not observed [19]. This result suggested that the translational inhibition of miR398 had an important role in the regulation of CSD expression under low Cu (II) condition.

Several miRNAs function in the maintenance of phytohormone signaling during exposure to abiotic stress. An miR168-mediated feedback regulatory loop regulates AGO1 homeostasis in ABA and abiotic stress responses in *Arabidopsis* [20]. *MIR168a*-overexpressing plants and *ago1* loss-of-function mutants showed ABA hypersensitivity and drought tolerance, while the *mir168a* mutants showed ABA hyposensitivity and drought hypersensitivity [20]. The promoter of *MIR168a* has four ABREs, suggesting that expression of *MIR168a* was directly induced by ABA. Although promoter activity of *AGO1* was induced by ABA, *AGO1* transcripts were negatively regulated by ABA-induced miR168a; and as a result, mRNA was maintained at a steady level during the stress response. These results suggest that a complex crosstalk exists between the global regulation of miRNA metabolism and ABA signaling functions to enable fine-tuning of abiotic stress response.

In another example of miRNA-phytohormone crosstalk in abiotic stress response, miR160 regulates the expression of *Auxin Response Factors* (*ARF10*, *ARF16* and *ARF17*) [21]. *Arabidopsis* plants expressing miR160-resistant *ARF10* not only showed an abnormal leaf shape but also showed hypersensitivity to ABA. Plants overexpressing miR160 showed hyposensitivity to ABA. In addition, other miRNAs targeting auxin signaling factors were also induced in response to abiotic stress [14]. Expression of miR393 was up-regulated by dehydration, salt and cold stresses and ABA [8]. A miR393 target gene, *TIR1*, an auxin receptor, is involved in the response to salt and oxidative stresses [22]. The promoter of miR167 contains ABREs, indicating their own regulation by ABA [14]. *ARF6* and *ARF8*, which are targeted by miR167, are regulators of female and male reproduction [23]. *TAS3*-siRNA also regulates auxin signaling [24]. It is thought that miRNAs targeting auxin signaling function as mediators that connect abiotic signaling with development [7,25]. These results suggest that multistep regulation by miRNAs is required for the correct adjustment of gene expression under abiotic stress.

Nutrient deficiency under abiotic stress is known to induce or suppress various miRNAs that regulate nutrient metabolism. Previous studies have demonstrated that the expression of miR395 was increased by sulfate starvation [26]. This specific miRNA suppresses ATP sulfurylases as target mRNAs, thus resulting in catalysis of the first step of inorganic sulfate assimilation [8]. A phosphate starvation-inducible miRNA (miR399) regulates Pi homeostasis by regulating the expression of *UBC24* mRNA encoding an ubiquitin-conjugating E2 enzyme [16,27]. miR399 functions as a positive regulator of Pi uptake and translocation. In addition to miR399 regulating *UBC24* expression, the cleavage activity of miR399 is suppressed by a long intergenic ncRNA, Induced by Phosphate Starvation 1 (IPS1) [28]. A detailed description is provided at a later point in this chapter.

2.2. Small Interfering RNAs (siRNAs)

Recent studies have reported that small interfering RNAs (siRNAs) function in stress responses. These siRNAs are generated from long double strand RNAs through various biological processes [29].

A specific type of siRNA that is involved in RNA-directed DNA methylation suppresses the activation of retrotransposons under heat stress [30]. The transcriptional gene silencing (TGS)-related heterochromatic siRNAs are generated from RNA Polymerase IV (PolIV)-derived transcripts in repetitive DNA sequences and heterochromatin. After heat stress, a *Copia*-type retrotransposon in *Arabidopsis*, named *ONSEN*, becomes transcriptionally active and has been shown to result in the synthesis of extrachromosomal DNA copies in the siRNA-mediated silencing deficient mutant *nrpdl* (*nrpdla*), which is the largest subcomponent of PolIV. Heat-induced expression and transgenerational retrotransposition of *ONSEN* were suppressed by siRNA-mediated silencing. It was also reported that abiotic stresses changed the genome-wide DNA methylation status across multiple generations [31].

Trans-acting siRNAs (ta-siRNA) can be classified as a specialized case of siRNAs in plants. These siRNAs are generated from dsRNAs, which are generated from miR173, miR390 and miR828 - cleaved lncRNAs [32,33]. The long antisense non-coding RNAs (lancRNAs) of these dsRNAs were synthesized by RNA-dependent RNA polymerase 6 (RDR6), those dsRNAs were positive candidates producing siRNAs [34]. The tight feedback regulation between miR390, *TAS3* ta-siRNA and ARF4, which is a target of *TAS3* ta-siRNA, was required for lateral root initiation [24]. Although ta-siRNA expression has not been found to change in response to abiotic stress in *Arabidopsis*, the expression of a *rice RDR6 homolog*, which produces dsRNAs from cleaved *TAS* RNAs, was induced by ABA in rice [35]. These results imply that *TAS3* ta-siRNA is involved in dynamic changes of root architecture during exposure to abiotic stress [36].

Borsani *et al.* 2005 reported that natural antisense transcript small interfering RNA (nat-siRNAs) were generated from dsRNAs produced from natural *cis*-antisense gene pairs of *Δ1-Pyrroline-5-Carboxylate Dehydrogenase* (*P5CDH*) and a high-salinity-stress inducible gene of unknown function (*SRO5*) during high-salinity stress [37]. Recent genome-wide analysis reported an accumulation of sRNAs in their overlapping region, suggesting the occurrence of an RNA interference event [38]. However, the biological process of generating nat-siRNAs is not completely understood at this time [39,40].

3. Long Non-Coding RNAs (lncRNAs)

Genome-wide tiling arrays and high-throughput sequencing have identified a vast amount of lncRNAs in plants [4,41–45]. These lncRNAs may represent alternatively spliced forms of known genes [46], products of antisense RNAs [4,38,47,48], double stranded RNAs [49], retained introns [46,50], short open reading frame [34,51,52], RNA polymerase III-derived RNAs [53] and RNA decoys mimicking miRNA targets [28]. In this manuscript, we classified the lncRNAs into long antisense non-coding RNAs (lancRNAs) and long intergenic non-coding RNAs (lincRNAs) based on its genomic locations (Figure 1; Table 1). These RNAs have various modifications that depend on each biosynthetic process.

3.1. Long Antisense Non-Coding RNAs (lancRNAs)

Over the past decade, genome-wide transcriptome analyses confirmed that approximately 30% of all annotated genes exhibited significant lancRNA expression in *Arabidopsis* [4,41]. These data regarding lancRNA expression are consistent with results from other organisms such as fly, human, and rice [54]. Expression profiles of lancRNAs in response to environmental stresses have been extensively characterized by an *Arabidopsis* tiling array analysis [4]. A certain type of lancRNAs belongs to a pair of fully overlapping sense–antisense transcripts (fSATs) in which the lancRNAs exist within the protein-coding gene regions in opposing orientations. The expression of these sense and antisense RNA transcripts are stress-responsive. On the other hand, partially overlapping sense–antisense transcripts (pSATs) do not exhibit synchronous expression patterns. These observations suggest that lancRNA were generated through multiple biosynthetic processes.

A type of lancRNA is co-expressed with sense protein-coding RNAs [4]. A large *Arabidopsis* tiling array analysis confirmed that more than 6000 lancRNAs were classified into the fSATs category. Interestingly, a significant linear correlation between the expression ratios (abiotic stress treated/untreated) of the sense transcripts and the ratios of the lancRNAs was observed in the fSATs. The *RD29A* and *CYP707A1* lancRNAs that were simultaneously accumulated with sense mRNAs, were accumulated by drought- and ABA treatments. Some of the lancRNAs that were identified contained complementary sequences to those of the sense mRNAs [4], indicating that lancRNA expression is dependent upon sense mRNAs.

Co-expression of sense RNA and lancRNA of a transgene was reported as a trigger for sense post-trancriptional gene silencing (S-PTGS) [55]. In the S-PTGS process, RNA-dependent RNA polymerase 6 (RDR6), one of six *Arabidopsis* RDRs, generates antisense RNAs from non-canonical sense RNAs of transgenes with aberrant features, such as non-cap structure of 5' end or poly(A) tail, to generate double stranded RNA (dsRNA) [56]. dsRNA-seq analysis showed that dsRNAs of more than 100 loci were reduced in *rdr6* in relative comparison to WT [57]. RNA-seq analysis of sRNAs also revealed that mutations of *ABH1* and *EIN5* (*XRN4*), which are involved in mRNA processing and mRNA degradation, respectively, affect the level of sRNAs mapped on the antisense strand of endogenous protein-coding genes [58]. The *xrn4* mutant was also screened as an enhancer of transgene silencing [59]. These results supported the observation that a certain type of lancRNA was generated from non-canonical sense RNA. Recent studies revealed that *Werner Exonuclease* (*WEX*), *Silencing Defective3* (*SDE3*), *DCL2*, *DCL4*, *Nuclear RNA Polymerase IVa* (*NRPD1a*), *RDR2* and *CLASSY1* were involved in S-PTGS and downstream PTGS [60–65]. Since S-PTGS-related siRNAs are generated from dsRNAs, these results also indicate that a certain type of lancRNA is synthesized from mRNA templates via a complex amplification pathway.

It is possible that the expression of lancRNA could serve as a functional link to the chromatin regulation of epigenetic silencing [48]. *Cold Induced Long Antisense Intragenic RNA* (*COOLAIR*) in the *FLC* locus is a well-characterized example of this in *Arabidopsis* [47,66]. Exposure of plants to low temperature treatment for a 2-week period resulted in a high level of *COOLAIR* expression. Several weeks after the induction of *COOLAIR*, the transcription of *FLC* was significantly decreased. During this period, tri-methylated histone H3 Lys27 (H3K27me3) levels progressively increased at a region around the transcription start site [67]. The level of H3K27me3 spreading the gene body was required to maintain the repression of *FLC* transcription after plants were returned to warm conditions [68]. *COOLAIR* has been suggested to be required for a plant homeodomain-polycomb repressive complex 2 (PHD-PRC2) located at a tightly localized nucleation region within *FLC*. Consequently, this results in an increase in H3K27me3 levels at the *FLC* locus [64]. Although further analysis is necessary to elucidate the role of *COOLAIR* and its epigenetic silencing of *FLC* during the short period of vernalization [69,70], a reporter gene that was fused with *COOLAIR* was shown and confirmed to be capable of causing cold-induced silencing [47].

The second class of lncRNA, *COLDAIR*, is transcribed from a region within the first intron of *FLC* on the sense strand [71]. The *COLDAIR* transcript has been shown to interact with PRC2 and its abundance also increased during vernalization. Reduction of *COLDAIR* transcript levels by RNAi confirmed that it is not required for the initial repression of *FLC* but is necessary for subsequent maintenance of repression. These results showed that the interaction of lncRNAs and the chromatin modification complex mediates cold-inducible epigenetic regulation. Based upon bioinformatic analyses comparing lancRNAs and chromatin status, it was recently hypothesized that a type of lancRNAs from the genome was repressed by cytosine methylation and H3K36me [48]. This prediction proposed that an interaction between chromatin regulation and a certain type of lancRNAs functions in genome regulation.

A recent strand-specific RNA-seq study showed that approximately 1300 ncRNA loci exist in the antisense strand of protein-coding genes in *Arabidopsis* [48]. The global ratio between sense and antisense tags in exons in this study was 0.01, which was similar to a previous strand-specific RNA-seq analysis using floral tissue [43,48]. A comparison between RNA-seq and genome-wide tiling array data showed that one-half to two-thirds of the sense–antisense transcripts were only represented in one experiment [48]. Since the expression levels of lancRNAs are low compared to sense transcripts, it is possible that this difference between the two techniques may reflect technical limitations inherent to these methods. Future transcriptome analyses are required to increase our understanding of non-canonical transcripts and to clarify the types of lancRNAs that function in molecular signals, RNA decoys, guides, and scaffolds [72].

3.2. Long Intergenic Non-Coding RNAs (lincRNAs)

Several transcriptome analyses have reported that more than 1000 lincRNA loci exist between protein-coding genes in *Arabidopsis* [4,41–45]. A part of these lincRNAs was transcribed from the methylated DNA regions by RNA polymerase IV and are thought to be positive candidates that generate into TGS-related heterochromatic siRNAs that guide DNA methylation [64].

Another type of lincRNA is transcribed by RNA Polymerase III. *In silico* genome sequence analysis predicted 20 novel ncRNA candidates [53]. A specific Pol III-derived ncRNA (*AtR18*) responded negatively to hypoxic stress and this regulation was evidently different from that of *U6* snRNA. Specifically, *AtR18* was not processed into a smaller fragment and no small open reading frames (sORFs) were included. Short interspersed elements (SINEs) and 7SL (signal recognition particle) RNA for protein trafficking are known as Pol III-derived RNA, with exception of the canonical functional RNA [73,74].

Many sORFs that have not been annotated as protein-coding genes have been identified as expressed genes during developmental or environmental conditions [51,52]. A specific sORF (*npc536*) has a large dynamic variation of expression across a wide range of tissue and hormonal, biotic, or abiotic treatment [34]. *npc536* exists in the antisense strand of a Golgi-transport complex related protein. *35S::npc536* transformants displayed heightened root growth under salt stress conditions. However, the biological function of this RNA is not still clearly understood at this time.

LincRNA is capable of regulating the cleavage activity of miRNA as a target mRNA decoy. *IPS1* is a lincRNA, which was found to be induced by phosphate starvation [28,75]. Although *Arabidopsis IPS1*, *AT4-1*, *AT4-2* and family members in other species share little sequence conservation with the predicted ORFs, high conservation was observed with a 22-nt sequence located in the 3' half of the transcript [75]. This observation suggests that the *IPS1* family contains the sequence that is targeted by miRNA and it does not function as a short peptide. The 22-nt sequence of *IPS1* is not perfectly complementary to miR399 due to a mismatch occurring at the 10–11th position [28]. Although miR399 was capable of hybridizing to *IPS1* transcript, there has been no indication of cleavage via miR399. On a whole plant level, phosphate starvation resulted in the induction of *IPS1*. As a result of the phosphate starvation, the miR399 target (*UBC24* mRNA) accumulated. miR399 has been characterized as a phosphate starvation inducible-miRNA and a positive regulator of Pi-uptake. Interestingly, two studies implied that miR399 may function as a long-distance signal from shoots to suppress *UBC24* expression in roots [76,77]. Taken together, it is reasonable to consider that the complex interactions between *IPS1*, miR399 and *UBC24* function to maintain spatial homeostasis of Pi. Computational analyses predicted the occurrence of these possible target mimics in rice and *Arabidopsis*, suggesting that RNA decoys (miRNA target mimic) are conserved and functional contributors to the regulation of RNA [78].

4. Conclusions and Perspectives

Whole genome transcriptome analyses of high-density microarrays and high throughput sequencing have expanded our understanding of a novel research area pertaining to networks, which govern abiotic stress responses. Despite the extensive body of published information regarding various types of ncRNAs, their association to the adaptation and response of plants to abiotic stress is not completely understood at this time. Recent advances in whole transcriptome analyses have enabled us to gain a greater understanding regarding the mechanisms of transcriptional and posttranscriptional regulation through ncRNAs, as well as sRNAs and lancRNAs. Functional analyses of miRNAs have deciphered and unraveled complex feedback loop regulations between miRNA and target genes. In addition, the characterized interaction between miR399, *IPS1* and *UBC24* has been shown to be a novel type of RNA interaction system in which a lincRNA reduced a cleavage activity of phosphate starvation-inducible miRNA and regulated a target mRNA. In addition, ncRNAs are also linked to chromatin regulation. The studies regarding epigenetics in the *FLC* locus enabled a hypothesis to be made which suggests that an interaction of lancRNA and chromatin modification complex mediates cold-inducible epigenetic regulation. Overall, these data have revealed a complex interaction between transcriptional regulators which function to fine-tune responses to various environmental stimuli. For the future, it will be essential to investigate the biological regulations of ncRNAs in order to enable scientists to completely elucidate the entire picture of gene regulation networks in stress response.

Acknowledgments

This work was supported by Grants-in-Aid for Scientific Research from the Ministry of Education, Culture, Sports, Science and Technology, Japan (Innovative Areas 25119724, Challenging Exploratory Research 24657041), Japan Science and Technology Agency (JST), Core Research for Evolutionary

Science and Technology (CREST) and grants from the RIKEN Research Institute, Japan, to Motoaki Seki.

References

1. Hirayama, T.; Shinozaki, K. Research on plant abiotic stress responses in the postgenome era: Past, present and future. *Plant J.* **2010**, *61*, 1041–1052.
2. Laporte, P.; Merchan, F.; Amor, B.B.; Wirth, S.; Crespi, M. Riboregulators in plant development. *Biochem. Soc. Trans.* **2007**, *35*, 1638–1642.
3. Rymarquis, L.A.; Kastenmayer, J.P.; Huttenhofer, A.G.; Green, P.J. Diamonds in the rough: mRNA-like non-coding RNAs. *Trends Plant Sci.* **2008**, *13*, 329–334.
4. Matsui, A.; Ishida, J.; Morosawa, T.; Mochizuki, Y.; Kaminuma, E.; Endo, T.A.; Okamoto, M.; Nambara, E.; Nakajima, M.; Kawashima, M.; *et al. Arabidopsis* transcriptome analysis under drought, cold, high-salinity and ABA treatment conditions using a tiling array. *Plant Cell Physiol.* **2008**, *49*, 1135–1149.
5. Mazzucotelli, E.; Mastrangelo, A.M.; Crosatti, C.; Guerra, D.; Stanca, A.M.; Cattivelli, L. Abiotic stress response in plants: When post-transcriptional and post-translational regulations control transcription. *Plant Sci.* **2008**, *174*, 420–431.
6. Khraiwesh, B.; Zhu, J.K.; Zhu, J. Role of miRNAs and siRNAs in biotic and abiotic stress responses of plants. *Biochim. Biophys. Acta* **2012**, *1819*, 137–148.
7. De Lima, J.C.; Loss-Morais, G.; Margis, R. microRNAs play critical roles during plant development and in response to abiotic stresses. *Genet. Mol. Biol.* **2012**, *35*, 1069–1077.
8. Sunkar, R.; Zhu, J.K. Novel and stress-regulated MicroRNAs and other small RNA from *Arabidopsis*. *Plant Cell* **2004**, *16*, 2001–2019.
9. Voinnet, O. Origin, biogenesis, and activity of plant microRNAs. *Cell* **2009**, *136*, 669–687.
10. Yamaguchi-Shinozaki, K.; Shinozaki, K. A novel *cis*-acting element in an *Arabidopsis* gene is involved in responsiveness. *Plant Cell* **1994**, *6*, 251–264.
11. Mundy, J.; Yamaguchi-Shinozaki, K.; Chua, N.H. Nuclear proteins bind conserved elements in the abscisic acid-responsive promoter of a rice *rab* gene. *Proc. Natl. Acad. Sci. USA* **1990**, *87*, 1406–1410.
12. Shang, Y.; Yan, L.; Liu, Z.Q.; Zheng, C.; Chao, M.; Xin, Q.; Wu, F.Q.; Wang, X.F.; Du, S.Y.; Jiang, T.; *et al.* The Mg-chelatase H subunit of *Arabidopsis* antagonizes a group of transcription repressors to relieve ABA-responsive genes of inhibition. *Plant Cell* **2010**, *22*, 1909–1935.
13. Zhou, X.; Wang, G.; Sutoh, K.; Zhu, J.K.; Zhang, W. Identification of cold-inducible microRNAs in plants by transcriptome analysis. *Biochim. Biophys. Acta* **2008**, *1779*, 780–788.
14. Liu, H.H.; Tian, X.; Li, Y.U.; Wu, C.A.; Zheng, C.C. Microarray-based analysis of stress-regulated microRNAs in *Arabidopsis* thaliana. *RNA* **2008**, *14*, 836–843.
15. Zhou, L.; Liu, Y.; Liu, Z.; Kong, D.; Duan, M.; Luo, L. Genome-wide identification and analysis of drought-responsive microRNAs in *Oryza sativa*. *J. Exp. Bot.* **2010**, *61*, 4157–4168.
16. Fuji, H.; Chiou, T.J.; Lin, S.I.; Aung, K.; Zhu, J.K. A miRNA involved in phosphate starvation response in *Arabidopsis*. *Curr. Biol.* **2005**, *15*, 2038–2043.

17. Sunkar, R.; Kapoor, A.; Zhu, J.K. Posttranscriptional induction of two Cu/Zn superoxide dismutase genes in *Arabidopsis* is mediated by downregulation of miR398 and important for oxidative stress tolerance. *Plant Cell* **2006**, *18*, 2051–2065.
18. Yamasaki, H.; Abdel-Ghany, S.E.; Cohu, C.M.; Kobayashi, Y.; Shikanai, T.; Pilon, M. Regulation of copper homeostasis by micro-RNA in *Arabidopsis*. *J. Biol. Chem.* **2007**, *282*, 16369–16378.
19. Brodersen, P.; Sakvarelidze-Achard, L.; Bruun-Rasmussen, M.; Dunoyer, P.; Yamamoto, Y.Y.; Sieburth, L.; Voinnet, O. Widespread translational inhibition by plant miRNAs and siRNAs. *Science* **2008**, *320*, 1185–1190.
20. Li, W.; Cui, X.; Meng, Z.; Huang, X.; Xie, Q.; Wu, H.; Jin, H.; Zhang, D.; Liang, W. Transcriptional regulation of *Arabidopsis* miR168a and ARGONAUTE1 homeostasis in abscisic acid and abiotic stress responses. *Plant Physiol.* **2013**, *158*, 1279–1292.
21. Liu, P.P.; Montgomery, T.A.; Fahlgren, N.; Kasschau, K.D.; Nonogaki, H.; Carrington, J.C. Repression of AUXIN RESPONSE FACTOR10 by microRNA160 is critical for seed germination and post-germination stages. *Plant J.* **2007**, *52*, 133–146.
22. Blomster, T.; Salojarvi, J.; Sipari, N.; Brosche, M.; Ahlfors, R.; Keinanen, M.; Overmyer, K.; Kangasjarvi, J. Apoplastic reactive oxygen species transiently decrease auxin signaling and cause stress-induced morphogenic response in *Arabidopsis*. *Plant Physiol.* **2011**, *157*, 1866–1883.
23. Wu, M.F.; Tian, Q.; Reed, J.W. *Arabidopsis* microRNA167 controls patterns of ARF6 and ARF8 expression, and regulates both female and male reproduction. *Development* **2006**, *133*, 4211–4218.
24. Marin, E.; Jouannet, V.; Herz, A.; Lokerse, A.S.; Weijers, D.; Vaucheret, H.; Nussaume, L.; Crespi, M.D.; Maizel, A. miR390, *Arabidopsis* TAS3 tasiRNAs, and their AUXIN RESPONSE FACTOR targets define an autoregulatory network quantitatively regulating lateral root growth. *Plant Cell* **2010**, *22*, 1104–1117.
25. Iglesias, M.J.; Terrile, M.C.; Bartoli, C.G.; D'Ippolito, S.; Casalongue, C.A. Auxin signaling participates in the adaptative response against oxidative stress and salinity by interacting with redox metabolism in *Arabidopsis*. *Plant Mol. Biol.* **2010**, *74*, 215–222.
26. Jones-Rhoades, M.W.; Bartel, D.P. Computational identification of plant microRNAs and their targets, including a stress-induced miRNA. *Mol. Cell.* **2004**, *14*, 787–799.
27. Aung, K.; Lin, S.I.; Wu, C.C.; Huang, Y.T.; Su, C.L.; Chiou, T.J. *pho2*, a phosphate overaccumulator, is caused by a nonsense mutation in a microRNA399 target gene. *Plant Physiol.* **2006**, *141*, 1000–1011.
28. Franco-Zorrilla, J.M.; Valli, A.; Todesco, M.; Mateos, I.; Puga, M.I.; Rubio-Somoza, I.; Leyva, A.; Weigel, D.; Garcia, J.A.; Paz-Ares, J. Target mimicry provides a new mechanism for regulation of microRNA activity. *Nat. Genet.* **2007**, *39*, 1033–1037.
29. Mallory, A.C.; Vaucheret, H. Functions of microRNAs and related small RNAs in plants. *Nat. Genet.* **2006**, *38*, S31–S36.
30. Ito, H.; Gaubert, H.; Bucher, E.; Mirouze, M.; Vaillant, I.; Paszkowski, J. An siRNA pathway prevents transgenerational retrotransposition in plants subjected to stress. *Nature* **2011**, *472*, 115–119.

31. Boyko, A.; Blevins, T.; Yao, Y.; Golubov, A.; Bilichak, A.; Ilnytskyy, Y.; Hollunder, J.; Meins, F., Jr.; Kovalchuk, I. Transgenerational adaptation of *Arabidopsis* to stress requires DNA methylation and the function of Dicer-like proteins. *PLoS One* **2010**, *5*, e9514.
32. Ronemus, M.; Vaughn, M.W.; Martienssen, R.A. MicroRNA-targeted and small interfering RNA-mediated mRNA degradation is regulated by argonaute, dicer, and RNA-dependent RNA polymerase in *Arabidopsis*. *Plant Cell* **2006**, *18*, 1559–1574.
33. Allen, E.; Xie, Z.; Gustafson, A.M.; Carrington, J.C. MicroRNA-directed phasing during *trans*-acting siRNA biogenesis in plants. *Cell* **2005**, *121*, 207–221.
34. Ben Amor, B.; Wirth, S.; Merchan, F.; Laporte, P.; d'Aubenton-Carafa, Y.; Hirsch, J.; Maizel, A.; Mallory, A.; Lucas, A.; Deragon J.M.; *et al.* Novel long non-protein coding RNAs involved in *Arabidopsis* differentiation and stress responses. *Genome Res.* **2009**, *19*, 57–69.
35. Yang, J.H.; Seo, H.H.; Han, S.J.; Yoon, E.K.; Yang, M.S.; Lee, W.S. Phytohormone abscisic acid control RNA-dependent RNA polymerase 6 gene expression and post-transcriptional gene silencing in rice cells. *Nucleic Acids Res.* **2008**, *36*, 1220–1226.
36. Bartels, D.; Sunkar, R. Drought and salt tolerance in plants. *Crit. Rev. Plant. Sci.* **2005**, *24*, 23–58.
37. Borsani, O.; Zhu, J.; Verslues, P.E.; Sunkar, R.; Zhu, J.K. Endogenous siRNAs derived from a pair of natural *cis*-antisense transcripts regulate salt tolerance in *Arabidopsis*. *Cell* **2005**, *123*, 1279–1291.
38. Zhang, X.; Lii, Y.; Wu, Z.; Polishko, A.; Zhang, H.; Chinnusamy, V.; Lonardi, S.; Zhu, J.K.; Liu, R.; Jin, H. Mechanisms of small RNA generation from *cis*-NATs in response to environmental and developmental cues. *Mol. Plant* **2013**, *6*, 704–715.
39. Henz, S.R., Cumbie, J.S.; Kasschau, K.D.; Lohmann, J.U.; Carrington, J.C.; Weigel, D.; Schmid, M. Distinct expression patterns of natural antisense transcripts in *Arabidopsis*. *Plant Physiol.* **2007**, *144*, 1247–1255.
40. Zhan, S.; Lukens, L. Protein-coding *cis*-natural antisense transcripts have high and broad expression in *Arabidopsis* thaliana. *Plant Physiol.* **2013**, *161*, 2171–2180.
41. Yamada, K.; Lim, J.; Dale, J.M.; Chen, H.; Shinn, P.; Palm, C.J.; Southwick, A.M.; Wu, H.C.; Kim, C.; Nguyen, M.; *et al.* Empirical analysis of transcriptional activity in the *Arabidopsis* genome. *Science* **2003**, *302*, 842–846.
42. Li, L.; Wang, X.F.; Stolc, V.; Li, X.Y.; Zhang, D.F.; Su, N.; Tongprasit, W.; Li, S.G.; Cheng, Z.K.; Wang, J.; *et al.* Genome-wide transcription analyses in rice using tiling microarrays. *Nat. Genet.* **2006**, *38*, 124–129.
43. Lister, R.; O'Malley, R.C.; Tonti-Filippini, J.; Gregory, B.D.; Berry, C.C.; Millar, A.H.; Ecker, J.R. Highly integrated single-base resolution maps of the epigenome in *Arabidopsis*. *Cell* **2008**, *133*, 523–536.
44. Laubinger, S.; Zeller, G.; Henz, S.R.; Sachsenberg, T.; Widmer, C.K.; Naouar, N.; Vuylsteke, M.; Scholkopf, B.; Ratsch, G.; Weigel, D. At-TAX: A whole genome tiling array resource for developmental expression analysis and transcript identification in *Arabidopsis thaliana*. *Genome Biol.* **2008**, *9*, R112.
45. Hazen, S.P.; Naef, F.; Quisel, T.; Gendron, J.M.; Chen, H.; Ecker, J.R.; Borevitz, J.O.; Kay, S.A. Exploring the transcriptional landscape of plant circadian rhythms using genome tiling arrays. *Genome Biol.* **2009**, *10*, R17.

46. Filichkin, S.A.; Priest, H.D.; Givan, S.A.; Shen, R.; Bryant, D.W.; Fox, S.E.; Wong, W.K.; Mockler, T.C. Genome-wide mapping of alternative splicing in *Arabidopsis thaliana*. *Genome Res.* **2010**, *20*, 45–58.
47. Swiezewski, S.; Liu, F.; Magusin, A.; Dean, C. Cold-induced silencing by long antisense transcripts of an *Arabidopsis* Polycomb target. *Nature* **2009**, *462*, 799–802.
48. Luo, C.; Sidote, D.J.; Zhang, Y.; Kerstetter, R.A.; Michael, T.P.; Lam, E. Integrative analysis of chromatin states in *Arabidopsis* identified potential regulatory mechanisms for natural antisense transcript production. *Plant J.* **2013**, *73*, 77–90.
49. Rajeswaran, R.; Aregger, M.; Zvereva, A.S.; Borah, B.K.; Gubaeva, E.G.; Pooggin, M.M. Sequencing of RDR6-dependent double-stranded RNAs reveals novel features of plant siRNA biogenesis. *Nucleic Acids Res.* **2012**, *40*, 6241–6254.
50. Ner-Gaon, H.; Halachmi, R.; Savaldi-Goldstein, S.; Rubin, E.; Ophir, R.; Fluhr, R. Intron retention is a major phenomenon in alternative splicing in *Arabidopsis*. *Plant J.* **2004**, *39*, 877–885.
51. Hanada, K.; Higuchi-Takeuchi, M.; Okamoto, M.; Yoshizumi, T.; Shimizu, M.; Nakaminami, K.; Nishi, R.; Ohashi, C.; Iida, K.; Tanaka, M.; *et al.* Small open reading frames associated with morphogenesis are hidden in plant genomes. *Proc. Natl. Acad. Sci. USA* **2013**, *110*, 2395–2400.
52. Moghe, G.D.; Lehti-Shiu, M.D.; Seddon, A.E.; Yin, S.; Chen, Y.; Juntawong, P.; Brandizzi, F.; Bailey-Serres, J.; Shiu, S.H. Characteristics and significance of intergenic polyadenylated RNA transcription in *Arabidopsis*. *Plant Physiol.* **2013**, *161*, 1210–1224.
53. Wu, J.; Okada, T.; Fukushima, T.; Tsudzuki, T.; Sugiura, M.; Yukawa, Y. A novel hypoxic stress-responsive long non-coding RNA transcribed by RNA polymerase III in *Arabidopsis*. *RNA Biol.* **2012**, *9*, 302–313.
54. Kapranov, P.; Willingham, A.T.; Gingeras, T.R. Genome-wide transcription and the implications for genomic organization. *Nat. Rev. Genet.* **2007**, *8*, 413–423.
55. Beclin, C.; Boutet, S.; Waterhouse, P.; Vaucheret, H. A branched pathway for transgene-induced RNA silencing in plants. *Curr. Biol.* **2002**, *12*, 684–688.
56. Yoshikawa, M.; Peragine, A.; Park, M.Y.; Poethig, R.S. A pathway for the biogenesis of *trans*-acting siRNAs in *Arabidopsis*. *Genes Dev.* **2005**, *19*, 2164–2175.
57. Zheng, Q.; Ryvkin, P.; Li, F.; Dragomir, I.; Valladares, O.; Yang, J.; Cao, K.; Wang, L.S.; Gregory, B.D. Genome-wide double-stranded RNA sequencing reveals the functional significance of base-paired RNAs in *Arabidopsis*. *PLoS Genet.* **2010**, *6*, e1001141.
58. Gregory, B.D.; O'Malley, R.C.; Lister, R.; Urich, M.A.; Tonti-Filippini, J.; Chen, H.; Millar, A.H.; Ecker, J.R. A link between RNA metabolism and silencing affecting *Arabidopsis* development. *Dev. Cell* **2008**, *14*, 854–866.
59. Gazzani, S.; Lawrenson, T.; Woodward, C.; Headon, D.; Sablowski, R. A link between mRNA turnover and RNA interference in *Arabidopsis*. *Science* **2004**, *306*, 1046–1048.
60. Dunoyer, P.; Himber, C.; Voinnet, O. DICER-LIKE 4 is required for RNA interference and produces the 21-nucleotide small interfering RNA component of the plant cell-to-cell silencing signal. *Nat. Genet.* **2005**, *37*, 1356–1360.
61. Xie, Z.; Allen, E.; Fahlgren, N.; Calamar, A.; Givan, S.A.; Carrington, J.C. Expression of *Arabidopsis* miRNA genes. *Plant Physiol.* **2005**, *138*, 2145–2154.

62. Jauvion, V.; Rivard, M.; Bouteiller, N.; Elmayan, T.; Vaucheret, H. RDR2 partially antagonizes the production of RDR6-dependent siRNA in sense transgene-mediated PTGS. *PLoS One* **2012**, *7*, e29785.
63. Glazov, E.; Phillips, K.; Budziszewski, G.J.; Schob, H.; Meins, F., Jr.; Levin, J.Z. A gene encoding an RNase D exonuclease-like protein is required for post-transcriptional silencing in *Arabidopsis*. *Plant J.* **2003**, *35*, 342–349.
64. Herr, A.J.; Jensen, M.B.; Dalmay, T.; Baulcombe, D.C. RNA polymerase IV directs silencing of endogenous DNA. *Science* **2005**, *308*, 118–120.
65. Dalmay, T.; Horsefield, R.; Braunstein, T.H.; Baulcombe, D.C. SDE3 encodes an RNA helicase required for post-transcriptional gene silencing in *Arabidopsis*. *EMBO J.* **2001**, *20*, 2069–2078.
66. Liu, F.; Marquardt, S.; Lister, C.; Swiezewski, S.; Dean, C. Targeted 3' processing of antisense transcripts triggers *Arabidopsis* FLC chromatin silencing. *Science* **2010**, *327*, 94–97.
67. Angel, A.; Song, J.; Dean, C.; Howard, M. A Polycomb-based switch underlying quantitative epigenetic memory. *Nature* **2011**, *476*, 105–108.
68. De Lucia, F.; Crevillen, P.; Jones, A.M.; Greb, T.; Dean, C. A PHD-polycomb repressive complex 2 triggers the epigenetic silencing of FLC during vernalization. *Proc. Natl. Acad. Sci. USA* **2008**, *105*, 16831–16836.
69. Helliwell, C.A.; Robertson, M.; Finnegan, E.J.; Buzas, D.M.; Dennis, E.S. Vernalization-repression of *Arabidopsis* FLC requires promoter sequences but not antisense transcripts. *PLoS One* **2011**, *6*, e2151.
70. Sun, Q.; Csorba, T.; Skourti-Stathaki, K.; Proudfoot, N.J.; Dean, C. R-loop stabilization represses antisense transcription at the *Arabidopsis* FLC locus. *Science* **2013**, *340*, 619–621.
71. Heo, J.B.; Sung, S. Vernalization-mediated epigenetic silencing by a long intronic noncoding RNA. *Science* **2011**, *331*, 76–79.
72. Wang, K.C.; Chang, H.Y. Molecular mechanisms of long noncoding RNAs. *Mol. Cell* **2011**, *43*, 904–914.
73. Kramerov, D.A.; Vassetzky, N.S. Short retroposons in eukaryotic genomes. *Int. Rev. Cytol.* **2005**, *247*, 165–221.
74. Yukawa, Y.; Felis, M.; Englert, M.; Stojanov, M.; Matousek, J.; Beier, H.; Sugiura, M. Plant 7SL RNA genes belong to type 4 of RNA polymerase III- dependent genes that are composed of mixed promoters. *Plant J.* **2005**, *43*, 97–106.
75. Shin, H.; Shin, H.S.; Chen, R.; Harrison, M.J. Loss of At4 function impacts phosphate distribution between the roots and the shoots during phosphate starvation. *Plant J.* **2006**, *45*, 712–726.
76. Lin, S.I.; Chiang, S.F.; Lin, W.Y.; Chen, J.W.; Tseng, C.Y.; Wu, P.C.; Chiou, T.J. Regulatory network of microRNA399 and PHO2 by systemic signaling. *Plant Physiol.* **2008**, *147*, 732–746.
77. Pant, B.D.; Buhtz, A.; Kehr, J.; Scheible, W.R. MicroRNA399 is a long-distance signal for the regulation of plant phosphate homeostasis. *Plant J.* **2008**, *53*, 731–738.
78. Wu, H.J.; Wang, Z.M.; Wang, M.; Wang, X.J. Wide-spread long non-coding RNAs (lncRNAs) as endogenous target mimics (eTMs) for microRNAs in plants. *Plant Physiol.* **2013**, *161*, 1875–1884.

2

How Can Plant DNA Viruses Evade siRNA-Directed DNA Methylation and Silencing?

Mikhail M. Pooggin

University of Basel, Department of Environmental Sciences, Botany, Schönbeinstrasse 6, Basel 4056, Switzerland; E-Mail: mikhail.pooggin@unibas.ch

Abstract: Plants infected with DNA viruses produce massive quantities of virus-derived, 24-nucleotide short interfering RNAs (siRNAs), which can potentially direct viral DNA methylation and transcriptional silencing. However, growing evidence indicates that the circular double-stranded DNA accumulating in the nucleus for Pol II-mediated transcription of viral genes is not methylated. Hence, DNA viruses most likely evade or suppress RNA-directed DNA methylation. This review describes the specialized mechanisms of replication and silencing evasion evolved by geminiviruses and pararetoviruses, which rescue viral DNA from repressive methylation and interfere with transcriptional and post-transcriptional silencing of viral genes.

Keywords: plant virus; DNA virus; geminivirus; pararetrovirus; silencing; siRNA; RNA-directed DNA methylation; cytosine methylation; silencing evasion; suppressor protein

1. Introduction

DNA viruses accumulate in the nuclei of infected plant cells as multiple circular minichromosomes. which resemble the host plant chromosomes in that the viral DNA is packaged into nucleosomes forming chromatin. Furthermore, viral minichromosomes are transcribed by the host Polymerase II (Pol II), which generates capped and polyadenylated viral RNAs, similar to mRNAs generated by Pol II from most plant protein-coding genes. Thus, viral minichromosomes must encounter the nuclear pathways that regulate host gene expression and chromatin states. However, DNA viruses have evolved specialized mechanisms of replication that differ from those replicating the plant chromosomes. These replication mechanisms can potentially rescue viral minichromosomes from repressive chromatin marks that silence certain plant genes and repetitive DNA elements in transcriptionally-inactive heterochromatic regions. Some of the repressive chromatin marks are established by the RNA-directed DNA methylation (RdDM) pathway. RdDM is a nuclear branch of the plant RNA silencing machinery that regulates gene expression and defends against invasive nucleic acids such as transposons, transgenes and viruses. The plant RNA silencing machinery generates 21, 22 and 24 nt small RNAs which are broadly classified into miRNAs and short interfering RNAs (siRNAs). These small RNAs serve as guide molecules for the silencing complexes that repress genes post-transcriptionally and/or transcriptionally in a sequence-specific manner. The transcriptional silencing through *de novo* DNA methylation is directed by 24-nt siRNAs, the most diverse and abundant class of plant small RNAs. Likewise, plant DNA viruses spawn massive quantities of viral 24-nt siRNAs which can potentially silence viral DNA. In this review, I will focus mainly on the nuclear events in life cycles of plant DNA viruses and describe the strategies of silencing evasion evolved by *Geminiviridae* (geminiviruses) and *Caulimoviridae* (pararetroviruses), the two major families of plant DNA viruses. The third DNA virus family, *Nanoviridae*, is discussed, because little is known about interactions of nanoviruses with the plant silencing system. Since they resemble geminiviruses in DNA replication mechanisms [1], the findings for geminiviruses could be extrapolated to nanoviruses. The post-transcriptional RNA silencing mechanisms which contribute to plant defenses against both RNA and DNA viruses, and the biogenesis and function of the three major classes viral siRNAs including 21-nt and 22-nt classes have been reviewed comprehensively [2–5]. Various silencing suppressor proteins encoded by plant viruses have also been reviewed [6,7], and I will focus only on those encoded by DNA viruses and describe emerging evidence that viral suppressor proteins may have effector functions in suppressing plant innate immunity [8].

2. Plant DNA Methylation

DNA methylation at cytosine nucleotides (5meC) is a reversible epigenetic mark that plays a key role in regulation of gene expression and chromatin states in most eukaryotes. Plants and mammals require cytosine methylation for proper development and genome defense against transposons [9,10]. In mammals, methylation occurs predominantly at symmetric CG sites and, following DNA replication, can be maintained by DNA METHYLTRASFERASE 1 (DNMT1). DNMT1 recognizes hemimethylated double-stranded DNA (dsDNA) with the help of methyl binding domain proteins and catalyzes methylation of symmetric cytosines on the newly-synthesized strand. Establishment of cytosine methylation on unmethylated dsDNA is catalyzed by *de novo* methyltransferases DNMT3a and DNMT3b. Furthermore, methylated dsDNA can be actively demethylated, which ensures dynamic regulation of chromatin states during development and in response to environmental cues. Generally, methylated DNA is repressed transcriptionally, because it is packed into heterochromatin inaccessible to RNA polymerases, whereas unmethylated DNA is present in open actively-transcribed euchromatin.

Figure 1. Models for maintenance methylation and RdDM at the plant genome loci. (Based mostly on the findings using the model plant *Arabidopsis*). The plant dsDNA associated with nucleosomes is depicted as solid lines and the methylated cytosines at one or both strands indicated with black lollypops. Following DNA replication, cytosine methylation at CG, CHG and CHH sites of the newly-synthesized strand (blue) is catalyzed by the maintenance methyltransferates MET1, CMT3, and CMT2, respectively, with the help of co-factors VIM and KYP that recognize hemimethylated dsDNA; CMT3 and CMT2 also bind the repressive histone methylation mark H3K9me2 indicated as grey lollypops. The RNA-directed DNA methylation (RdDM) pathway establishing methylation of dsDNA *de novo* is catalyzed by DRM2 that interacts with the DRD1-Pol V complex generating a scaffold transcript. The nascent scaffold transcript is targeted by the 24-nt siRNA-AGO4 complex. Following DRM2-catalyzed *de novo* methylation of both DNA strands, Pol IV with the help of SHH1 (binding H3K9me2) and CLSY1 initiates siRNA biogenesis. The Pol IV transcript is converted by RDR2 to dsRNA. The resulting dsRNA is processed by DCL3 into 24-nt siRNA duplexes. The duplexes are handed over to AGO4 to form the silencing complexes with a single-stranded siRNA guide. This completes an siRNA amplification loop that reinforces RdDM-mediated transcriptional silencing. The chromatin remodeler DDM1 facilitates the access of all the methyltransferases to dsDNA.

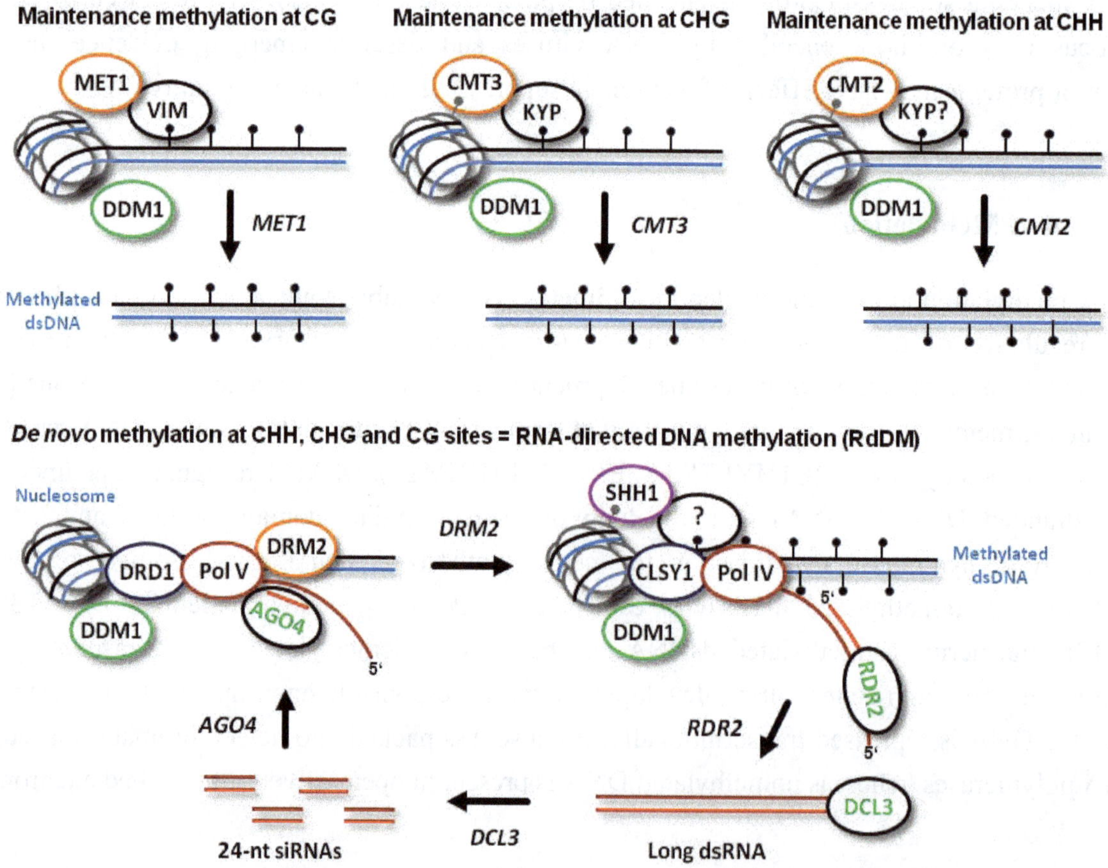

In flowering plants, cytosines in all possible sequence contexts can be methylated, including symmetric (CG and CHG, where H is A, C, or T) and asymmetric (CHH). *De novo* establishment of methylation at CG, CHG and CHH sites is catalyzed by DOMAINS REARRANGED

METHYLTRANSFERASE 2 (DRM2), the plant homolog of mammalian DNMT3a and DNMT3b, which requires 24-nt siRNA guide molecules and other components of the RdDM pathway (Figure 1; see below for more details). Following DNA replication, symmetric CG methylation is maintained by DNA METHYLTRANSFERASE 1 (MET1), the plant homolog of mammalian DNMT1, which recognizes hemimethylated dsDNA with the help of CG-specific methyl binding proteins VARIANT IN METHYLATION 1 (VIM1), VIM2 and VIM3 [11] (Figure 1). Symmetric CHG methylation is maintained by CHROMOMETHYLASE 3 (CMT3), a plant-specific methyltransferase that recognizes dimethylated histone 3 tails at lysine 9 (H3K9m2) on the nucleosomes (Figure 1). In this process, CHG methylation at the template strand is recognized by the H3K9m2 methyltransferase KRYPTONITE (KYP), which can bind methylated cytosines in both CHG and CHH context [12]. Thus, CHG methylation is maintained through a reinforcing loop of DNA and histone (H3K9) methylation. Recently, a homolog of CMT3, CMT2, has been implicated in maintenance methylation at CHH sites [13]. Like CMT3, CMT2 is recruited through direct recognition of the methylated histone H3K9me2 and does not require siRNA guides or other components of RdDM (previously thought to be the only pathway maintaining CHH methylation). Furthermore, indirect recognition of the hemimethylated DNA by CMT2 may also require KYP that binds methylated CHH sites (Figure 1).

Both maintenance methylation and RdDM are facilitated by a chromatin remodeler DEFFICIENT IN DNA METYLATION 1 (DDM1). Indeed, 70% of CG, CHG and CHH methylation is lost in *ddm1* mutant plants. It is believed that maintenance methylation does not take place on naked dsDNA immediately following passage of the DNA replication fork, and that cytosine methylation occurs in a nucleosomal context involving both core and linker histones [14]. DDM1 remodels heterochromatin by removing the repressive linker histone H1 [13]. Obviously, all the DNA methyltransferases need the access to DNA, which can be facilitated by DDM1 (Figure 1). Together, DDM1 and RdDM synergize to maintain all the cytosine methylation in the plant genome [13].

Other factors required for normal DNA methylation include those that have direct or indirect impact on the levels of *S*-adenosyl-L-methionine (SAM), the donor of methyl groups.

3. Mechanism of RNA-Directed DNA Methylation (RdDM)

RdDM is mediated by two plant-specific DNA-dependent RNA polymerases, Pol IV and Pol V: Pol IV functions to initiate siRNA biogenesis, while Pol V generates scaffold transcripts that recruit downstream RdDM factors [15]. Both Pol IV and Pol V are plant-specific enzymes that have evolved from Pol II and share several core Pol II subunits. However, little is known about promoters and other regulatory elements driving transcription at the RdDM loci; the transcripts generated by Pol V and Pol IV were not precisely mapped.

The model depicted in Figure 1 (based mostly on the findings using the model plant *Arabidopsis*) states that Pol V scaffold transcripts are produced at DNA loci to be methylated *de novo*. The nascent scaffold transcript is targeted by an ARGONAUTE 4 (AGO4) protein complex containing a 24-nt siRNA guide molecule via complementary interaction of the siRNA and the scaffold RNA. AGO4 belongs to a family comprising ten members, most of which possess catalytic activity required for sequence-specific cleavage of their target RNAs and subsequent gene silencing at both transcriptional and post-transcriptional levels [16,17]. Besides catalyzing cleavage of the nascent Pol V transcript,

AGO4 interacts with Pol V itself. Together, these interactions are required for recruitment of the methyltransferase DRM2 (or its homolog DRM1) and for subsequent *de novo* methylation of both DNA strands (Figure 1). Other factors that facilitate Pol V transcription and DRM2 recruitment include DEFECTIVE IN RNA-DIRECTED DNA METHYLATION 1 (DRD1), DEFECTIVE IN MERISTEM SILENCING 3 (DMS3) and RNA-DIRECTED DNA METHYLATION 1 (RDM1). These proteins form a complex proposed to unwind dsDNA in front of Pol V (via a putative DNA translocase/ATPase activity of DRD1) and to mediate recruitment of DRM2 to the AGO4-bound scaffold transcript. Following AGO4-catalyzed cleavage of the scaffold transcript, the released siRNA-AGO4 complex may bind the complementary DNA and thereby define the region to be methylated by DRM2 [15]. Other members of the nuclear AGO clade, AGO6 and AGO9, which display tissue specific expression, might also function in RdDM together with, or in place of AGO4 [18].

The biogenesis of 24-nt siRNAs at the RdDM loci is initiated by Pol IV transcription. Pol IV transcripts are then converted to double-stranded RNA (dsRNA) by RNA-DEPENDENT RNA-POLYMERASE 2 (RDR2) (Figure 1). RDR2 belongs to a family with at least three functional enzymes involved in the biogenesis of distinct classes of endogenous plant siRNAs and viral secondary siRNAs. Thus, RDR6 generates dsRNA precursors of plant trans-acting siRNAs (tasiRNAs), which have been precisely mapped [19,20], while RDR1 and RDR6 together are involved in the biogenesis of secondary siRNAs derived from RNA viruses [21–23]. RDR2-dependent dsRNA precursors of 24-nt siRNAs have not been mapped, and it is presumed that RDR2 converts to dsRNA a complete Pol IV transcript, or generates Okazaki-like fragments on the nascent Pol IV transcript [15]. Notably, RDR2 and Pol IV form a complex, and RDR2 has no activity in the absence of Pol IV [24]. Together, Pol IV and RDR2 are required for the biogenesis of virtually all endogenous plant 24-nt siRNAs.

Pol IV is localized at the target loci through interaction with SAWADEE HOMEODOMAIN HOMOLOG 1 (SHH1) that recognizes H3K9me2 [25]. Furthermore, Pol IV occupancy at actively-transcribed, siRNA-generating loci may also require methyl binding protein activity, because Pol IV is believed to transcribe methylated DNA following *de novo* methylation (Figure 1). Pol IV transcription of methylated DNA at the RdDM loci would amplify 24-nt siRNAs to reinforce silencing *in cis*, maintain methylation following replication, and enable *de novo* methylation of homologous DNA loci *in trans*.

De novo methylation might also occur at some Pol II loci via targeting of nascent Pol II transcripts by 24-nt siRNAs [26–28]. In fact, such events might trigger *de novo* methylation and transcriptional silencing of active long-terminal repeat (LTR) retrotransposons (whose genomic RNA is generated by Pol II) following their transposition at new loci.

RdDM and Pol IV activity require CLASSY 1 (CLSY1), a putative ATP-dependent nucleic acid translocase predicted to evict nucleosomes and unwind dsDNA (Figure 1). As discussed above, the chromatin remodeler DDM1 might also facilitate RdDM by removing the repressive histone H1. Establishment of other repressive histone modifications at RdDM loci is catalyzed by a Jumonji domain protein JMJ14 and HISTONE DEACETYLASE 6 (HDA6). JMJ14 demethylates histone H3 lysine 4, thus removing the mark associated with active chromatin. Likewise, HDA6 removes acetyl groups from histone lysines (*i.e.*, active chromatin marks), which is a prerequisite for their subsequent methylation creating the repressive marks such as H3K9me2 [15].

The final step in the biogenesis of endogenous 24-nt siRNAs is accomplished by DICER-LIKE 3 (DCL3), an RNase III-like enzyme that belongs to a family of four prototype members [29]. DCL3 catalyzes processing of RDR2-dependent dsRNA into 24-nt siRNA duplexes (Figure 1). These duplexes are then methylated at the 3'-terminal nucleotides' hydroxyls by HUA ENHANCER 1 (HEN1) and sorted by AGO4, AGO6, or AGO9 to form the silencing complexes containing a single-stranded 24-nt siRNA guide molecule [18]. Either strand of the siRNA duplex can get incorporated into the AGO complex, which enables targeting of both sense and antisense transcripts, potentially generated at the RdDM loci.

4. DNA Demethylation

DNA demethylation can occur passively through several rounds of DNA replication in the absence of efficient maintenance methylation, or actively through enzymatic activities. In plants, DNA glycosylases have been implicated in active removal of 5meC from DNA [9,30]. These include DEMETER (DME) which controls imprinting in reproductive tissues, REPRESSOR OF SILENCING 1 (ROS1) initially identified as suppressor of transcriptional silencing of a plant promoter-driven transgene, and two DEMETER-LIKE enzymes (DML2 and DML3) which, together with ROS1, counteract excessive methylation at several hundred loci across the genome [31–33]. The DNA glycosylases can remove repressive cytosine methylation marks in all the sequence contexts without the need for DNA replication and thereby release transcriptional silencing. However, it is not clear what provides sequence specificity for these enzymes. Animals apparently lack 5meC DNA glycosylases and demethylation involves excision of de-aminated and/or oxidized derivatives of 5meC [34].

A crosstalk between demethylation and *de novo* methylation pathways has been recently illustrated by the finding that ROS1 expression is controlled by the RdDM pathway and mutations in Pol IV and Pol V cause transcriptional silencing at the ROS1 target loci [35].

5. Replication Modes of Geminiviruses

The family *Geminiviridae* comprises circular single-stranded DNA (ssDNA) viruses with 2.5–3.2 kb genomes [36]. The *Begomovirus* genus contains monopartite or bipartite geminiviruses with an additional circular ssDNA component of similar size (DNA-B). Viral ssDNA is encapsidated by viral coat protein in twinned (geminate) virions. The life cycle of geminiviruses, their replication and gene expression strategies have been comprehensively reviewed [37,38]. According to the current model (Figure 2), following insect injection into a plant cell, the viral particle is targeted via a coat protein-based nuclear localization signal to the nucleus, where viral ssDNA is released into nucleoplasm. The circular ssDNA is then converted to circular dsDNA by the host DNA polymerase and other components of the DNA repair machinery. In genus *Begomovirus*, the complementary strand synthesis is primed by an RNA primer [39]. By contrast, in genus *Mastrevirus*, a nested set of complementary strand DNA primers with major species ranging from 78 to 88 nts were found to be associated with virion-derived ssDNA [40].

Figure 2. Models for RCR and RDR modes of geminivirus DNA replication. (**a**) RCR. The viral circular ssDNA is released from the virion (yellow) into the nucleus. The host DNA polymerase synthesizes the complementary strand, yielding circular covalently-closed dsDNA. This dsDNA serves as a template for bidirectional transcription of the early leftward (Rep) and the late rightward (coat protein) genes. Viral mRNAs are transported to the cytoplasm. Following translation, Rep moves to the nucleus to initiate replication of the viral dsDNA by a rolling circle replication (RCR) mechanism. Rep (in yellow) nicks the virion strand in the origin of replication and recruits the host DNA polymerase to extend 3'-end of the cleaved virion strand on the complementary strand template. As the extension progresses, the polymerase complex, associated with Rep covalently linked to the 5'-end of the virion strand, displaces the virion strand. After one or more rounds of replication on the circular complementary strand template, Rep nicks and religates the displaced virion strand extended by one or more copies of the newly-synthesized virion strand and thereby releases one or more copies of circular ssDNA. The resulting circles can re-enter the replication cycle or get packaged into virions; (**b**) The circular covalently-closed dsDNA is invaded by a short viral DNA primer. The primer is extended by the host DNA polymerase on the circular viral template strand. After (or during) one or more rounds of replication, the newly-synthesized linear ssDNA gets fully or partially converted to linear dsDNA by the same (or another) DNA polymerase complex. Thus, RDR generates a heterogeneous population of linear dsDNAs. The long linear dsDNAs that harbor two or more origins of replication are transcribed by Pol II in both orientations to generate viral mRNAs. Following translation, Rep initiates replication of the long linear dsDNA with two or more origins of replication. The replicational release of ssDNA from the multimeric linear dsDNA generates circular ssDNA that can re-enter the replication cycle or get packaged.

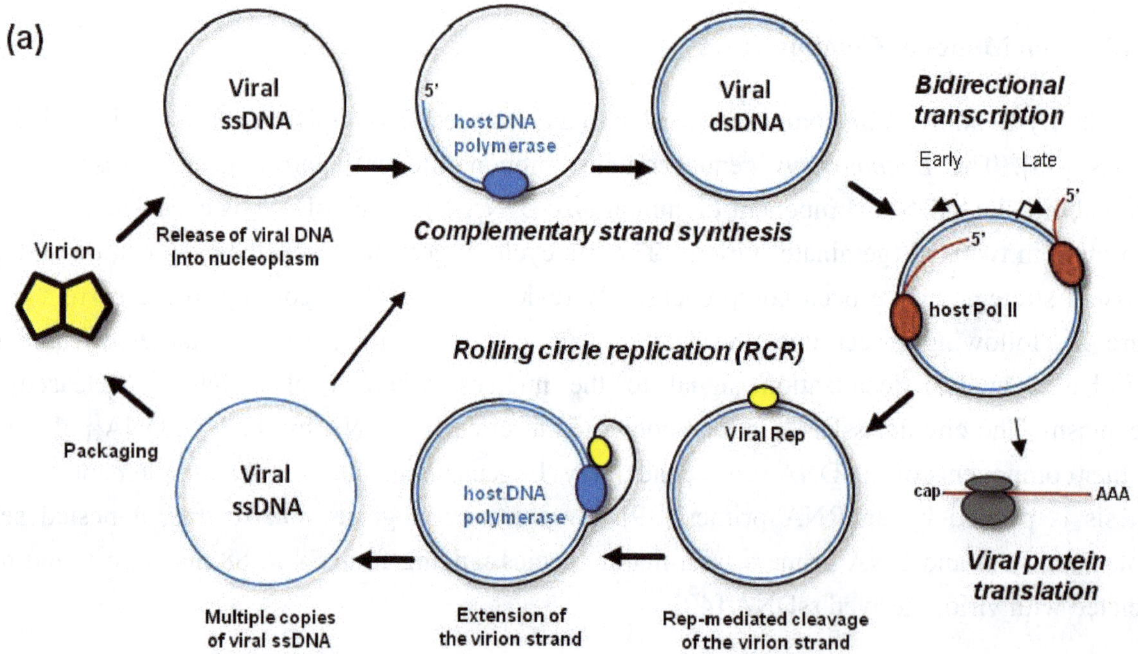

Figure 2. *Cont.*

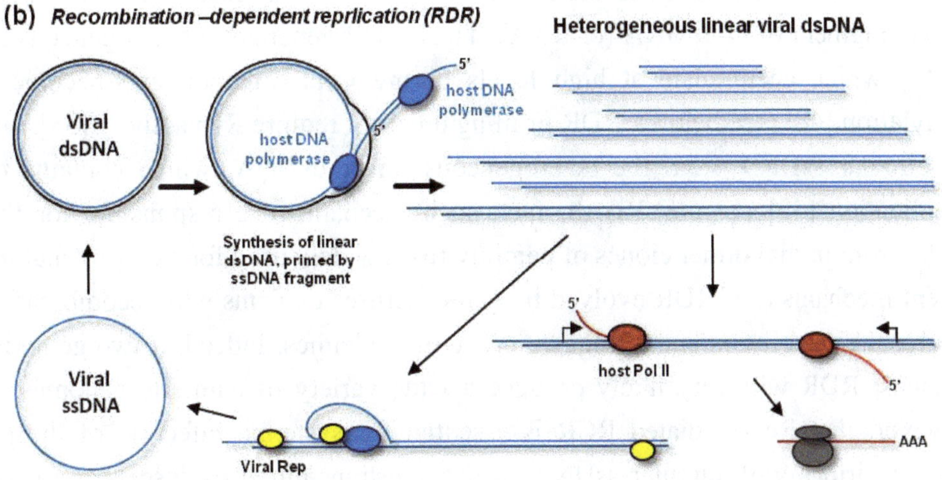

Following complementary strand synthesis, the resulting covalently-closed circular dsDNA gets associated with nucleosomes [41,42] and transcribed by the host Pol II (Figure 2A). Pol II transcribes the viral minichromosome in the leftward orientation to generate mRNAs for viral replication-initiator protein (Rep) and other proteins assisting replication and transcription. At a later stage, the minichromosome is transcribed by Pol II in the rightward orientation to generate mRNA for coat protein [37,43]. In begomovirus-infected plants, the number of nucleosomes per viral minichromosome is varying between 11 and 12, presumably representing transcriptionally active states, and 13, representing inactive state [42].

After production by the cytoplasmic ribosomes, the viral Rep protein moves to the nucleus to initiate rolling circle replication (RCR) of the viral dsDNA that had given rise to the Rep mRNA. Rep is the only viral protein essential for the RCR mechanism generating multiple copies of circular ssDNA. Rep initiates RCR by nicking the virion strand of dsDNA in a conserved nonanucleotide sequence of the replication origin and by recruiting the host DNA polymerase complex. The polymerase uses the circular complementary strand as a template to extend 3'-end of the cleaved virion strand. During this process, the virion strand with Rep covalently linked to its 5'-end is displaced from the template strand (Figure 2A). Rep helicase activity has also been implicated in a post-initiation phase of RCR [44]. After one or more rounds of RCR, Rep (being associated with the polymerase complex) nicks and ligates the displaced virion strand extended with one or more copies of the newly-synthesized virion strand, and thereby releases circular ssDNA from the complex. Thus, multiple circles of viral ssDNA are synthesized on one complementary ssDNA circle (Figure 2A). These circles can re-enter the replication cycle, or get packaged into virions at later stages of infection, when viral coat protein is accumulated. As a result of RCR, multiple copies of viral minichromosomes accumulate in the initially-infected nucleus and eventually in the nuclei of other cells that are infected by cell-to-cell and long-distance movement of viral particles.

In addition to RCR, geminiviruses can replicate their dsDNA by a recombination-dependent replication (RDR) mechanism [38,45–47]. According to a model shown in Figure 2B, RDR is initiated by a viral ssDNA fragment that invades a homologous region of the circular dsDNA with the help of the host recombination enzymes. Then the host DNA polymerase extends the invaded ssDNA on a template strand. During (or after) one or more rounds of the extension on the circular template, the resulting linear ssDNA is converted to dsDNA by the DNA polymerase complex primed by a short complementary fragment of viral DNA (or RNA). Thus, RDR generates a heterogeneous population of linear dsDNAs, which accumulate at high levels during viral infection and become targeted for cytosine methylation [48] (see below). RDR priming does not require Rep activity [38]. However, Rep may release circular ssDNA from the heterogeneous linear dsDNA, which contains two or more origins of replication [49] (Figure 2B). In fact, such mechanism is responsible for the release of circular ssDNA from partial dimer clones of geminiviruses widely used for experimental inoculations.

The efficient mechanism of RDR evolved by geminiviruses explains why recombination is a major driving force for their evolution and a frequent cause of epidemics. Indeed, if two geminiviruses enter the same nucleus, RDR will very likely produce a wide variety of chimeric genomes. It should be stressed, however, that Rep-mediated RCR is essential for systemic infection of the plant and for formation of the virions with circular ssDNA, which are transmitted by insects from plant to plant. Thus, both modes of replication are required for robust infection and spread of geminiviruses. The following section describes how RCR and RDR help geminiviruses evade repressive cytosine methylation and transcriptional silencing.

6. Evasion of Maintenance Methylation and RdDM by Geminiviruses

It has been proposed that cytosine methylation is one of the major host defense mechanisms against geminiviruses and therefore these viruses have evolved different suppressor proteins to interfere with repressive methylation and transcriptional silencing of viral DNA [50]. Here I argue that geminiviruses can evade repressive methylation simply via efficient Rep-dependent replication as has been suggested earlier [48,51].

Experimental evidence based on bisulfite treatment of total DNA from geminivirus-infected plants, followed by PCR amplification and sequencing of the virion strand, shows that 50% to 99% (depending on the virus or the host used) of viral DNA is not methylated [52,53]. Note that technical biases of the bisulfite sequencing method may have prevented correct evaluation of the percentage of 5meC in viral DNA, as discussed by Paprotka *et al.* [48]. Interestingly, methylated cytosines were not randomly distributed between the viral molecules, but concentrated in a small fraction of densely methylated molecules [53]. Hence, a large fraction of viral DNA is not methylated at all. Since the most abundant form of viral DNA is circular ssDNA that gets encapsidated into virions, the above findings imply that this form is not methylated and therefore maintenance methylation does not occur during Rep-mediated RCR. As discussed above, maintenance methylation likely occurs in a nucleosomal context (Figure 1). During the first round of RCR the nucleosomes are removed from the replicating viral DNA and their formation is prevented by continuous rounds of replication displacing newly-synthesized ssDNA (Figure 2A). Moreover, the latter ssDNA is only transiently associated with the template strand, thus preventing an access of the hemimethylated dsDNA-binding proteins required for recruitment of methyltransferases (Figure 1). Likewise, the RDR mechanism generating heterogeneous linear dsDNA on a circular dsDNA template (Figure 2B) is not compatible with maintenance methylation. The variable levels of cytosine methylation detected by bisulfite sequencing in all the sequence contexts [52,53] likely reflect the amounts of *de novo* methylated viral dsDNA in circular or linear forms, both containing the virion strand. Using the bisulfite sequencing approach to evaluate a methylation status of the complementary strand (*i.e.*, RCR template) revealed 36% to 45% of methylation in all the sequence contexts [51]. Hence, a large fraction of viral dsDNA is also not methylated. Taking the above findings and considerations together, detectable methylation of geminiviral DNA is established *de novo*, possibly through RdDM using viral 24-nt siRNA guides. Potential targets of RdDM could be both circular dsDNA and heterogeneous linear dsDNA, which undergo transcription (Figure 2), because RdDM requires on-going transcription at the endogenous target loci (Figure 1).

The lack of maintenance methylation during geminivirus replication is further supported by the findings that geminiviral clones methylated *in vitro* gave rise to unmethylated dsDNA progeny in plant protoplasts, although viral DNA replication was inhibited compared to unmethylated controls [54,55]. These findings illustrate the repressive nature of cytosine methylation, likely inhibiting initial transcription of viral genes. However, more importantly, they demonstrate the ability of geminiviruses to resurrect viral DNA from repressive methylation by evading maintenance methylation during replication.

The methylation status of viral dsDNA in the above-described protoplast studies was evaluated by treatment of total DNA with methylation sensitive enzymes, followed by Southern blot hybridization with virus-specific probes. This approach did not reveal any substantial methylation of viral circular dsDNA in plants infected with different geminiviruses [48,54,56]. The conflicting results obtained with two different methods can be explained by the inability of PCR-based bisulfite sequencing to discriminate between different forms of viral DNA. To resolve this problem, more advanced methods have been applied, using treatment of total DNA with methylation-dependent enzyme McrBC, followed by 1-D or 2-D gel separation and Southern blot analysis or detection with 5meC-specific antibodies [48]. This study has confirmed that circular dsDNA, the template for both replication and transcription, is not methylated. The only viral DNA form that possessed detectable cytosine methylation is heterogeneous linear dsDNA, the product of RDR. Therefore, the extremely variable levels of DNA methylation detected by bisulfite sequencing, ranging for wild-type geminiviruses from 1.25% to 50%–60% [52,53], may reflect the amounts of heterogeneous linear dsDNA accumulated in the respective virus-host systems. The highest methylation level (88%) was reported for an intergenic region of the curtovirus *Beet curly top virus* (BCTV) mutant lacking an L2 gene [52]. *Arabidopsis* plants recover from this mutant virus infection and accumulate very low levels of highly methylated viral DNA. The residual replication of this defective virus in recovered tissues may proceed mainly by RDR that generates linear dsDNA, the target for methylation. Another explanation is that the BCTV L2 protein acts an active suppressor of cytosine methylation [57] (discussed below).

It has been reported that plants deficient in cytosine methylation exhibit increased sensitivity to geminivirus infection [52]. Thus, enhanced disease symptoms were observed for the begomovirus *Cabbage leaf curl virus* (CaLCuV) and the curtovirus BCTV in *Arabidopsis* mutants lacking core components of maintenance methylation or RdDM, which included DRM1/2, Pol IV/V, DDM1, MET1, CMT3, KYP, DCL3, or AGO4. However, mutant plants lacking RDR2, which is also required for RdDM (Figure 1B), did not display enhanced symptoms. Moreover, the mutant plants displaying enhanced symptoms accumulated the wild-type levels of viral DNA [52]. Hence, viral DNA replication is not "de-repressed" in the absence of core components of RdDM or maintenance methylation pathways and, in wild type plants, repressive cytosine methylation can be effectively evaded, likely by

Rep-mediated replication of viral DNA. Consistent with this notion, another study did not reveal increased titres or enhanced symptoms of CaLCuV in *Arabidopsis* mutants lacking Pol IV, RDR2, DCL3, or AGO4 [58]. Notably, viral 24-nt siRNAs were normally produced in all these mutants, except *dcl3*, indicating that the biogenesis of viral 24-nt siRNAs does not require the RdDM components essential for production of dsRNA precursors of endogenous 24-nt siRNAs (see below).

7. Suppression of Cytosine Methylation and Transcriptional Silencing by Geminiviral Proteins

Geminiviral proteins implicated in suppression of cytosine methylation and transcriptional silencing include AC2/AL2/C2/L2 homologs encoded by *Begomovirus* and *Curtovirus* genera and betaC1 encoded by betasatellites associated with some begomoviruses. Since these proteins are not conserved in all genera of *Geminiviridae*, other viral protein(s) may suppress transcriptional silencing (see below) or, as argued here, all geminiviruses should be able to evade cytosine methylation through Rep-dependent replication.

The begomovirus AC2/AL2/C2 gene encodes a transcriptional activator (TrAP) required for activation of late viral genes in the nucleus [37,43]. This protein was also shown to suppress post-transcriptional silencing and its nuclear localization was required for this activity [59,60]. Notably, the suppressor activity of AC2 from two Old World begomoviruses correlated with upregulation of a common subset of host genes including WERNER-LIKE EXONUCLEASE 1 (WEL1), which may act as negative regulators of RNA silencing [60]. Thus, AC2 appears to suppress post-transcriptional silencing indirectly via transcriptional activation of the host genes. Interestingly, the *WEL1* gene that codes for a putative silencing suppressor [60] seats in a transcriptionally-silent locus containing seven *WEL1* paralogs in head-to-tail orientation [60]. Transcriptional silencing of this repetitive DNA locus, likely associated with repressive chromatin marks, might be reversed by the viral TrAP activity. It should be mentioned that in addition to its antisilencing function, *WEL1* may also function in viral DNA replication, because it encodes a putative 3'–5' exonuclease [60] which resembles the WERNER exonuclease involved in DNA replication, recombination and repair.

It has been demonstrated that the begomovirus CaLCuV AL2 and the curtovirus BCTV L2 can reverse transcriptional silencing at transgenic and some endogenous loci repressed by cytosine methylation [57]. The reversal of silencing correlated with partial reduction of non-CG methylation at the respective loci as well as with genome-wide reduction in CHG methylation. In this process, CaLCuV AL2 did not require the C-terminal transcriptional activation domain [57]. This is in contrast to a homologous AC2 protein from the Old Word begomovirus *Mungbean yellow mosaic virus*, which needs this domain to activate the host genes and suppress post-transcriptional silencing [60]. Surprisingly, both AL2 from the New World begomoviruses (CaLCuV and TGMV) and L2 from the curtovirus BCTV reverse transcriptional silencing and cytosine methylation by a mechanism that does not require their nuclear localization. These proteins interact with and inactivate ADENOSINE KINASE (ADK), a cytoplasmic enzyme involved in the methyl cycle producing SAM, the donor of methyl groups [50]. Curiously, AL2/L2-mediated inactivation of ADK was required for suppression of both transcriptional [57] and post-transcriptional [61] silencing.

A different mechanism of silencing suppression was reported for C2 of the curtovirus *Beet severe curly top virus*. This protein interacts with SAM DECARBOXYLASE 1 and thereby interferes with the methyl cycle and DNA methylation [62]. Given that different mechanisms were reported for the closely homologous proteins such as C2 and L2 from curtoviruses as well as AC2 and AL2 from begomoviruses, further research should clarify their activities in the nucleus and the cytoplasm.

Some monopartite begomoviruses are associated with betasatellites that enhance disease symptoms [63]. Betasatellites code for a single protein, betaC1, reported to act as a suppressor of post-transcriptional silencing [64,65]. Like begomoviral AC2/C2, betaC1 from a betasatellite of *Tomato yellow leaf curl China virus* (TYLCCNV) is a nuclear protein and its nuclear localization is required for silencing suppression [64]. The same betaC1 protein was also reported to reverse transcriptional silencing and CHG methylation through inactivation of S-ADENOSYL HOMOCYSTEINE HYDROLASE (SAHH), a methyl cycle enzyme required for synthesis of the methyl donor SAM [53]. Curiously, betaC1 requires an intact nuclear localization signal for the cytoplasmic interaction with SAHH. In the presence of betasatellite, cytosine methylation of the TYLCCNV virion strand was reduced from 5.4% to 1.25%. Since TYLCCNV infection or expression of the TYLCCNV C2 protein failed to reverse transcriptional silencing or cytosine methylation at endogenous loci, betaC1 was proposed to functionally substitute for a loss-of-function mutation in the TYLCCNV C2 gene [53]. However, very low methylation of TYLCCNV DNA in the absence of betasatellite (5.4%) would argue against the absolute necessity for a geminivirus to possess a suppressor of cytosine methylation.

Enhanced symptoms of geminiviral disease in the presence of betasatellite as well as in the methylation-deficient mutant plants described above could be explained by possible involvement of hypomethylation of the host genome in anti-viral defense responses. The symptom severity could be proportional to the expression levels of host defense genes which are induced though demethylation in response to viral infection. The activities of certain geminiviral viral proteins might be recognized by the immune receptors from NUCLEOTIDE BINDING-LEUCINE RICH REPEAT (NB-LRR) family, which induce expression of defense genes in response to both non-viral and viral pathogens [8]. It remains to be investigated if the immune responses to viral infection require active demethylation of the host genome, triggered by recognition of viral proteins.

It has been reported that begomoviral Rep has the ability to reverse transcriptional silencing and reduce CG methylation at endogenous loci, possibly through Rep-mediated downregulation of *MET1* [66]. However, in addition to *MET1*, the transcript levels of *CMT3* and *ROS1* (but not *DRM2*) were also downregulated by transient expression of Rep or by geminivirus infection. It is unclear how the downregulation of the maintenance methyltransferases and the demethylase together would reduce CG methylation and reverse transcription silencing at the endogenous loci, and whether these effects of Rep are important for viral infection. The ability of geminiviral Rep to modify cell cycle and trigger host DNA reduplication [37,67,68] may explain the reduced levels of cytosine methylation at endogenous loci in the Rep transgenic plants upon induction of Rep expression [66]. As described above, Rep-mediated replication of TYLCCNV failed to suppress transcriptional silencing or cytosine methylation [53]. Moreover, geminivirus infection could induce transcriptional silencing of transgenes containing cognate geminiviral sequences, which correlated with hypermethylation of these sequences at CG, CHG and CHH sites [51,69]. This process of virus-induced transgene silencing did not affect geminivirus symptom development or viral DNA accumulation [51]. Thus, while Rep-mediated replication rescues viral DNA from repressive methylation, Rep activity does not prevent *de novo* methylation and silencing of the target transgenes.

8. Plant Recovery from Geminiviral Infection and RdDM

Recovery from virus disease symptoms correlating with reduced viral titres has been observed for RNA and DNA viruses in certain host plants. For RNA viruses that spawn massive quantities of 21-nt and/or 22-nt siRNAs [2–5], the post-transcriptional RNA silencing pathway appears to mediate plant recovery [70]. For geminiviruses that spawn 24-nt siRNAs in addition to 21- and 22-nt siRNAs [58,71–73], both post-transcriptional and transcriptional silencing pathways have been implicated in recovery [72].

Evidence for a role of transcriptional silencing and cytosine methylation in the recovery process comes from several studies. Thus, plant recovery from infections with cassava mosaic begomoviruses correlated with increased accumulation of viral siRNAs of all the size-classes [73], suggesting that viral 24-nt siRNA may direct transcriptional silencing and thereby contribute to recovery. Recovery of pepper plants from begomovirus infection was associated with much low titres of both viral DNA and siRNAs in the youngest recovered leaves, compared to those in the severely infected old leaves. The levels of viral DNA methylation in all the contexts were elevated from ca. 10% in the old leaves to ca. 20% in the youngest leaves [72]. The inverse correlation between 5meC levels and 24-nt siRNA quantities in the respective leaves raises a question whether cytosine methylation is established through the action of viral 24-nt siRNAs. In another study, the abundance of begomovirus-derived siRNAs was also negatively correlated with plant recovery and positively correlated with viral titre [56].

The involvement of RdDM in plant recovery was deduced from the observation that contrary to wild-type plants, the mutant plants lacking AGO4 could not recover from infection with the BCTV mutant lacking L2. This correlated with lower 5meC levels of viral DNA in the disease-displaying *ago4* mutant plants (20%) than in the recovered wild type plants (80%) [52]. However, no difference in cytosine methylation of the wild type BCTV in *ago4* mutant plants *versus* wild-type plants was observed (18% in both cases). Since an AGO4-siRNA complex is a major effector of RdDM (Figure 1), it remains unclear which mechanism mediates cytosine methylation of geminiviral DNA and whether this mechanism requires viral 24-nt siRNAs.

Plant recovery from geminivirus infection can also be triggered by transient or stable expression of inverted-repeat transgenes that generate dsRNA cognate to the geminivirus intergenic region [74,75]. However, it is unclear whether dsRNA-derived 24-nt siRNAs or dsRNA itself contributed to the recovery and whether *de novo* methylation of the viral DNA plays a role in this process. Interestingly, the intergenic region of geminiviruses is a poor source of siRNAs (see below). Thus, targeting this naturally-protected region by artificial dsRNA could help the plant recover from the viral disease.

9. Genetic Requirements for the Biogenesis of Geminiviral siRNAs

Geminivirus-infected plants produce abundant virus-derived 21-, 22- and 24-nt siRNAs. As evaluated by deep sequencing, a sub-population of viral siRNAs can vary from ca. 1%–3% to 30%–50% of the total small RNA population in infected plants [76–78]. Thus, despite a tiny size of the geminivirus genome, the quantity of viral siRNAs in some virus-host systems (e.g., CaLCuV-infected *Arabidopsis* [78]) is comparable to a combined quantity of siRNAs and miRNAs expressed from the plant genome. It should be noted that the percentage of viral siRNAs in a total sRNA population is lower for those geminiviruses that are strictly limited to phloem tissues. Taking into account the dilution factor, geminivirus-infected phloem cells must produce massive amounts of viral siRNAs.

Figure 3. Models for the biogenesis of geminiviral and pararetorviral siRNAs. (**a**) The biogenesis of geminiviral siRNAs is initiated by bi-directional readthrough transcription beyond the poly(A) signals that normally terminate transcription of the viral leftward genes (in begomoviruses, AC1/Rep, AC4, AC2/TrAP and AC3) and the rightward genes (in begomoviruses, AV2 and AV1/CP). The resulting sense and antisense readthrough transcripts (dotted lines) anneal to the complementary viral mRNAs (solid lines with arrowheads) and to each other (in the intergenic region between the transcription start sites). This creates dsRNAs spanning the entire circular viral genome. Every DCL digests these dsRNAs into siRNAs of different sizes, with DCL3 (24-nt), DCL4 (21-nt) and DCL2 (22-nt) being favored (in that order); (**b**) Pol II transcribes both the discontinuous and the covalently-closed dsDNA forms of pararetrovial dsDNA. Abrupt termination of Pol II transcription at the unrepaired minus-strand DNA gap (Met-tRNA gap), results in production of aberrant 8S RNA lacking poly(A) tail (Leader RNA). This RNA forms a viroid-like secondary structure which can be converted by Pol II to dsRNA. The resulting dsRNA serves as a decoy to engage all the four DCLs in massive production of 21-, 22-, and 24-nt vsRNAs. Pol II-mediated transcription of the covalently-closed circular dsDNA generates pgRNA covering the entire genome as well as antisense transcript(s) (red dotted line). The 35S pgRNA promoter was reported to drive transcription not only in the forward but also in the reverse orientation [79] (indicated with bent lines with arrowheads). The pgRNA and antisense Pol II transcripts form low-abundance dsRNA spanning the entire virus genome. This dsRNA is diced by the four DCLs to generate viral 21, 22 and 24-nt siRNAs.

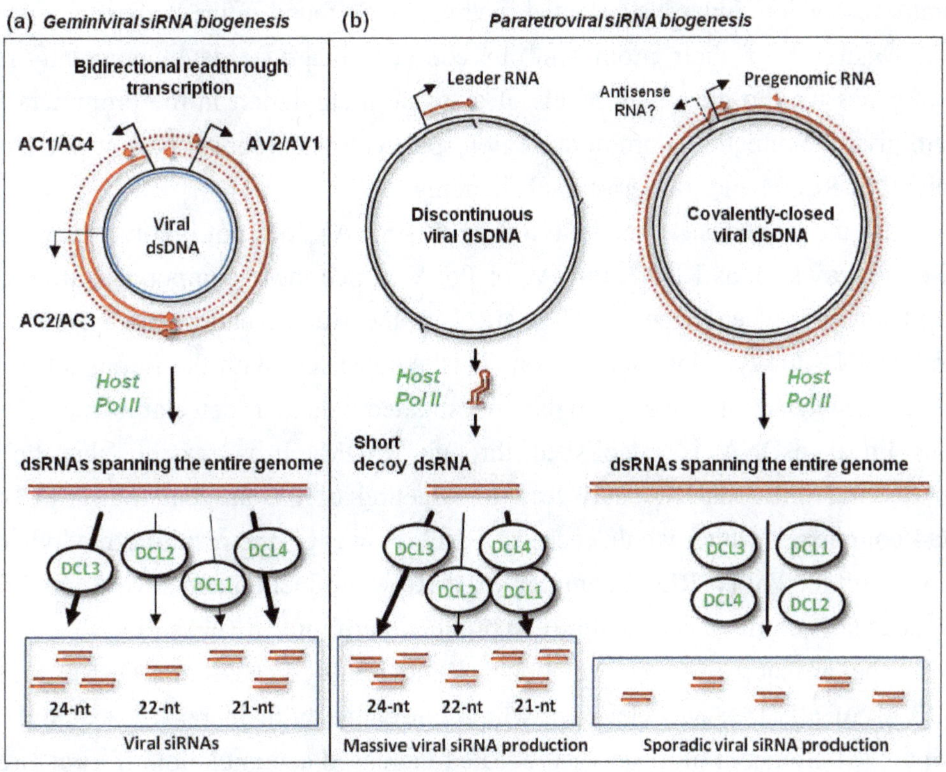

Although the hotspots of viral siRNA production are not equally distributed along the virus genome, unique (non-redundant) siRNAs of each size-class cover the entire circular viral DNA in both sense and antisense polarities, as demonstrated for CaLCuV DNA-A and DNA-B [78]. Based on genetic evidence combined with small RNA deep sequencing and blot hybridization, the biogenesis of CaLCuV siRNAs is mediated by all the four plant DCLs, but does not require RDR1, RDR2, RDR6, Pol IV, or Pol V [58,78]. The precursors of viral siRNAs are likely produced by Pol II-mediated bi-directional readthrough transcription of viral circular dsDNA far beyond the poly(A) signals (Figure 3A; further discussed in [4,78]). Such readthrough transcripts of sense and antisense polarities can potentially form dsRNA substrates for DCLs. Interestingly, the intergenic region harboring bidirectional promoter elements between the transcription start sites is a poor source of viral siRNAs [77,78]. This implies that the readthrough transcripts of each polarity are preferentially associated with more abundant viral mRNAs to form dsRNA. The resulting dsRNAs covering the leftward and rightward genes as well as less abundant dsRNAs covering the intergenic region are then processed by each of the four DCLs to generate 21-nt (DCL4 and DCL1), 22-nt (DCL2) and 24-nt (DCL3) siRNA duplexes [58,78] (Figure 3A). Both strands of these siRNA duplexes are then methylated at the 3'-terminal nucleotide's hydroxyl by HEN1 [58] and presumably sorted by AGO proteins to form silencing complexes. By using CaLCuV as a vector for virus-induced gene silencing (VIGS) targeting a host gene, it was demonstrated that viral siRNA generated by each DCL has the ability to knock down target mRNA accumulation [58]. Furthermore, CaLCuV-VIGS targeting an enhancer region of 35S promoter-driven transgene could induce transcriptional silencing of the transgene in virus-infected plants [78]. Whether viral 24-nt siRNAs get associated with AGO4 to direct *de novo* methylation and transcriptional silencing remains to be investigated.

The most abundant viral 24-nt siRNAs, which can potentially direct *de novo* methylation of viral DNA, map to the coding regions of the geminivirus genome [77,78], where cytosine methylation may not affect viral transcription. Indeed, substantial methylation is found in the bodies of active *Arabidopsis* genes, *i.e.*, downstream of their promoters. In contrast, inactive, developmentally-regulated and tissue-specific genes tend to have high levels of cytosine methylation in the promoters [9]. As noted above the geminiviral bidirectional promoter region spawns low amounts of 24-nt siRNAs, which may not be sufficient for RdDM and transcriptional silencing.

Taken together, the biogenesis of geminiviral 24-nt siRNAs does not involve the core components of the RdDM pathway such as RDR2, Pol IV, or Pol V. Since these components are required for the biogenesis and function of endogenous 24-nt siRNAs, the RdDM pathway may not be effective in targeting viral dsDNA for cytosine methylation. This is consistent with the findings that circular viral dsDNA is not methylated. It remains to be investigated whether detectable methylation of viral heterogeneous linear dsDNA is established through RdDM. It is feasible that Pol II-mediated transcription of viral linear dsDNA may lead to targeting of the nascent transcript by viral 24-nt siRNA-AGO4 complexes, which would recruit DRM2. However, the resulting methylated DNA may not be able to recruit the Pol IV-RDR2 complex for dsRNA production and siRNA amplification, since this complex does not contribute substantially to production of viral siRNAs [58,78].

Notably, in the absence of three functional RDRs (RDR1, RDR2, RDR6), accumulation of CaLCuV siRNAs of all sizes was elevated, which correlated with increased accumulation of some viral transcripts [78]. Southern blot analysis revealed increased accumulation of viral circular ssDNA, but not circular dsDNA. This implies that plant RDR activities may repress both Pol II-mediated siRNA production via readthrough transcription and Rep-dependent production of circular ssDNA. Interestingly, a distinct plant siRNA-generating pathway has been implicated in recombination-dependent DNA repair [80]. It is tempting to speculate that viral siRNAs accumulating in the nucleus may facilitate viral DNA replication, e.g., by serving as primers for the host DNA polymerase.

10. Pararetrovirus Replication and Evasion of Transcriptional and Post-Transcriptional Silencing

Plants do not host retroviruses, but their genomes are populated by LTR retrotransposons whose transcriptional activity is repressed by RdDM. Only episomal pararetroviruses that do not obligatorily integrate into the host genome can replicate and spread in plants. The family *Caulimoviridae* comprises several genera of pararetroviruses with circular dsDNA gemomes of 7.4 to 8 kbp [81,82]. Like retroviruses, the pararetroviruses replicate via reverse transcription. The pararetroviral reverse transcriptase (RT) possesses RNA-dependent and DNA-dependent DNA polymerase activities and RNaseH activity, but lacks an integrase activity [81]. Nonetheless, some plant pararetroviruses have managed to integrate into the host genomes and form complex repetitive integration loci. Some of them, e.g., endogenous *Banana streak virus* and *Petunia vein clearing virus* (PVCV), can be released from the genome upon stress and cause disease [83,84].

The genomic DNA of episomal pararetroviruses is encapsidated in icosahedral or bacilliform virions and transmitted from plant to plant by insect vectors [81]. Like in geminiviruses, a nuclear localization signal of pararetroviral coat protein promotes delivery of viral DNA into the nucleus. The virion-associated circular dsDNA has at least one gap (discontinuity) in each strand, the remnants from reverse transcription of viral pregenomic RNA (pgRNA) in the cytoplasm [85]. These gaps are sealed in the nucleus by the host DNA repair machinery and the resulting covalently-closed circular dsDNA gets associated with nucleosomes to form a viral minichromosome, the template for Pol II transcription (Figure 4). Pol II generates a capped and polyadenylated pgRNA that covers the entire virus genome and has a terminal redundancy, owing to the recognition of the poly(A) signal located at a short distance downstream of the transcription start site only on a second encounter. In some genera, Pol II transcription also generates a subgenomic RNA, the mRNA for P6/TAV protein. This multifunctional protein is involved in formation of dense inclusion bodies in the cytoplasm, translation reinitiation and suppression of plant defenses (see below).

The pgRNA harboring all the viral ORFs serves as an mRNA for polycistronic translation of viral proteins (including coat protein and RT) and as a template for reverse transcription. Following translation in the cytoplasm, the pgRNA is reverse transcribed by viral RT enzymatic activities with the help of coat protein. The resulting open-circular dsDNA with gaps at both strands can be delivered into the nucleus by coat protein, or get incorporated into a mature virion, which can re-infect the same nucleus or move out of the cell (Figure 4). As a result of multiple rounds of replication as well as cell-to-cell and long-distance movement of virions, the infected cells' nuclei accumulate multiple copies of viral minichromosomes.

Figure 4. Model for pararetrovirus replication. Viral circular dsDNA from the virion (in yellow) is released into the nucleus. The gaps at both DNA strands are sealed by the host

DNA repair machinery. The resulting covalently-closed dsDNA serves as a template for Pol II transcription generating viral pregenomic RNA (pgRNA). The capped and polyadenylated pgRNA is transported to the cytoplasm for translation of viral proteins including the reverse transcriptase (RT), and for subsequent reverse transcription catalyzed by RT. The resulting dsDNA with discontinuities at both strands can get packaged into a new virion or targeted to the nucleus for the next round of replication.

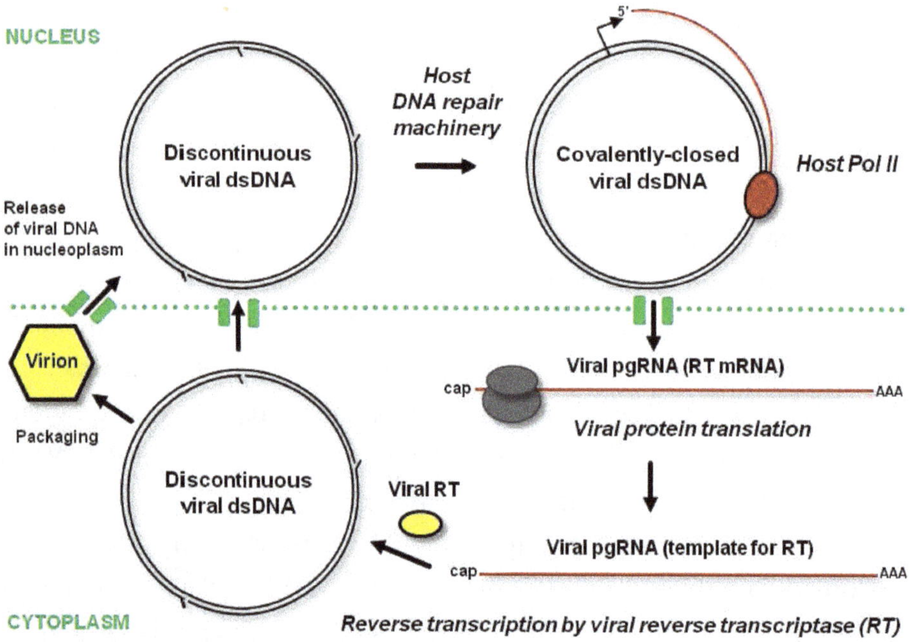

The cytoplasmic step of viral replication through pgRNA should effectively protect viral DNA from maintenance methylation and RdDM. However, covalently-closed circular dsDNA, which is transcribed in the nucleus, can potentially be methylated *de novo* by the RdDM machinery charged with viral 24-nt siRNAs. If this is the case, even inefficient transcription of viral minichromosomes with the repressive marks will generate pgRNA, and the next round of pgRNA translation and reverse transcription will produce unmethylated viral dsDNA.

Deep-sequencing analysis of small RNAs from *Arabidopsis* plants infected with *Cauliflower mosaic virus* (CaMV), a type member of genus *Caulimovirus*, has demonstrated that 21-, 22- and 24-nt viral siRNAs accumulate in massive quantities comparable to the entire complement of endogenous plant siRNA and miRNAs [86]. Moreover, massive production of all size-classes of viral siRNAs of both sense and antisense polarities is largely restricted to a 600 bp non-coding region of the CaMV genome, between the pgRNA transcription start site and the reverse transcription primer binding site. Other genomic sequences spawn much less abundant siRNAs of each size-class and polarity. Given that Pol II-mediated transcription of the CaMV genome generating pgRNA and P6 mRNA is mono-directional, the precursors of viral siRNAs covering the entire genome in both polarities are likely generated by antisense transcription driven by cryptic promoter(s) on viral DNA (Figure 3B). Alternatively, host RDR activities may convert viral RNAs into dsRNA. However, genetic evidence combined with siRNA deep sequencing and blot hybridization ruled out this hypothesis. Indeed, the biogenesis of viral siRNAs from both hot and cold regions does not require RDR1, RDR2, or RDR6 [58,86]. Furthermore, Pol V and Pol IV do not contribute to CaMV siRNA production. Hence, both sense and antisense strands of dsRNA precursors of viral siRNAs are likely generated by Pol II. The resulting dsRNAs are then processed by each of the four Dicers, which generate 21-nt (DCL1 and DCL4), 22-nt (DCL2) and 24-nt (DCL3) siRNAs [58,86]. DCL1, which normally generates plant miRNAs, produces a larger fraction of viral 21-nt siRNAs than DCL4 [58,86]. DCL4 is a primary dicer generating 21-nt siRNAs from RNA viruses [87] but its activity is inhibited by CaMV P6/TAV protein [88,89] (further discussed below).

The 600 bp non-coding region of CaMV genome generating the majority of viral siRNAs was proposed to produce a decoy dsRNA that would engage all the four DCLs and available AGOs in production and sorting of viral siRNAs [86] (Figure 3B). Such decoy strategy would protect other regions from silencing at both transcriptional and post-transcriptional levels. Indeed, the upstream pgRNA promoter elements and the downstream coding sequences spawn only small amounts of viral siRNAs that would have to compete with abundant, decoy dsRNA-derived siRNAs for AGOs to form silencing complexes. Consistent with the decoy model, immuno-precipitation with AGO-specific antibodies revealed that AGO1 is associated with 21-nt siRNAs from the non-coding region but not other regions of CaMV genome [86]. Surprisingly, only a tiny fraction of abundant 24-nt siRNAs from the non-coding region was associated with AGO4. AGO4 complexes in the nucleus are likely saturated with endogenous 24-nt siRNAs and only a small pool of free AGO4 is available. If the non-coding region becomes *de novo* methylated through the action of detectable silencing complexes, transcriptional activity of the upstream promoter will not be affected. At the post-transcriptional level, the 600 nt non-coding leader sequence of pgRNA folds into a stable secondary structure bypassed by ribosomes to initiate translation [90–92], which may not be accessible for 21-nt siRNA-AGO1 complexes. Taken together, the decoy strategy evolved by CaMV [86] and possibly other pararetroviruses with a similar configuration of the non-coding region elements and structures [93] would help the virus evade silencing at both transcriptional and post-transcriptional levels.

Like in the case of geminiviruses, pararetrovirus infection can induce silencing of transgenes sharing homology with the virus. In CaMV-infected plants, the transgenes driven by the CaMV 35S pgRNA promoter were silenced at the transcriptional levels, whereas those with the CaMV 3'UTR sequences at the post-transcriptional level [94,95]. Notably, CaMV replication and viral transcript accumulation were not affected by ongoing silencing of the transgenes [94]. Thus, CaMV can indeed evade both transcriptional and posttranscriptional silencing as argued above. It remains to be investigated if silencing of homologous transgenes is directed by viral siRNAs.

Some host plants can recover from pararetrovirus disease symptoms, but abundant viral dsDNA can still persist in the recovered tissues. The recovery of kohlrabi plants from CaMV infection was preceded by overaccumulation of covalently-closed viral dsDNA in the nucleus, followed by arrest of reverse transcription [96]. Interestingly, overall transcription of viral dsDNA in the nucleus (evaluated by a "nuclear run-on" method) did not change after the transition to recovery, but accumulation of polyadenylated viral transcripts was strongly reduced. This implicates post-transcriptional silencing in the recovery process. Notably, covalently-closed viral dsDNA was not found to be methylated before or after recovery [96]. The mechanisms underlying the overaccumulation of viral minichromosomes before recovery and the posttranscriptional degradation of viral RNAs remain to be further investigated.

Endogenous pararetroviruses integrated in the host genomes are likely repressed by cytosine methylation and histone modifications. These repressive marks can potentially be established *de novo* by RdDM and efficiently maintained following plant DNA replication. The integrated copies of PVCV in the petunia genome were found to be associated with repressive H3K9me2 marks [84]. In this case, accumulation of 21–24 nt viral siRNAs was barely detectable, and only disease induction could boost viral siRNA production. Hence, the released episomal virus spawns much more abundant siRNAs than the integrated copies. Deep-sequencing of siRNAs combined with cytosine methylation analysis should clarify whether the infectious copies of integrated pararetroviral DNA are densely methylated and whether cytosine methylation is established and maintained by RdDM. In the case of an endogenous tomato pararetrovirus, which cannot be released as episomal virus, the integrated viral sequences were found to be methylated at CHG and CHH sites and virus-derived 21–24 nt siRNAs accumulated at detectable levels [97].

11. Suppression of Plant Defenses by Pararetroviral Proteins

CaMV P6/TAV protein has been implicated in suppression of the plant defenses based on RNA silencing [88,89,98] and innate immunity [8,99]. Since this protein has no homologs in several genera of *Caulimoviridae* [81], plant pararetroviruses have to rely on other strategies to suppress or evade plant defenses. The tungrovirus *Rice tungro bacilliform virus* possesses a P4 gene of unknown function, which is missing in closely related badnaviruses. Like CaMV P6 gene, the P4 gene is located downstream of the RT gene and expressed from a separate mRNA [81]. These similarities suggest that P4 may have been acquired by a badnavirus to cope with plant defenses in a new host.

The mechanism of silencing suppression by CaMV P6/TAV has been extensively investigated [88,89,98]. According to the current model, P6 interferes with amplification of secondary siRNAs by blocking DCL4-mediated processing of RDR6-dependent dsRNAs. Curiously, nuclear import of P6 was required for P6-mediated suppression of endogenous tasiRNA biogenesis, which presumably occurs in the cytoplasm, and for P6 interaction with DOUBLE-STRANDED RNA BINDING 4 (DRB4), a partner of DCL4 [88]. It remains to be demonstrated if these activities of CaMV P6 are also required for suppression of antiviral silencing. Indirect evidence supporting this hypothesis is that only a fraction of CaMV 21-nt siRNAs is produced by DCL4 and the biogenesis of the DCL4-dependent fraction of viral 21-nt siRNAs does not require RDR6 activity [58,86]. However, presumptive RDR6-dependent precursors of viral secondary siRNAs, which should be stabilized by the P6 action, could not be detected in CaMV-infected plants. In contrast the RDR6-dependent dsRNA precursors of plant tasiRNAs are readily detectable in both CaMV-infected and P6 transgenic plants [19,20,58,89]. Thus, CaMV infection or P6 expression does not interfere with RDR6 activity, but viral mRNAs (and their degradation products) appear to be poor substrates for RDR6. Similar findings have been reported for the geminivirus CaLCuV [78]. Thus, DNA viruses have evolved to protect their mRNAs from RDR activity that would amplify and spread antiviral siRNAs. Likewise, most of the plant genes controlled by miRNAs do not spawn RDR6-dependent secondary siRNAs.

The caulimovirus P6/TAV is a multifunctional protein harboring the domains implicated in interactions with ribosomal proteins and translation initiation factors, in binding RNA (single and double-stranded), in formation of inclusion bodies, and in hypersensitive immune responses [100–102]. The domain responsible for suppression of RNA silencing has not been identified yet.

It has been hypothesized that P6/TAV and suppressor proteins of other plant viruses can interfere with the innate immune responses, which restrict growth of non-viral pathogens [8]. In resistant hosts, CaMV P6 triggers hypersensitive responses and its avirulence domain recognized by the immune system has been mapped. Notably, this P6 domain is also required for CaMV virulence in susceptible hosts [101]. By analogy with effector proteins of non-viral pathogens, a primary function for P6 is to suppress basal immune responses. In resistant hosts, P6 effector activity is recognized by the immune receptors of the NB-LRR family, which triggers hypersensitive response and programmed cell death restricting viral infection [8]. This hypothesis is supported by the finding that CaMV P6 expression in transgenic plants promotes growth of a bacterial pathogen [99]. The mechanism of P6 interference with the immune responses remains to be investigated.

12. Concluding Remarks

Unlike animals, land plants do not host "true" dsDNA viruses whose replication mechanisms generate dsDNA genome copies without a ssDNA or RNA intermediate, or "true" retroviruses with a provirus stage of replication that involves viral DNA integration in the host genome. This exclusion is likely because the land plants have evolved the mechanisms of siRNA-directed *de novo* methylation of all cytosines (RdDM) and maintenance methylation at both CG and non-CG sites. These mechanisms establish and maintain cytosine methylation in all sequence contexts of the plant genome and thereby effectively repress unwanted transcription in the nucleus. This repressive methylation system is evaded by ssDNA viruses which can resurrect their dsDNA forms from cytosine methylation by Rep-dependent replication generating unmethylated ssDNA. Likewise, pararetroviruses that omit a host genome integration step can thereby evade the transcriptional silencing reinforced by a concert action of maintenance methylation and RdDM-dependent amplification of siRNAs. Furthermore, episomal pararetroviruses can evade repressive methylation by constant delivery of multiple unmethylated copies of circular dsDNA to the nucleus from the cytoplasm where pgRNA is reverse transcribed. Having these replication strategies, pararetroviruses and ssDNA viruses have not been under a strong pressure in land plants to evolve suppressors of cytosine methylation.

Acknowledgments

I thank Thomas Boller for supporting research of MMP group at the University of Basel and Rajendran Rajeswaran for fruitful collaboration and critical reading of the manuscript. Special thanks to the anonymous reviewer for constructive criticism and ideas. The work of MMP group is being funded by Swiss National Science Foundation (grant 31003A_143882/1), Indo-Swiss Collaboration in Biotechnology, and European Collaboration in Science and Technology (COST grant SER No. C09.0176 and COST Action FA0806).

References

1. Gronenborn, B. Nanoviruses: Genome organisation and protein function. *Vet. Microbiol.* **2004**, *98*, 103–109.

2. Llave, C. Virus-derived small interfering RNAs at the core of plant-virus interactions. *Trends Plant Sci.* **2010**, *15*, 701–707.
3. Pantaleo, V. Plant RNA silencing in viral defence. *Adv. Exp. Med. Biol.* **2011**, *722*, 39–58.
4. Rajeswaran, R.; Pooggin, M.M. Role of Virus-Derived Small RNAs in Plant Antiviral Defense: Insights from DNA Viruses. In *MicroRNAs in Plant Development and Stress Response*; Sunkar, R., Ed.; Springer-Verlag: Berlin Heidelberg, Germany, 2012; pp. 261–289.
5. Szittya, G.; Burgyán, J. RNA interference-mediated intrinsic antiviral immunity in plants. *Curr. Top. Microbiol. Immunol.* **2013**, *371*, 153–181.
6. Burgyán, J.; Havelda, Z. Viral suppressors of RNA silencing. *Trends Plant Sci.* **2011**, *16*, 265–272.
7. Incarbone, M.; Dunoyer, P. RNA silencing and its suppression: Novel insights from in planta analyses. *Trends Plant Sci.* **2013**, doi:10.1038/nrg1555.
8. Zvereva, AS.; Pooggin, M.M. Silencing and innate immunity in plant defense against viral and non-viral pathogens. *Viruses* **2012**, *4*, 2578–2597.
9. Law, J.A.; Jacobsen, S.E. Establishing, maintaining and modifying DNA methylation patterns in plants and animals. *Nat. Rev. Genet.* **2010**, *11*, 204–220.
10. Klose, R.J.; Bird, A.P. Genomic DNA methylation: The mark and its mediators. *Trends Biochem. Sci.* **2006**, *31*, 89–97.
11. Woo, H.R.; Dittmer, T.A.; Richards, E.J. Three SRA-domain methylcytosine-binding proteins cooperate to maintain global CpG methylation and epigenetic silencing in *Arabidopsis*. *PLoS Genet.* **2008**, *4*, e1000156.
12. Johnson, L.M.; Bostick, M.; Zhang, X.; Kraft, E.; Henderson, I.; Callis, J.; Jacobsen, S.E. The SRA methyl-cytosine-binding domain links DNA and histone methylation. *Curr. Biol.* **2007**, *17*, 379–384.
13. Zemach, A.; Kim, M.Y.; Hsieh, P.H.; Coleman-Derr, D.; Eshed-Williams, L.; Thao, K.; Harmer, S.L.; Zilberman, D. The *Arabidopsis* nucleosome remodeler DDM1 allows DNA methyltransferases to access H1-containing heterochromatin. *Cell* **2013**, *153*, 193–205.
14. Pikaard, C.S. Methylating the DNA of the most repressed: Special access required. *Mol. Cell* **2013**, *49*, 1021–1022.
15. Pikaard, C.S.; Haag, J.R.; Pontes, O.M.; Blevins, T.; Cocklin, R. A transcription fork model for Pol IV and Pol V-dependent RNA-Directed DNA methylation. *Cold Spring Harb. Symp. Quant. Biol.* **2013**, doi:10.1101/sqb.2013.77.014803.
16. Mallory, A.; Vaucheret, H. Form, function, and regulation of ARGONAUTE proteins. *Plant Cell* **2010**, *22*, 3879–3889.
17. Carbonell, A.; Fahlgren, N.; Garcia-Ruiz, H.; Gilbert, K.B.; Montgomery, T.A.; Nguyen, T.; Cuperus, J.T.; Carrington, J.C. Functional analysis of three *Arabidopsis* ARGONAUTES using slicer-defective mutants. *Plant Cell* **2012**, *24*, 3613–3629.
18. Havecker, E.R.; Wallbridge, L.M.; Hardcastle, T.J.; Bush, M.S.; Kelly, K.A.; Dunn, R.M.; Schwach, F.; Doonan, J.H.; Baulcombe, D.C. The *Arabidopsis* RNA-directed DNA methylation

argonautes functionally diverge based on their expression and interaction with target loci. *Plant Cell* **2010**, *22*, 321–334.

19. Rajeswaran, R.; Pooggin, M.M. RDR6-mediated synthesis of complementary RNA is terminated by miRNA stably bound to template RNA. *Nucleic Acids Res.* **2012**, *40*, 594–599.
20. Rajeswaran, R.; Aregger, M.; Zvereva, A.S.; Borah, B.K.; Gubaeva, E.G.; Pooggin, M.M. Sequencing of RDR6-dependent double-stranded RNAs reveals novel features of plant siRNA biogenesis. *Nucleic Acids Res.* **2012**, *40*, 6241–6254.
21. Wang, X.B.; Wu, Q.; Ito, T.; Cillo, F.; Li, W.X.; Chen. X.; Yu, J.L.; Ding, S.W. RNAi-mediated viral immunity requires amplification of virus-derived siRNAs in *Arabidopsis thaliana. Proc. Natl. Acad. Sci. USA* **2010**, *107*, 484–489.
22. Wang, X.B.; Jovel, J.; Udomporn, P.; Wang, Y.; Wu, Q.; Li, W.X.; Gasciolli, V.; Vaucheret, H.; Ding, S.W. The 21-Nucleotide, but Not 22-Nucleotide, viral secondary small interfering RNAs direct potent antiviral defense by two cooperative argonautes in *Arabidopsis thaliana. Plant Cell* **2011**, *23*, 1625–1638.
23. Garcia-Ruiz, H.; Takeda, A.; Chapman, E.J.; Sullivan, C.M.; Fahlgren, N.; Brempelis, K.J.; Carrington, J.C. *Arabidopsis* RNA-dependent RNA polymerases and dicer-like proteins in antiviral defense and small interfering RNA biogenesis during Turnip Mosaic Virus infection. *Plant Cell* **2010**, *22*, 481–496.
24. Haag, J.R.; Ream, T.S.; Marasco, M.; Nicora, C.D.; Norbeck, A.D.; Pasa-Tolic, L.; Pikaard, C.S. *In vitro* transcription activities of Pol IV, Pol V, and RDR2 reveal coupling of Pol IV and RDR2 for dsRNA synthesis in plant RNA silencing. *Mol. Cell* **2012**, *48*, 811–818.
25. Law, J.A.; Du, J.; Hale, C.J.; Feng, S.; Krajewski, K.; Palanca, A.M.; Strahl, B.D.; Patel, D.J.; Jacobsen, S.E. Polymerase IV occupancy at RNA-directed DNA methylation sites requires SHH1. *Nature* **2013**, doi:10.1038/nature12178.
26. Zheng, B.; Wang, Z.; Li, S.; Yu, B.; Liu, J.Y.; Chen, X. Intergenic transcription by RNA polymerase II coordinates Pol IV and Pol V in siRNA-directed transcriptional gene silencing in Arabidopsis. *Genes Dev.* **2009**, *23*, 2850–2860.
27. Stroud, H.; Greenberg, M.V.; Feng, S.; Bernatavichute, Y.V.; Jacobsen, S.E. Comprehensive analysis of silencing mutants reveals complex regulation of the *Arabidopsis* methylome. *Cell* **2013**, *152*, 352–364.
28. You, W.; Lorkovic, Z.J.; Matzke, A.J.; Matzke, M. Interplay among RNA polymerases II, IV and V in RNA-directed DNA methylation at a low copy transgene locus in *Arabidopsis thaliana. Plant Mol. Biol.* **2013**, *82*, 85–96.
29. Schauer, S.E.; Jacobsen, S.E.; Meinke, D.W.; Ray A. DICER-LIKE1: Blind men and elephants in *Arabidopsis* development. *Trends Plant Sci.* **2002**, *7*, 487–491.
30. Furner, I.J.; Matzke, M. Methylation and demethylation of the *Arabidopsis* genome. *Curr. Opin. Plant Biol.* **2011**, *14*, 137–141.
31. Penterman, J.; Zilberman, D.; Huh, J.H.; Ballinger, T.; Henikoff, S.; Fischer, R.L. DNA demethylation in the *Arabidopsis* genome. *Proc. Natl. Acad. Sci. USA* **2007**, *104*, 6752–6757.
32. Zhu, J.; Kapoor, A.; Sridhar, V.V.; Agius, F.; Zhu, J.K. The DNA glycosylase/lyase ROS1 functions in pruning DNA methylation patterns in *Arabidopsis*. *Curr. Biol.* **2007**, *17*, 54–59.

33. Ortega-Galisteo, A.P.; Morales-Ruiz, T.; Ariza, R.R.; Roldán-Arjona, T. *Arabidopsis* DEMETER-LIKE proteins DML2 and DML3 are required for appropriate distribution of DNA methylation marks. *Plant Mol. Biol.* **2008**, *67*, 671–681.
34. Bhutani, N.; Burns, D.M.; Blau, H.M. DNA demethylation dynamics. *Cell* **2011**, *146*, 866–872.
35. Li, X.; Qian, W.; Zhao, Y.; Wang, C.; Shen, J.; Zhu, J.K.; Gong, Z. Antisilencing role of the RNA-directed DNA methylation pathway and a histone acetyltransferase in *Arabidopsis*. *Proc. Natl. Acad. Sci. USA* **2012**, *109*, 11425–11430.
36. Fauquet, C.M.; Briddon, R.W.; Brown, J.K.; Moriones, E.; Stanley, J.; Zerbini, M.; Zhou, X. Geminivirus strain demarcation and nomenclature. *Arch. Virol.* **2008**, *153*, 783–821.
37. Hanley-Bowdoin, L.; Settlage, S.B.; Orozco, B.M.; Nagar, S.; Robertson, D. Geminiviruses: Models for plant DNA replication, transcription, and cell cycle regulation. *Crit. Rev. Plant Sci.* **1999**, *18*, 71–106.
38. Jeske, H. Geminiviruses. *Curr. Top. Microbiol. Immunol.* **2009**, *331*, 185–226.
39. Saunders, K.; Lucy, A.; Stanley, J. RNA-Primed complementary-sense DNA synthesis of the geminivirus African cassava mosaic virus. *Nucleic Acids Res.* **1992**, *20*, 6311–6315.
40. Andersen, M.T.; Richardson, K.A.; Harbison, S.A.; Morris, B.A. Nucleotide sequence of the geminivirus chloris striate mosaic virus. *Virology* **1988**, *164*, 443–449.
41. Pilartz, M.; Jeske, H. Abutilon mosaic geminivirus double-stranded DNA is packed into minichromosomes. *Virology* **1992**, *189*, 800–802.
42. Pilartz, M.; Jeske, H. Mapping of abutilon mosaic geminivirus minichromosomes. *J. Virol.* **2003**, *77*, 10808–10818.
43. Shivaprasad, P.V.; Akbergenov, R.; Trinks, D.; Rajeswaran, R.; Veluthambi, K.; Hohn, T.; Pooggin, M.M. Promoters, transcripts, and regulatory proteins of Mungbean yellow mosaic geminivirus. *J. Virol.* **2005**, *79*, 8149–8163.
44. Singh, D.K.; Malik, P.S.; Choudhury, N.R.; Mukherjee, S.K. MYMIV replication initiator protein (Rep): Roles at the initiation and elongation steps of MYMIV DNA replication. *Virology* **2008**, *380*, 75–83.
45. Jeske, H.; Lütgemeier, M.; Preiss, W. DNA forms indicate rolling circle and recombination-dependent replication of Abutilon mosaic virus. *EMBO J.* **2001**, *20*, 6158–6167.
46. Preiss, W.; Jeske, H. Multitasking in replication is common among geminiviruses. *J. Virol.* **2003**, *77*, 2972–2980.
47. Erdmann, J.B.; Shepherd, D.N.; Martin, D.P.; Varsani, A.; Rybicki, E.P.; Jeske, H. Replicative intermediates of maize streak virus found during leaf development. *J. Gen. Virol.* **2010**, *91*, 1077–1081.
48. Paprotka, T.; Deuschle, K.; Metzler, V.; Jeske, H. Conformation-selective methylation of geminivirus DNA. *J. Virol.* **2011**, *85*, 12001–12012.
49. Stenger, D.C.; Revington, G.N.; Stevenson, M.C.; Bisaro, D.M. Replication release of geminivirus genomes from tandemly repeated copies: Evidence for a rolling circle replication of a plant viral DNA. *Proc. Natl. Acad. Sci. USA* **1991**, *88*, 8029–8033.
50. Raja, P.; Wolf, J.N.; Bisaro, D.M. RNA silencing directed against geminiviruses: Post-Transcriptional and epigenetic components. *Biochim. Biophys. Acta* **2010**, *1799*, 337–351.

51. Bian, X.Y.; Rasheed, M.S.; Seemanpillai, M.J.; Rezaian, M.A. Analysis of silencing escape of tomato leaf curl virus: An evaluation of the role of DNA methylation. *Mol. Plant-Microbe Interact.* **2006**, *19*, 614–624.
52. Raja, P.; Sanville, B.C.; Buchmann, R.C.; Bisaro, D.M. Viral genome methylation as an epigenetic defense against geminiviruses. *J. Virol.* **2008**, *82*, 8997–9007.
53. Yang, X.; Xie, Y.; Raja, P.; Li, S.; Wolf, J.N.; Shen, Q.; Bisaro, D.M.; Zhou, X. Suppression of methylation-mediated transcriptional gene silencing by βC1-SAHH protein interaction during geminivirus-betasatellite infection. *PLoS Pathog.* **2011**, *7*, e1002329.
54. Brough, C.L.; Gardiner, W.E.; Inamdar, N.M.; Zhang, X.Y.; Ehrlich, M.; Bisaro, D.M. DNA methylation inhibits propagation of tomato golden mosaic virus DNA in transfected protoplasts. *Plant Mol. Biol.* **1992**, *18*, 703–712.
55. Ermak, G.; Paszkowski, U.; Wohlmuth, M.; Mittelsten Scheid, O.; Paszkowski, J. Cytosine methylation inhibits replication of African cassava mosaic virus by two distinct mechanisms. *Nucleic Acids Res.* **1993**, *21*, 3445–3450.
56. Hagen, C.; Rojas, M.R.; Kon, T.; Gilbertson, R.L. Recovery from Cucurbit leaf crumple virus (family *Geminiviridae*, genus *Begomovirus*) infection is an adaptive antiviral response associated with changes in viral small RNAs. *Phytopathology* **2008**, *98*, 1029–1037.
57. Buchmann, R.C.; Asad, S.; Wolf, J.N.; Mohannath, G.; Bisaro, D.M. Geminivirus AL2 and L2 proteins suppress transcriptional gene silencing and cause genome-wide reductions in cytosine methylation. *J. Virol.* **2009**, *83*, 5005–5013.
58. Blevins, T.; Rajeswaran, R.; Shivaprasad, P.V.; Beknazariants, D.; Si-Ammour, A.; Park, H.S.; Vazquez, F.; Robertson, D.; Meins, F., Jr.; Hohn, T.; *et al.* Four plant Dicers mediate viral small RNA biogenesis and DNA virus induced silencing. *Nucleic Acids Res.* **2006**, *34*, 6233–6246.
59. Dong, X.; van Wezel, R.; Stanley, J.; Hong, Y. Functional characterization of the nuclear localization signal for a suppressor of posttranscriptional gene silencing. *J. Virol.* **2003**, *77*, 7026–7033.
60. Trinks, D.; Rajeswaran, R.; Shivaprasad, P.V.; Akbergenov, R.; Oakeley, E.J.; Veluthambi, K.; Hohn, T.; Pooggin, M.M. Suppression of RNA silencing by a geminivirus nuclear protein, AC2, correlates with transactivation of host genes. *J. Virol.* **2005**, *79*, 2517–2527.
61. Wang, H.; Buckley, K.J.; Yang, X.; Buchmann, R.C.; Bisaro, D.M. Adenosine kinase inhibition and suppression of RNA silencing by geminivirus AL2 and L2 proteins. *J. Virol.* **2005**, *79*, 7410–7418.
62. Zhang, Z.; Chen, H.; Huang, X.; Xia, R.; Zhao, Q.; Lai, J.; Teng, K.; Li, Y.; Liang, L.; Du, Q.; *et al.* BSCTV C2 attenuates the degradation of SAMDC1 to suppress DNA methylation-mediated gene silencing in Arabidopsis. *Plant Cell* **2011**, *23*, 273–288.
63. Briddon, R.W.; Stanley, J. Subviral agents associated with plant single-stranded DNA viruses. *Virology* **2006**, *344*, 198–210.
64. Cui, X.; Li, G.; Wang, D.; Hu, D.; Zhou, X. A Begomovirus DNAbeta-encoded protein binds DNA, functions as a suppressor of RNA silencing, and targets the cell nucleus. *J. Virol.* **2005**, *79*, 10764–10775.

65. Eini, O.; Dogra, S.C.; Dry, I.B.; Randles, J.W. Silencing suppressor activity of a begomovirus DNA β encoded protein and its effect on heterologous helper virus replication. *Virus Res.* **2012**, *167*, 97–101.
66. Rodríguez-Negrete, E.; Lozano-Durán, R.; Piedra-Aguilera, A.; Cruzado, L.; Bejarano, E.R.; Castillo, A.G. Geminivirus Rep protein interferes with the plant DNA methylation machinery and suppresses transcriptional gene silencing. *New Phytol.* **2013**, *199*, 464–475.
67. Nagar, S.; Hanley-Bowdoin, L.; Robertson, D. Host DNA replication is induced by geminivirus infection of differentiated plant cells. *Plant Cell* **2002**, *14*, 2995–3007.
68. Kittelmann, K.; Rau, P.; Gronenborn, B.; Jeske, H. Plant geminivirus rep protein induces rereplication in fission yeast. *J. Virol.* **2009**, *83*, 6769–6778.
69. Seemanpillai, M.; Dry, I.; Randles, J.; Rezaian, A. Transcriptional silencing of geminiviral promoter-driven transgenes following homologous virus infection. *Mol. Plant Microbe Interact.* **2003**, *16*, 429–438.
70. Ratcliff, F.G.; MacFarlane, S.A.; Baulcombe, D.C. Gene silencing without DNA. RNA-mediated cross-protection between viruses. *Plant Cell* **1999**, *11*, 1207–1216.
71. Akbergenov, R.; Si-Ammour, A.; Blevins, T.; Amin, I.; Kutter, C.; Vanderschuren, H.; Zhang, P.; Gruissem, W.; Meins, F., Jr.; Hohn, T.; *et al.* Molecular characterization of geminivirus-derived small RNAs in different plant species. *Nucleic Acids Res.* **2006**, *34*, 462–471.
72. Rodríguez-Negrete, E.A.; Carrillo-Tripp, J.; Rivera-Bustamante, R.F. RNA silencing against geminivirus: Complementary action of posttranscriptional gene silencing and transcriptional gene silencing in host recovery. *J. Virol.* **2009**, *83*, 1332–1340.
73. Chellappan, P.; Vanitharani, R.; Pita, J.; Fauquet, C.M. Short interfering RNA accumulation correlates with host recovery in DNA virus-infected hosts, and gene silencing targets specific viral sequences. *J. Virol.* **2004**, *78*, 7465–7477.
74. Pooggin, M.; Shivaprasad, P.V.; Veluthambi, K.; Hohn, T. RNAi targeting of DNA virus in plants. *Nat. Biotechnol.* **2003**, *21*, 131–132.
75. Vanderschuren, H.; Akbergenov, R.; Pooggin, M.M.; Hohn, T.; Gruissem, W.; Zhang, P. Transgenic cassava resistance to African cassava mosaic virus is enhanced by viral DNA-A bidirectional promoter-derived siRNAs. *Plant Mol. Biol.* **2007**, *64*, 549–557.
76. Donaire, L.; Wang, Y.; Gonzalez-Ibeas, D.; Mayer, K.F.; Aranda, M.A.; Llave, C. Deep-sequencing of plant viral small RNAs reveals effective and widespread targeting of viral genomes. *Virology* **2009**, *392*, 203–214.
77. Yang, X.; Wang, Y.; Guo, W.; Xie, Y.; Xie, Q.; Fan, L.; Zhou, X. Characterization of small interfering RNAs derived from the geminivirus/betasatellite complex using deep sequencing. *PLoS One* **2011**, *6*, e16928.
78. Aregger, M.; Borah, B.K.; Seguin, J.; Rajeswaran, R.; Gubaeva, E.G.; Zvereva, A.S.; Windels, D.; Vazquez, F.; Blevins, T.; Farinelli, L.; *et al.* Primary and secondary siRNAs in geminivirus-induced gene silencing. *PLoS Pathog.* **2012**, *8*, e1002941.
79. Xie, M.; He, Y.; Gan, S. Bidirectionalization of polar promoters in plants. *Nat. Biotechnol.* **2001**, *19*, 677–679.

80. Wei, W.; Ba, Z.; Gao, M.; Wu, Y.; Ma, Y.; Amiard, S.; White, C.I.; Rendtlew Danielsen, J.M.; Yang, Y.G.; Qi, Y. A role for small RNAs in DNA double-strand break repair. *Cell* **2012**, *149*, 101–112.
81. Hull, R. Caulimoviridae (Plant Pararetroviruses) In *eLS*; John Wiley & Sons Ltd: Chichester, UK, 2007.
82. Rothnie, H.M.; Chapdelaine, Y.; Hohn, T. Pararetroviruses and retroviruses: A comparative review of viral structure and gene expression strategies. *Adv. Virus Res.* **1994**, *44*, 1–67.
83. Chabannes, M.; Baurens, F.C.; Duroy, P.O.; Bocs, S.; Vernerey, M.S.; Rodier-Goud, M.; Barbe, V.; Gayral, P.; Iskra-Caruana, M.L. Three infectious viral species lying in wait in the banana genome. *J. Virol.* **2013**, doi:10.1128/JVI.00899-13.
84. Noreen, F.; Akbergenov, R.; Hohn, T.; Richert-Pöggeler, K.R. Distinct expression of endogenous Petunia vein clearing virus and the DNA transposon dTph1 in two *Petunia hybrida* lines is correlated with differences in histone modification and siRNA production. *Plant J.* **2007**, *50*, 219–229.
85. Pfeiffer, P.; Hohn, T. Involvement of reverse transcription in the replication of cauliflower mosaic virus: A detailed model and test of some aspects. *Cell* **1983**, *33*, 781–789.
86. Blevins, T.; Rajeswaran, R.; Aregger, M.; Borah, B.K.; Schepetilnikov, M.; Baerlocher, L.; Farinelli, L.; Meins, F., Jr.; Hohn, T.; Pooggin, M.M. Massive production of small RNAs from a non-coding region of Cauliflower mosaic virus in plant defense and viral counter-defense. *Nucleic Acids Res.* **2011**, *39*, 5003–5014.
87. Deleris, A.; Gallego-Bartolome, J.; Bao, J.; Kasschau, K.D.; Carrington, J.C.; Voinnet, O. Hierarchical action and inhibition of plant dicer-like proteins in antiviral defense. *Science* **2006**, *313*, 68–71.
88. Haas, G.; Azevedo, J.; Moissiard, G.; Geldreich, A.; Himber, C.; Bureau, M.; Fukuhara, T.; Keller, M.; Voinnet, O. Nuclear import of CaMV P6 is required for infection and suppression of the RNA silencing factor DRB4. *EMBO J.* **2008**, *27*, 2102–2112.
89. Shivaprasad, P.V.; Rajeswaran, R.; Blevins, T.; Schoelz, J.; Meins, F., Jr.; Hohn, T.; Pooggin, M.M. The CaMV transactivator/viroplasmin interferes with RDR6-dependent trans-acting and secondary siRNA pathways in Arabidopsis. *Nucleic Acids Res.* **2008**, *36*, 5896–5909.
90. Pooggin, M.M.; Futterer, J.; Skryabin, K.G.; Hohn, T. Ribosome shunt is essential for infectivity of cauliflower mosaic virus. *Proc. Natl. Acad. Sci. USA* **2001**, *98*, 886–891.
91. Pooggin, M.M.; Fütterer, J.; Hohn, T. Cross-species functionality of pararetroviral elements driving ribosome shunting. *PLoS One* **2008**, *3*, e1650.
92. Pooggin, M.M.; Rajeswaran, R.; Schepetilnikov, M.V.; Ryabova, L.A. Short ORF-dependent ribosome shunting operates in an RNA picorna-like virus and a DNA pararetrovirus that cause rice tungro disease. *PLoS Pathog.* **2012**, *8*, e1002568.
93. Pooggin, M.M.; Fütterer, J.; Skryabin, K.G.; Hohn, T. A short open reading frame terminating in front of a stable hairpin is the conserved feature in pregenomic RNA leaders of plant pararetroviruses. *J. Gen. Virol.* **1999**, *80*, 2217–2228.
94. Al-Kaff, N.S.; Covey, S.N.; Kreike, M.M.; Page, A.M.; Pinder, R.; Dale, P.J. Transcriptional and posttranscriptional plant gene silencing in response to a pathogen. *Science* **1998**, *279*, 2113–2115.

95. Al-Kaff. N.S.; Kreike, M.M.; Covey, S.N.; Pitcher, R.; Page, A.M.; Dale, P.J. Plants rendered herbicide-susceptible by cauliflower mosaic virus-elicited suppression of a 35S promoter-regulated transgene. *Nat. Biotechnol.* **2000**, *18*, 995–999.
96. Covey, S.N.; Al-Kaff, N.S.; Lángara, A.; Turner, D.S. Plants combat infection by gene silencing. *Nature* **1997**, *385*, 781–782.
97. Staginnus, C.; Gregor, W.; Mette, M.F.; Teo, C.H.; Borroto-Fernández, E.G.; Machado, M.L.; Matzke, M.; Schwarzacher, T. Endogenous pararetroviral sequences in tomato (*Solanum lycopersicum*) and related species. *BMC Plant. Biol.* **2007**, *7*, 24.
98. Love, A.J.; Laird, J.; Holt, J.; Hamilton, A.J.; Sadanandom, A.; Milner, J.J. Cauliflower mosaic virus protein P6 is a suppressor of RNA silencing. *J. Gen. Virol.* **2007**, *88*, 3439–3444.
99. Love, A.J.; Geri, C.; Laird, J.; Carr, C.; Yun, B.W.; Loake. G.J.; Tada, Y.; Sadanandom, A.; Milner, J.J. Cauliflower mosaic virus protein P6 inhibits signaling responses to salicylic acid and regulates innate immunity. *PLoS One* **2012**, *7*, e47535.
100. Ryabova, L.A.; Pooggin, M.M.; Hohn, T. Viral strategies of translation initiation: Ribosomal shunt and reinitiation. *Prog. Nucleic Acid Res. Mol. Biol.* **2002**, *72*, 1–39.
101. Kobayashi, K.; Hohn, T. The avirulence domain of Cauliflower mosaic virus transactivator/viroplasmin is a determinant of viral virulence in susceptible hosts. *Mol. Plant Microbe Interact.* **2004**, *17*, 475–483.
102. Schepetilnikov, M.; Kobayashi, K.; Geldreich, A.; Caranta, C.; Robaglia, C.; Keller, M.; Ryabova, L.A. Viral factor TAV recruits TOR/S6K1 signalling to activate reinitiation after long ORF translation. *EMBO J.* **2011**, *30*, 1343–1356.

3

Up-Regulation of microRNA* Strands by their Target Transcripts

Sung-Min Kang [1], Ji-Woong Choi [1], Su-Hyung Hong [1] and Heon-Jin Lee [1,2,*]

[1] Department of Oral Microbiology, School of Dentistry, Kyungpook National University, Daegu 700-412, Korea; E-Mails: dkdkdk43@naver.com (S.-M.K.); cjwboyboy@naver.com (J.-W.C.); hongsu@knu.ac.kr (S.-H.H.)

[2] Brain Science and Engineering Institute, Kyungpook National University, Daegu 700-412, Korea

* Author to whom correspondence should be addressed; E-Mail: heonlee@knu.ac.kr;

Abstract: During microRNA (miRNA) biogenesis, one strand of a 21–23 nucleotide RNA duplex is preferentially selected for entry into an RNA-induced silencing complex (RISC). The other strand, known as the miRNA* species, is typically thought to be degraded. Previous studies have provided miRNA* selection models, but it remains unclear how the dominance of one arm arises during the biogenesis of miRNA. Using miRNA sponge-like methods, we cloned four tandem target sequences (artificial target) of miR-7b* and then measured miR-7b* expression levels after transfection of the artificial target. miR-7b* levels were found to significantly increase after transfection of the artificial target. We postulate that the abundance of target transcripts drives miRNA arm selection.

Keywords: microRNA (miRNA); pre-miRNA; RISC; Drosha; Dicer; Argonaute (Ago)

---

1. Introduction

MicroRNAs (miRNAs) are noncoding 21–23 nucleotide (nt) strands and constitute an evolutionarily conserved class of pleiotropically acting small RNAs. miRNAs usually control posttranscriptional processes, such as, sequence-specific interactions with 3' untranslated regions (UTRs) of cognate mRNA targets in animals [1]. Nucleotides at positions 2–8 are considered to be important for pairing with target messenger RNAs and are referred to as "seed" sequences [2,3].

A miRNA gene is first transcribed into a primary miRNA (pri-miRNA), then processed into a ~70 nt hairpin precursor miRNA (pre-miRNA) by the RNase III enzyme Drosha and double-stranded RNA-binding domain protein, DGCR8 [4]. Pre-miRNA is then cleaved to generate the ~22 nt miRNA:miRNA* duplex by another RNase III enzyme, Dicer [5]. One strand of the duplex is loaded onto Argonaute (Ago) protein to produce RNA-induced silencing complex (RISC). Furthermore, it has been suggested that, whereas all Ago proteins participate in the stabilization of mature miRNA, only Ago2 (which has endonuclease activity) cleaves the miRNA* strand and activates RISC [6–9]. This suggestion led to the observation that Ago1 facilitated RISC-mediated translational repression and Ago2-RISC led to target mRNA cleavage [8,10]. However, Ago proteins have also been demonstrated to have dual roles, for example, Ago proteins increased the abundance of mature miRNAs, and decreased miRNA expression was observed in a cell line from an Ago2 knockout mouse [11].

The nomenclature originated because one arm, the miRNA, of the RNA duplex preferentially accumulates and the opposite arm, miRNA*, degrades. Another nomenclature often used is miR-3p/miR-5p, which refers to the direction of the mature miRNA strand. 3p and 5p miRNAs usually exhibit partial complementary overlap and have different target genes, despite being produced from the same transcript [12]. However, it remains unclear how dominance of one arm arises during the biogenesis of miRNA. Previous models suggest that the choice of the dominant miRNA arm is based on two mechanisms, that is, on the thermodynamic stability and structural properties of the processed duplex [13,14], or on energy-independent protein-mediated selection by Ago2, an endonuclease that cleaves complementary siRNA strands to facilitate RISC loading of the siRNA strand [6].

However, in a recent study, it was suggested the hairpin arm that makes dominant miRNA differs in different tissues, at different times of development, and between species [15]. In human gastric cancer, miRNA hairpin arm (3p or 5p) selection exhibits different tissue expression preferences in healthy and tumor tissues [16]. Furthermore, some miRNA precursors are processed to produce significant amounts of mature miRNAs from both arms and both miRNAs might regulate target transcripts [17]. These findings suggest the existence of another mechanism for controlling the selection of mature miRNAs.

We expressed an artificial target of miR-7b* that normally presents less than its mature miRNA, miR-7b in order to investigate the effect of the target mRNA on miRNA*. The expression of artificial targets can force the accumulation of miR-7b* rather than miR-7b, which suggests that target abundance might be a critical prerequisite of miRNA* strand stabilization.

2. Results and Discussion

The importance of the influences of target mRNA abundance and turnover rates on miRNA activity have been discussed [18,19]. Recently, Chatterjee *et al.* suggested the term "target-mediated miRNA protection (TMMP)", and showed that target mRNAs in *C. elegans* can protect their cognate miRNAs from degradation *in vivo* [20]. However, little is known about the decay of miRNA*, the other arm of the same hairpin precursor. We hypothesized that target mRNAs of high abundance may drive miRNA arm selection, and in a previous study using miRNA sponge-like methods [21], we cloned multiple target sequences of miR-7b and miR-7b* (miRNA artificial targets; Figure 1B). It has been reported that hyperosmolar stimulation induces miR-7b in the hypothalamus and that the neuronal marker Fos expression is inhibited by miR-7b [22]. Usually, miR-7b is dominant (approximately eight fold higher than miR-7b*) in AtT-20 mouse pituitary cells (Figure 1A). However, in the present study, both qRT-PCR and Northern blotting clearly showed dramatic elevated expression of endogenous miR-7b* by the miR-7b* artificial target. The seed-sequence mutated artificial targets (miR-7b* mutation I and II) reduced the miR-7b* up-regulating effect (Figure 2A,B), suggesting that the up-regulation of miR-7b* occurred in a sequence specific manner (see Figure 1B for artificial and seed mutated targets). Furthermore, luciferase assays showed that miR-7b* mimic oligonucleotide strongly suppressed luciferase activity by binding to its artificial target with a perfect complementary match: this effect was reduced by the seed mutated artificial targets (Figure 1C). We observed the same effects for miR-338-3p (dominant or guide strand) and for miR-338-5p (miR-338* or passenger strand) after artificial targets transfection (Figure 3). Interestingly, when the artificial targets of the dominant strands (miR-7b and miR-338-3p) were transfected, the elevation effect was not as great as that of the artificial target of the non-dominant strands (miR-7b* and miR-338-5p) (Figures 2A,B and 3C). This suggests there might be a certain threshold for this differential regulation or some unknown mechanism that overrides miRNA arm selection.

Figure 1. The expression patterns of miR-7b and miR-7b*. (**A**) Basal levels of miR-7b and miR-7b* expression in AtT-20 cells were analyzed by qRT-PCR; (**B**) Design of miRNA artificial targets. *Renilla* (hR*luc*) containing miRNA artificial targets were constructed by inserting multiple miRNA binding sites into the 3' UTR of a hR*luc* reporter gene driven by a T7 promoter. The figure shows the nucleotide sequence of the miRNA artificial targets: the red letters are a mutated sequence; (**C**) Luciferase assay using the miRNA artificial target reporter constructs. COS-7 cells were co-transfected with the miR-7b* mimic oligonucleiotide and miRNA artificial target (1. miR-7b, 2. miR-7b*, 3. miR-7b* mutation I, 4. miR-7b* mutation II, 5. miR-338-3p). *Renilla* luciferase activity was normalized against *firefly* luciferase activity and fold changes are compared to miR-338-3p. The results shown are the means of three independent transfections (error bars indicate standard deviations).

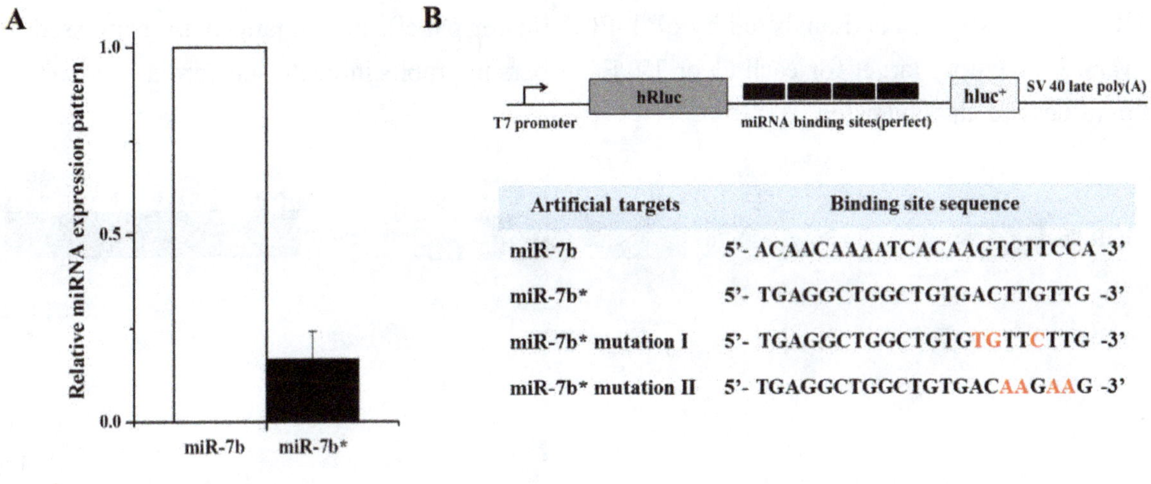

Figure 1. *Cont.*

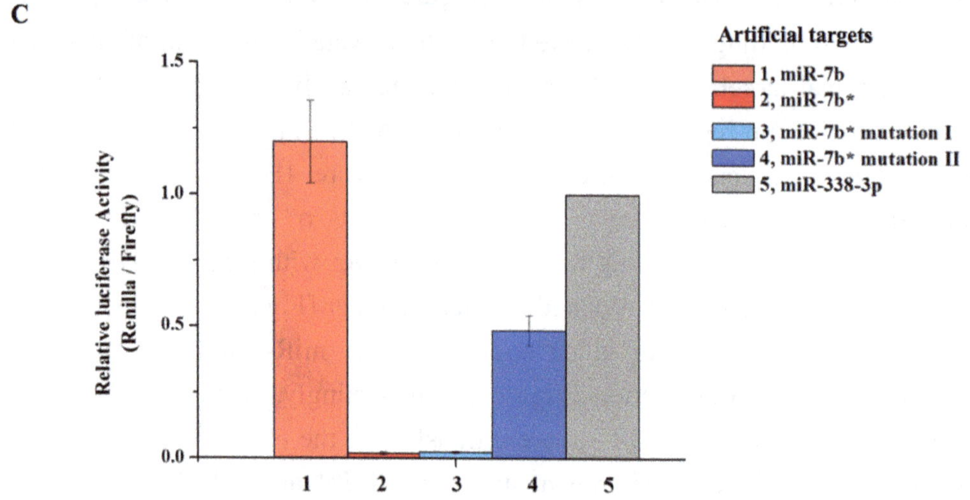

Figure 2. Changed expression levels of miR-7b and miR-7b* after transfection with multi-transcript artificial targets. (**A**) 24 h after transfection of miRNA artificial targets (1 and 1'. null, 2 and 2'. empty vector, 3 and 3'. miR-7b*, 4 and 4'. miR-7b* mutation I, 5 and 5'. miR-7b* mutation II), AtT-20 total RNA was extracted and analyzed by northern blotting using probes specific for miR-7b and miR-7b*, respectively. Lane 3' shows that miR-7b* expression was significantly increased following transfection of the miR-7b* artificial target. In contrast, miR-7b expression was similar in non-transfected and miR-7b* artificial targets transfected samples; (**B**) Twenty-four hours after various artificial targets (1 and 1'. null, 2 and 2'. empty vector, 3 and 3'. miR-7b, 4 and 4'. miR-7b*, 5 and 5'. miR-7b* mutation I, 6 and 6'. miR-7b* mutation II) were transfected, miR-7b (left) and miR-7b* (right) expression levels were analyzed by qRT-PCR. miRNA expression levels were normalize to U6. Fold changes are expressed *versus* empty target of each (2 or 2'); (**C**) Twenty-four hours after various artificial targets (1 and 1'. empty vector, 2 and 2'. miR-7b*, 3 and 3'. miR-7b* mutation I, 4 and 4'. miR-7b* mutation II) were transfected, AtT-20 lysates were immunoprecipitated (IP) using anti-Ago2 antibody. IPs were analyzed by immunoblotting with the same anti-Ago2 antibody (upper panel). Co-immunoprecipitated RNA was extracted and analyzed by qRT-PCR (lower panel). Fold changes are expressed *versus* the empty target for each (1 or 1'). Error bars in graphs indicate standard deviations; *p*-values are also indicated.

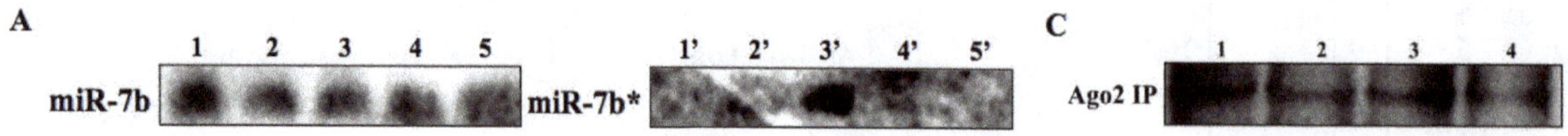

Figure 2. *Cont.*

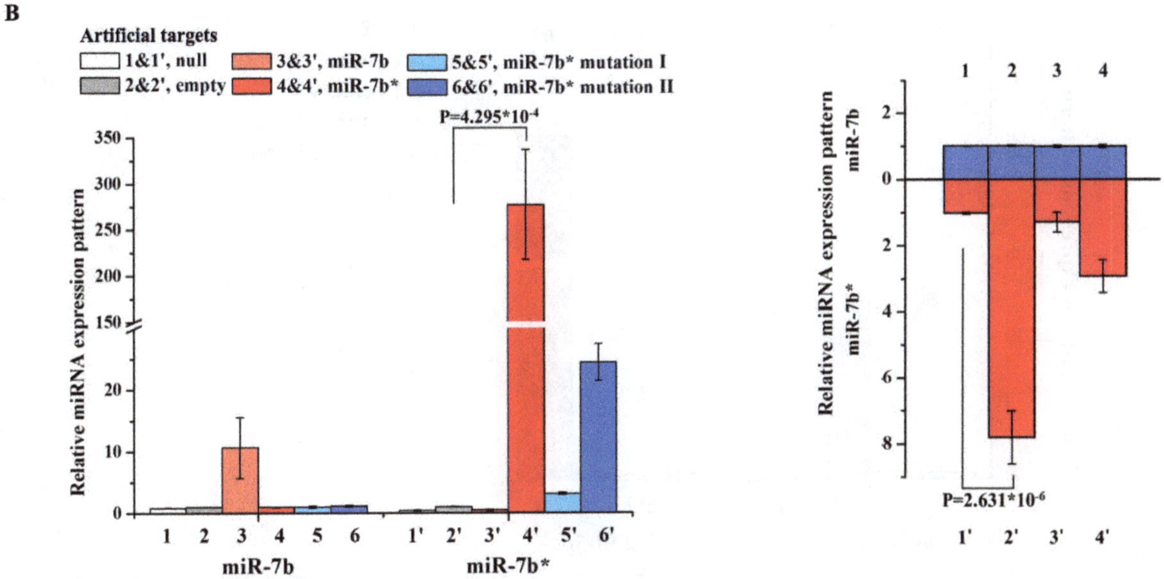

Figure 3. miR-338-3p and miR-338-5p expressions. (**A**) miR-338-3p and miR-338-5p expression were analyzed by qRT-PCR of the total RNA extracted from AtT-20 cells. miR-338-3p levels were about twice as high as miR-338-5p levels in AtT-20 cells; (**B**) Schematic representation of miRNA artificial targets. Construct map of the miR-338-3p and miR-338-5p artificial targets with miR-338-3p and miR-338-5p artificial target sequences below; (**C**) Both miR-338-3p and miR-338-5p expression was up-regulated by each of artificial targets. However, the change of miR-338-3p was not as great as miR-338-5p (miR-338*) by their artificial target. The expression levels of miR-338-5p were significantly elevated by miR-338-5p artificial target transfection but not by transfection of the miR-338-3p artificial target or empty vector (1 and 1'. null, 2 and 2'. empty vector, 3 and 3'. miR-338-3p, 4 and 4'. miR-338-5p). Fold changes are expressed *versus* the empty target for each (2 or 2'). Error bars in graphs indicate standard deviations; *p*-values are also indicated.

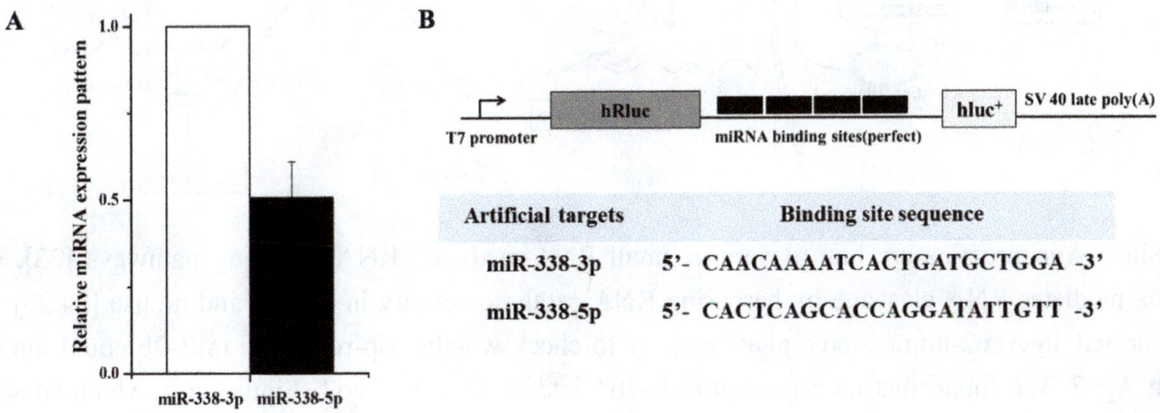

Figure 3. *Cont.*

C

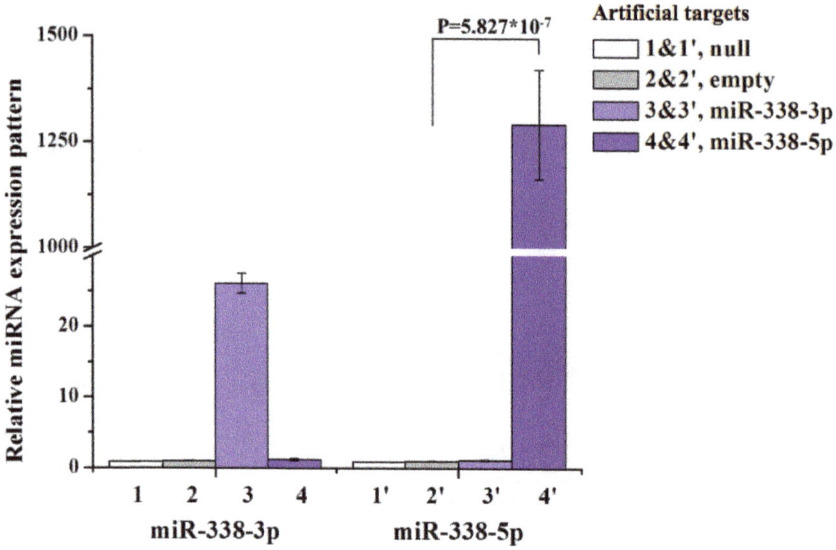

Figure 4. Schematic diagram of miRNA arm selection. Pri-miRNA is transcribed by RNA polymerase II and processed into pre-miRNA by Drosha/DGCR8 in the nucleus. Then pre-miRNA is processed into a duplex form of mRNA that is unwound during RISC assembly in the cytoplasm. Selection of the miRNA* sequence is determined by target transcript abundances. When miRNA* target transcripts are sufficiently abundant the miRNA* sequence is stabilized and acts on target transcripts in the same manner as miRNA.

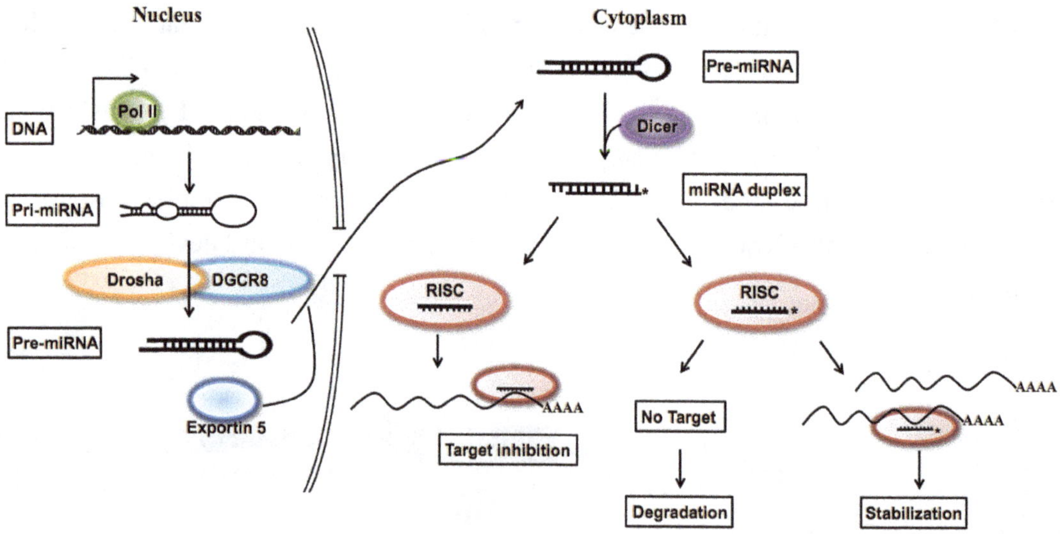

Since Ago proteins are key players in small RNA-mediated RNA silencing pathways [23], and Ago2 mediates RNA cleavage by harboring RNA catalytic activity in human and mouse [24,25], we performed an Ago2-immunoprecipitation assay to check whether up-regulated miR-7b could interact with Ago2. We found that up-regulated miR-7b* interacted with Ago2 (Figure 2C), which suggests that miR-7b* may have the same inhibitory function as mature miRNAs, similar to the regulatory activity of miRNA* function without degradation [17]. We tried to examine the effect of TMMP on miR-7b* by using endogenous miR-7b* target transcripts, but failed to identify targets of miR-7b experimentally due to the difficulty of predicting miRNA targets. Based on our results, we postulate mature miRNA arm selection is influenced by the abundance of miRNA target transcripts, and that selection may occur as duplex miRNAs, incorporated into the RISC complex (Figure 4), unwind from RISC assembly [26].

3. Experimental Section

3.1. Cell Culture

COS-7 cells (a monkey kidney fibroblast cell line) were obtained from the Korea Cell Line Bank (Seoul). COS-7 cells were cultured in Dulbecco's Modified Eagle's Medium (DMEM/High Glucose) (Hyclone, Logan, UT, USA) containing 10% fetal bovine serum (FBS) (Hyclone, Logan, UT, USA) and 1× antibiotic-antimycotic (GIBCO, Grand Island, NY, USA), and then incubated at 37 °C in a 5% CO_2 atmosphere. AtT-20 cells (a mouse pituitary tumor cell line) were obtained from the American Type Culture Collection (ATCC, CCL-89), and were grown in Kaighn's Modification of Ham's F-12 (F-12K) medium (Hyclone) supplemented with 15% horse serum, 2.5% FBS and 1× antibiotic-antimycotic. These cells were used in the non-luciferase assays.

3.2. Construction of Artificial miRNA Target

We annealed, ligated, and cloned oligonucleotides for miRNA binding sites with four tandem target sequences (Figure 1B) into psiCEHCK2 vector (Promega, Madison, WI, USA) downstream of the *Renilla* luciferase (hR*luc*) coding sequence.

3.3. Luciferase Assays

COS-7 cells were plated the day before transfection and transfected in triplicate with Lipofectamine 2000 (Invitrogen, Carlsbad, CA, USA) and 800 ng of various artificial target plasmids and 25 nM of miR-7b* mimic oligonucleotide (Bioneer, Daejeon, Korea). All assays were performed 24 h after transfection using the dual luciferase assay (Promega, Madison, WI, USA), according to the manufacturer's protocol. All experiments were performed in triplicate.

3.4. Isolation of miRNA and Quantitative Real-Time PCR (qRT-PCR) Analysis

Total RNAs were extracted from the artificial target transfected AtT-20 cells using QIAzol (Qiagen, Valencia, CA, USA) using a modification of the manufacturer's instructions, and then treated with DNaseI (Ambion, Foster City, CA, USA). qRT-PCR of miRNAs was conducted on an ABI 7500 real-time PCR system using TaqMan Universal PCR Master Mix, miRNA Expression Assay primer, and probe sets (Applied Biosystems, Foster City, CA, USA). U6 RNA (a small nuclear RNA) was used as an internal cDNA loading control. Threshold cycle times (*Ct*) were obtained and relative gene expressions were calculated using the comparative cycle time method.

3.5. Immunoprecipitation

For the immunoprecipitation of endogenous Ago2, AtT-20 cells were grown on 10 cm dishes and harvested at 24 h after miRNA transfections. Cells were then incubated with lysis buffer for 20 min on ice, homogenized, and centrifuged at 12,000 rpm for 20 min at 4 °C. Supernatants were incubated with anti-Ago2 antibody (Sigma Aldrich, St. Louis, MO, USA) with constant rotation for one day at 4 °C. Then, 20 μL of protein G Sepharose® beads (Sigma Aldrich, St. Louis, MO, USA) were added and incubated with rotation for 4 h at 4 °C. Beads were then washed three times with lysis buffer.

3.6. Isolation of Ago2-Associated miRNA

To measure amounts of RISC-associated miRNAs, cell lysates were prepared from AtT-20 cells after the transfection of miRNA artificial targets. Ago2-miRNA complex was immunoprecipitated from lysates using anti-Ago2 antibody and total RNA was isolated from immunoprecipitates using QIAzol reagent (Qiagen, Valencia, CA, USA). The miRNAs levels were measured by quantitative RT-PCR and normalized against U6 levels in cell lysates.

3.7. Northern Blot Analysis for miRNA

Ten μg aliquots of AtT-20 total RNA isolated from AtT-20 cells using QIAzol (Qiagen, Valencia, CA, USA), according to the manufacturer's instructions, was separated on 15% TBE-urea gels (Invitrogen, Carlsbad, CA, USA) and electro-transferred to Nylon+ membranes (Invitrogen, Carlsbad, CA, USA). Hybridizations were carried out in North2South® hybridization buffer (Invitrogen) at 37 °C using miR-7b* probe (Bioneer, Daejeon, Korea), or at 42 °C using miR-7b LNA (Locked Nucleic Acid) probe (Exiqon, Vedbaek, Denmark).

3.8. Statistical Analysis

All data are presented as means ± standard deviations (SD). Significant variation analysis was conducted to calculate the parametric two-tailed non-paired *t*-test. All analyses were performed using Origin 8.0 (OriginLab, Northampton, MA, USA), and *p*-values of ≤ 0.05 were considered statistically significant.

4. Conclusions

Despite extensive study of miRNA, it remains largely unclear how one miRNA arm becomes less dominant (often referred as miRNA*) during the miRNA maturation process. In this study, we introduced an artificial target of miR-7b* in order to check miR-7b* stability. Transfection of the miR-7b* artificial target led to a dramatic up-regulation of miR-7b*, but did not have much effect on miR-7b, the dominant sequence of miR-7b hairpin precursor. A similar phenomenon was observed in miR-338-3p and miR-338-5p (miR-338*). Therefore, we postulate that selection of the miRNA arm might be decided by the mechanism "target-mediated miRNA protection (TMMP)" and TMMP is probably more selective to miRNA* strands.

Acknowledgments

The authors sincerely thank Scott Young at the NIH (Bethesda, MD, USA) for his critical reading of the manuscript. This research was supported by the Basic Science Research Program of the Korean National Research Foundation (NRF) funded by the Korean Ministry of Education, Science and Technology (Grant nos. 2011-0028240 and 2011-0014086).

References

1. Bartel, D.P. MicroRNAs: Genomics, biogenesis, mechanism, and function. *Cell* **2004**, *116*, 281–297.
2. Kim, Y.K.; Heo, I.; Kim, V.N. Modifications of small RNAs and their associated proteins. *Cell* **2010**, *143*, 703–709.
3. Lewis, B.P.; Burge, C.B.; Bartel, D.P. Conserved seed pairing, often flanked by adenosines, indicates that thousands of human genes are microRNA targets. *Cell* **2005**, *120*, 15–20.
4. Bushati, N.; Cohen, S.M. microRNA Functions. *Annu. Rev. Cell Dev. Biol.* **2007**, *23*, 175–205.
5. Lee, Y.; Jeon, K.; Lee, J.T.; Kim, S.; Kim, V.N. MicroRNA maturation: Stepwise processing and subcellular localization. *EMBO J.* **2002**, *21*, 4663–4670.
6. Rand, T.A.; Petersen, S.; Du, F.; Wang, X. Argonaute2 cleaves the anti-guide strand of siRNA during RISC activation. *Cell* **2005**, *123*, 621–629.
7. Matranga, C.; Tomari, Y.; Shin, C.; Bartel, D.P.; Zamore, P.D. Passenger-strand cleavage facilitates assembly of siRNA into Ago2-containing RNAi enzyme complexes. *Cell* **2005**, *123*, 607–620.
8. Wang, H.W.; Noland, C.; Siridechadilok, B.; Taylor, D.W.; Ma, E.; Felderer, K.; Doudna, J.A.; Nogales, E. Structural insights into RNA processing by the human RISC-loading complex. *Nat. Struct. Mol. Biol.* **2009**, *16*, 1148–1153.
9. Leuschner, P.J.; Ameres, S.L.; Kueng, S.; Martinez, J. Cleavage of the siRNA passenger strand during RISC assembly in human cells. *EMBO Rep.* **2006**, *7*, 314–320.
10. Iwasaki, S.; Kawamata, T.; Tomari, Y. Drosophila argonaute1 and argonaute2 employ distinct mechanisms for translational repression. *Mol. Cell.* **2009**, *34*, 58–67.
11. Diederichs, S.; Haber, D.A. Dual role for argonautes in microRNA processing and posttranscriptional regulation of microRNA expression. *Cell* **2007**, *131*, 1097–1108.
12. Bhayani, M.K.; Calin, G.A.; Lai, S.Y. Functional relevance of miRNA sequences in human disease. *Mutat. Res.* **2012**, *731*, 14–19.
13. Khvorova, A.; Reynolds, A.; Jayasena, S.D. Functional siRNAs and miRNAs exhibit strand bias. *Cell* **2003**, *115*, 209–216.
14. Schwarz, D.S.; Hutvagner, G.; Du, T.; Xu, Z.; Aronin, N.; Zamore, P.D. Asymmetry in the assembly of the RNAi enzyme complex. *Cell* **2003**, *115*, 199–208.
15. Griffiths-Jones, S.; Hui, J.H.; Marco, A.; Ronshaugen, M. MicroRNA evolution by arm switching. *EMBO Rep* **2011**, *12*, 172–177.
16. Li, S.C.; Liao, Y.L.; Ho, M.R.; Tsai, K.W.; Lai, C.H.; Lin, W.C. miRNA arm selection and isomiR distribution in gastric cancer. *BMC Genomics Electron. Resour.* **2012**, *13*, S13.

17. Okamura, K.; Phillips, M.D.; Tyler, D.M.; Duan, H.; Chou, Y.T.; Lai, E.C. The regulatory activity of microRNA* species has substantial influence on microRNA and 3' UTR evolution. *Nat. Struct. Mol. Biol.* **2008**, *15*, 354–363.
18. Arvey, A.; Larsson, E.; Sander, C.; Leslie, C.S.; Marks, D.S. Target mRNA abundance dilutes microRNA and siRNA activity. *Mol. Syst. Biol.* **2010**, *6*, 363.
19. Larsson, E.; Sander, C.; Marks, D. mRNA turnover rate limits siRNA and microRNA efficacy. *Mol. Syst. Biol.* **2010**, *6*, 433.
20. Chatterjee, S.; Fasler, M.; Bussing, I.; Grosshans, H. Target-mediated protection of endogenous microRNAs in *C. elegans*. *Dev. Cell* **2011**, *20*, 388–396.
21. Ebert, M.S.; Neilson, J.R.; Sharp, P.A. MicroRNA sponges: Competitive inhibitors of small RNAs in mammalian cells. *Nat. Methods* **2007**, *4*, 721–726.
22. Lee, H.J.; Palkovits, M.; Young, W.S., 3rd. miR-7b, a microRNA up-regulated in the hypothalamus after chronic hyperosmolar stimulation, inhibits Fos translation. *Proc. Natl. Acad. Sci. USA* **2006**, *103*, 15669–15674.
23. Hutvagner, G.; Simard, M.J. Argonaute proteins: Key players in RNA silencing. *Nat. Rev. Mol. Cell Biol.* **2008**, *9*, 22–32.
24. Meister, G.; Landthaler, M.; Patkaniowska, A.; Dorsett, Y.; Teng, G.; Tuschl, T. Human Argonaute2 mediates RNA cleavage targeted by miRNAs and siRNAs. *Mol. Cell* **2004**, *15*, 185–197.
25. Liu, J.; Carmell, M.A.; Rivas, F.V.; Marsden, C.G.; Thomson, J.M.; Song, J.J.; Hammond, S.M.; Joshua-Tor, L.; Hannon, G.J. Argonaute2 is the catalytic engine of mammalian RNAi. *Science* **2004**, *305*, 1437–1441.
26. Kwak, P.B.; Tomari, Y. The N domain of Argonaute drives duplex unwinding during RISC assembly. *Nat. Struct. Mol. Biol.* **2012**, *19*, 145–151.

4

Non-Coding RNAs: The "Dark Matter" of Cardiovascular Pathophysiology

Claudio Iaconetti, Clarice Gareri, Alberto Polimeni and Ciro Indolfi *

Division of Cardiology, Magna Graecia University, URT Consiglio Nazionale delle Ricerche (CNR), Catanzaro 88100, Italy; E-Mails: iaconetticlaudio@unicz.it (C.I.); c.gareri@unicz.it (C.G.); polimeni@unicz.it (A.P.)

* Author to whom correspondence should be addressed; E-Mail: indolfi@unicz.it;

Abstract: Large-scale analyses of mammalian transcriptomes have identified a significant number of different RNA molecules that are not translated into protein. In fact, the use of new sequencing technologies has identified that most of the genome is transcribed, producing a heterogeneous population of RNAs which do not encode for proteins (ncRNAs). Emerging data suggest that these transcripts influence the development of cardiovascular disease. The best characterized non-coding RNA family is represented by short highly conserved RNA molecules, termed microRNAs (miRNAs), which mediate a process of mRNA silencing through transcript degradation or translational repression. These microRNAs (miRNAs) are expressed in cardiovascular tissues and play key roles in many cardiovascular pathologies, such as coronary artery disease (CAD) and heart failure (HF). Potential links between other ncRNAs, like long non-coding RNA, and cardiovascular disease are intriguing but the functions of these transcripts are largely unknown. Thus, the functional characterization of ncRNAs is essential to improve the overall understanding of cellular processes involved in cardiovascular diseases in order to define new therapeutic strategies. This review outlines the current knowledge of the different ncRNA classes and summarizes their role in cardiovascular development and disease.

Keywords: non-coding RNA; microRNA; long non-coding RNA; vascular development; vascular disease; heart pathophysiology

---

1. Introduction

Many studies have recently focused on understanding RNA metabolism and its implication in development and disease processes. Genomic tiling arrays and RNA-Sequencing have showed that the human genome is dynamically transcribed and leads to the production of a complex world of RNA molecules of which only a small fraction is translated into proteins [1]. In fact, application of high-throughput sequencing technologies in the analysis of mammalian transcriptomes, revealed a wide spectrum of RNA molecules that do not encode protein, termed non-coding RNAs (ncRNAs) [2]. For many years the role of these molecules remained unknown, so ncRNAs were called the "Dark Matter" of biology. To date many studies have been carried out on these molecules, especially on microRNAs, partially clarifying their roles. However many mechanisms and functions of different classes of ncRNA still remain unknown. Emerging evidence indicates that the non-coding portion of the genome is critical in the regulation of multiple biological processes, such as differentiation, development, post-transcriptional regulation of gene expression and epigenetic regulation [3–5]. Recently, many classes of ncRNA have been described to be associated with human disease [6]. Cardiovascular disease is a major cause of mortality and hospitalization worldwide [7], and the work of multiple research groups has been devoted to determine the molecular mechanism underlying heart and vascular disease. Recent studies indicate that altered ncRNA expression and function have been strongly implicated in cardiovascular disease such as myocardial infarction, cardiac hypertrophy and coronary artery disease [8–10]. The transcriptome of a cell contains different types of ncRNA that can be divided into two principal classes (Table 1): structural and regulatory ncRNAs. Structural ncRNAs include RNA molecules that are usually constitutively expressed such as ribosomal and transfer RNAs. Regulatory ncRNAs can be classified into three major classes based on transcript size: small (small ncRNAs), medium and long non-coding RNAs (lncRNAs) [6]. The most studied class of small ncRNAs in cardiovascular research is the microRNAs (miRNAs). MiRNAs are endogenous, single-stranded molecules consisting of approximately 20–22 nucleotides that regulate their target genes by reducing mRNA stability and/or translation [11]. Changes in microRNA expression lead to changes in gene function. This dysregulation of miRNA expression appears to play a significant role in the onset and progression of cardiovascular diseases [12]. Despite the progress in defining the role of microRNAs in cardiac and vascular biology, the complex network of ncRNAs and their interaction with different states of cardiovascular development and disease is still unknown. This is related to the multiple diversity of biogenesis, expression and functional properties of different classes of ncRNAs. Among these, the long non-coding RNAs (lncRNAs) are apparently the most numerous and functionally different [13]. LncRNAs are broadly classified as transcripts longer than 200 nucleotides and some of them are preferentially expressed in specific tissues [14]. Thus it is becoming increasingly clear that lncRNAs can regulate numerous molecular mechanisms. Recently, lncRNAs have emerged as new players in cardiovascular development and disease demonstrating potential roles in different cellular processes [15,16]. However, the characteristics and functions of the overwhelming majority of these lncRNAs are currently unknown. Accordingly, the functional characterization of lncRNAs is essential to advance our comprehensive understanding of cellular processes underlying cardiovascular development and disease. In the present review, emerging roles of ncRNAs in cardiovascular pathophysiology are discussed. Particular focus will be on the evaluation of biological roles of microRNAs and lncRNAs in vascular as well as cardiac disorders. Moreover, the focus of this review is to provide an overview of the current state of knowledge of molecular processes implicated in differentiation and cardiovascular development, which are related to the function of ncRNAs.

Table 1. Classes of non-coding RNAs (ncRNAs).

Non-coding RNAs	Symbol	Functions
	Structural ncRNAs	
Transfer RNA	tRNA	mRNA translation
Ribosomal RNA	rRNA	mRNA translation
	Regulatory ncRNA Short ncRNA	
Micro RNAs	miRNA	post-transcriptional regulators
PIWI-interacting RNA	piRNA	DNA methylation, transposon repression
Short interfering RNA	siRNA	RNA interference
	Medium ncRNA	
Small nucleolar RNAs	snoRNA	RNA modification, rRNA processing
Promoter upstream transcripts	PROMPTs	Associated with chromatin changes
Transcription initiation RNAs	tiRNAs	Epigenetic regulation
	Long ncRNAs	
Long intergenic ncRNA	lincRNAs	Epigenetic regulators of transcription
Enhancer-like ncRNA	eRNA	Transcriptional gene activation
Transcribed ultraconserved regions	T-UCRs	Regulation of miRNA and mRNA levels
Natural antisense transcripts	NATs	mRNA stability
Promoter-associated long RNAs	PALRs	chromatin changes
Pseudogenes	None	microRNA decoys

2. An Overview on the Main Methods to Analyze the ncRNAs Expression

Each ncRNA has expression levels that are tissue- or stage-specific. In recent years several methods have been developed to study ncRNA expression. A common approach is Real-time PCR, which is employed mainly to analyze microRNA expression levels but can be used also for studies on long ncRNA [17]. Also many approaches based on immunoprecipitation assays have been developed in recent years (e.g., RNA immunoprecipitation or RIP, Cross-linking and immunoprecipitation or CLIP, RNA-chromatin immunoprecipitation or RNA-ChIP) [18–20]. RNA-IP was developed to identify ncRNAs, especially ncRNA, that interact with a specific protein. The basic principal behind all immunoprecipitation approaches is the same. Using a specific antibody it is possible to isolate a ncRNA-protein complex, then a cDNA library is constructed and the ncRNA is sequenced. Unfortunately, for any of these immunoprecipitation-based approaches the results are influenced by the specificity and affinity of the antibodies. Moreover, these methods (Real-time PCR or IP) allow evaluation of the expression of a few specific molecules but do not permit the discovery of new ncRNAs or provide an overview of all ncRNAs. Recently, advances in technology enabled the development of new genome-wide screening methods to study ncRNAs and their targets. Among these the most commonly used are microarray analysis and RNA sequencing. These technologies are very accurate and permit large-scale analysis of ncRNAs. In particular, the microarray [21–23] approach offers various platforms allowing the study of microRNAs and mRNAs targets, although to date there are only a few chips to analyze long non-codingRNA. Analysis with traditional microarrays is limited to detecting the presence or the absence of known ncRNAs and it is incapable of identifying new molecules or revealing different splicing variants. To get around this problem a new approach has been defined: tiling arrays. Unlike traditional microarrays, these platforms permit identification of new ncRNAs in a selected DNA region without prior knowledge of their precise location. For instance, Rinn *et al.* used this approach to study lncRNAs expressed in the region of HOX genes in humans [24]. The RNA sequencing or RNAseq [25] refers to the use of high-throughput sequencing technologies to get information about a sample's RNA content. This approach permits information to be obtained on differential expression of the interest gene, microRNA or long ncRNA. RNAseq is very sensitive in detecting less-abundant transcript and it can reveal alternatively spliced isoforms. Moreover, sequencing the entire transcriptome has been widely used to discover new non coding molecules. However, given the time and the cost related to the downstream analysis of the data generated by RNA sequencing, microarrays remain the first choice in many applications.

3. Functions of Non Coding RNAs

Although the function of most lncRNAs remains unknown, it has become clear that these molecules are intimately involved in many biological processes. LncRNAs can regulate gene expression programs through a variety of mechanisms, such as epigenetic modifications of DNA, alternative splicing, post-transcriptional gene regulation and mRNA stability and translation [5,26,27]. Given their established roles in transcriptional regulation, lncRNAs play a key role in several cellular events including proliferation, migration, apoptosis and development [21,28]. LncRNAs are now known to regulate the expression of protein-coding genes: they can positively or negatively control the expression of their target genes. Several lncRNAs are involved with *in cis* inactivation of larger genomic regions by epigenetic mechanisms. Kcnq1ot1 is a regulatory non coding antisense RNA that regulates epigenetic gene silencing in an imprinted gene cluster *in cis* [29]. This lncRNA specifically interacts with nearby genes in embryonic tissues causing transcriptional gene silencing. More recently, it was found that lncRNAs can act *in cis* to regulate expression of neighboring genes during cardiomyocyte differentiation [30]. Notably, many lncRNAs are now known to regulate the expression of genes by a *trans* mechanism. One example of a lncRNA that acts *in trans* is AK143260, termed Braveheart (Bvht) that specifically promotes activation of a core gene regulatory network to direct cardiovascular lineage commitment [15]. So far, several other functions have been attributed to lncRNAs. These molecules can act as scaffolds bringing together multiple proteins to form ribonucleoprotein complexes. For example, Miao-Chih Tsai *et al.* showed that a long non coding transcript, termed HOTAIR (HOX Antisense Intergenic RNA), acts as a scaffold for Polycomb Repressive Complex 2 (PRC2) and LSD1/CoREST/REST complex [31]. In addition to their role in chromatin regulation, lncRNAs can also function as molecular "decoys" of transcription factors and other regulatory proteins. PANDA (P21 associated ncRNA DNA damage activated) is an example of a lncRNA with decoy functionality. In fact, PANDA interacts with the transcription factor NF-YA to limit expression of pro-apoptotic genes [32]. Finally, the presence of a complex network of interactions between lncRNAs and miRNAs is becoming increasingly clear. In fact, lncRNAs may exert their biological activity through their ability to act as endogenous decoys for miRNAs. For example, a muscle-specific long noncoding RNA, linc-MD1, could interact with two specific miRNAs, miR-133 and miR-135, and promote muscle differentiation by acting as a competing endogenous RNA (ceRNA) in mouse and human myoblasts [33]. Another lncRNA which has been identified in association with microRNAs is the pseudogene PTENP1 [34]. Similar to Linc-MD1, PTENP1 mRNA acts as a decoy for miRNAs that directly target the tumor suppressor protein PTEN. Accordingly, PTENP1 reduces down-regulation of PTEN messenger RNA. Recent reports also show that stable circular lncRNAs (circRNAs) can act as molecular decoys of microRNAs [35,36]. Taken together, these observations suggest that lncRNAs could have profound effects on several molecular mechanisms. Nevertheless, lncRNAs are poorly conserved among species resulting in an additional degree of complexity in the definition of their functions. Despite rapid progress in lncRNA discovery, evidence of physiologic function for lncRNAs remains poor and further investigation is necessary.

Figure 1. Role of non-coding RNAs in Vascular Development and Disease.

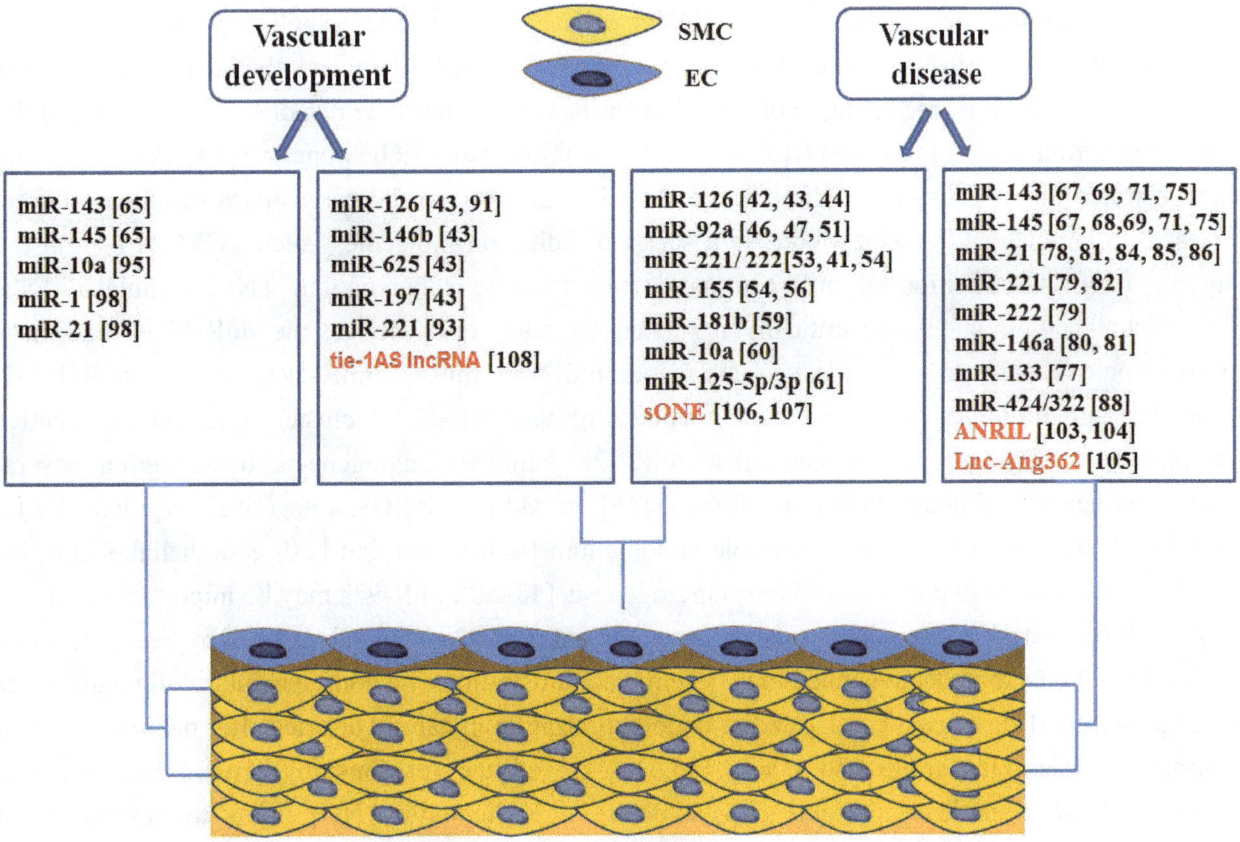

4. Roles of ncRNAs in Vascular Biology and Disease

The vessel wall is composed of endothelial cells (ECs) and smooth muscle cells (SMCs) that play central roles in vascular biology and disease. In fact, these cells can undergo profound changes in phenotype during vascular injury and remodeling; these changes are correlated with pathologies such as atherosclerosis and proliferative thickening of the vessel known as restenosis. Atherosclerosis is a chronic inflammatory disease of the arterial wall and is the major cause of death in western countries [37]. It is a complex process involving multiple cell types and the interactions of many different molecular pathways. The events that lead to the formation of atherosclerotic lesions include modification of endothelial cell function, monocyte adherence and entry into vessel wall, phenotypic modulation of smooth muscle cell, and platelet adhesion and aggregation [38]. Phenotypic modulation of smooth muscle cells is, also, crucial in the neointimal lesion formation after stent implantation [39]. Numerous ncRNAs, especially microRNAs, have been shown to govern these processes during vascular disease. In fact, miRNA control endothelial cell and vascular smooth muscle cell biology, and thereby regulate the progression of vascular disease, such as atherosclerosis and restenosis. Current evidence also suggests that other ncRNA classes, such as lnc-RNA molecules play a critical role in endothelial and smooth muscle cell function. Figure 1 summarizes the role of ncRNA classes in different cells of the vessel wall.

4.1. microRNAs in Endothelial Biology and Dysfunction

In endothelial cells (ECs) the action of specific miRNAs is important for vascular signaling and function. Different studies indicate that the major miRNA-regulating enzymes, Dicer and Drosha, are essential for angiogenic functions of endothelial cells [40,41]. The endothelial-specific miR-126 is the most abundant miRNA found in adult ECs and it is involved in endothelial dysfunction and inflammation [42]. It is interesting to observe how miR-126 regulates the response of ECs to VEGF by inhibiting sprout-related protein SPRED1, a negative inhibitor of VEGF signaling [43]. Another group demonstrated that VCAM-1 is a direct target of miR-126 [44]. In the early phase of atherosclerotic disease, inflammatory cytokines increase a series of adhesion molecules, such as VCAM-1, on the surface of ECs. Inhibition of miR-126 increases leukocyte adherence in TNFα-stimulated ECs. Endothelial cell functions are critically regulated by other microRNAs: the miR-17-92 cluster, a polycistronic miRNA gene that produces six mature miRNAs: miR-17, miR-18a, miR-19a, miR-19b-1, miR-20a, and miR-92a [45]. Individual members of the miR-17-92 cluster, function as negative regulators of angiogenesis. In particular, miR-92a inhibited angiogenesis by targeting several functional genes, including integrin α5 (ITGa5) [46]. In addition, miR-92a negatively regulates KLF2 and KLF4 expression in athero-susceptible endothelium [47]. Given that both endothelial KLF4 and KLF2 are implicated in protection against atherogenesis [48–50], miR-92a may be important in arterial disease. Moreover, we have recently analyzed the effect of miR-92a in endothelial cell by loss-of-function studies [51]. Our group demonstrated that systemic administration of a complementary oligonucleotide (antagomiR-92a) significantly enhanced re-endothelialization in carotid arteries after balloon injury or arterial stenting. Our group and others [46,51] showed the relationship between miR-92a and endothelial nitric oxide synthase (eNOS) expression. Nitric oxide (NO) limits the formation of neointimal hyperplasia in animal models of arterial injury to a large part by inhibiting vascular smooth muscle cell proliferation [52]. Accordingly, the functional consequences of the miR-92a inhibition are an increase in NO bioavailability and an antiproliferative effect on SMCs [51]. A further example of negative correlation between microRNAs and eNOS activity is represented by miR-221 and miR-222. These microRNAs are highly expressed in ECs and exhibit anti-angiogenic effects [53]. Notably, over-expression of miR-221 and miR-222 indirectly reduces the expression of eNOS [33]. miR221 and miR-222 directly target c-kit, the receptor for stem cell factor (SCF), which plays a key role in endothelial cell migration [53]. Recently it has been shown that the miR-221 and miR-222 are negatively correlated with the expression of Ets-1 [54] that regulates the expression of several inflammatory molecules in the endothelial cell during vascular inflammation [55]. Another miRNA which has been identified in endothelial cells is miR-155. Similar to miR-221 and miR-222, miR-155 directly targets ETS-1 in ECs [54]. Also, miR-155 down-regulates eNOS expression through decreasing eNOS mRNA stability by binding its 3'-UTR [56]. Given their role in regulating endothelial cell biology, miR-221, miR 222 and miR-155 represent possible therapeutic targets in the inflammatory response of endothelial cells during the initial stage of atherosclerosis. Several other groups provide additional examples of the intersection between microRNAs and endothelial cell activation and dysfunction. In response to inflammatory stimuli, the nuclear factor-KappaB (NF-κB) signaling pathway is activated leading to the expression of multiple pro-inflammatory genes in ECs [57]. In fact, in Apolipoprotein E (ApoE)-deficient mice, endothelial cell-specific inhibition of NF-κB resulted in reduced development of atherosclerosis [58]. Two endothelial-specific microRNAs, miR-10a and miR-181b, inhibit the activation of the NF-κB signaling pathway in ECs. Recently, miR-181b has been identified as a key player in vascular inflammatory disease. miR-181b expression is reduced in response to TNF-a in the vascular endothelium, whereas its over-expression inhibits TNF-α-induced NF-κB-responsive targets gene such as VCAM-1 and E-selectin [59]. Moreover, miR-181b targets importin-α3, a critical protein in NF-κB nuclear translocation and activation. miR-10a directly inhibits mitogen-activated kinase kinase kinase 7 (MAP3K7) and beta-transducin repeat-containing gene (β-TRC) [60]. These molecules are essential in promoting IκBα degradation, an inhibitor of NF-κB activation. Inhibition of miR-10a enhances the NF-κB-dependent expression of adhesion molecules in ECs. Other specific microRNAs that regulate endothelial cell function have been described. For example, miR-125a-5p and miR-125b-5p have been identified as negative regulators of ET-1 [61], a potent vasoconstrictive and mitogen peptide that plays multiple roles in the progression of vascular disorder [62]. Taken together, the results described above indicate that several microRNAs play an essential role in endothelial pathophysiology. Accordingly, the identification of specific microRNAs involved in biological processes, such as angiogenesis and inflammation, could lead to the definition of new strategies to treat vascular diseases. Given that the same microRNAs may have opposite effects in different biological contexts, further studies are necessary to clarify their roles in endothelial dysfunction. For example, identification of signaling pathways which modulate the activity of microRNAs is critical for development of microRNA-based therapeutic strategies.

4.2. microRNAs in Phenotypic Switching of VSMCs

SMCs within adult animals retain remarkable plasticity and can undergo profound and reversible changes in phenotype, a process referred to as phenotypic switching [63]. SMCs play a role during all phases of the atherogenic process as well as in proliferative disease [64–66]. Several microRNAs are implicated in VSMC phenotypic switching in response to vascular injury or atherosclerotic disease.

4.2.1. miR-143 and miR-145 Play a Role in the Regulation of Phenotype of VSMCs in Response to Injury

miRNA-143 and -145 are considered the master regulators of contractile phenotype by promoting contractile protein expression [67]. The expression levels of miR-143/145 are down-regulated in injured or atherosclerotic vessels and are associated with the phenotypic switch from a contractile/quiescent to a synthetic/proliferative phenotype. A recent study reported that adenovirus-mediated over-expression of miR-145 could partially restore down-regulation of SMC marker genes and neointima formation following balloon injury of the rat carotid artery [68]. In addition, miR-143/145-knockout mice present morphological changes in the aorta, due to an incomplete differentiation of VSMCs [69]. Several growth factors promote phenotypic switching of VSMCs during vascular disease [63]. The growth factor, PDGF-BB, is a critical regulator of VSMCs phenotype in vessel injury. Indeed, PDGF stimulation increases migration and proliferation of SMCs *in vitro* and *in vivo* [70]. It has been shown that PDGF can reduce miR-145 and miR-143 expression through Src and p53 activity [71] and promote formation of podosomes. miR-143 and miR-145 directly target key regulators of podosome formation, such as PDGF receptor α (PDGF-Rα), protein kinase Cε (PKCε) and fascin. Another cytochines that participate in the phenotypic control of VSMCs are Transforming Growth Factor beta (TGFβ) and bone morphogenetic proteins (BMPs). Unlike PDGF, the TGF-family of growth factors has been shown to promote contractile phenotype [72–74]. Induction of miR-143 and miR-145 by TGF-β or BMP4 leads to down-regulation of KLF4 expression and activation of contractile genes [75]. It is interesting to note that the expression of miR-143 and miR-145 is regulated by multiple growth factor signaling pathways promoting phenotypic modulation of SMC. Therefore, because miR-143 and mir-145 may be important modulators of vascular disorder, further studies are necessary to define the regulatory mechanisms of their expression in response to vascular injury or atherosclerotic disease.

4.2.2. miR-133 Is a Negative Regulator of SP-1 and Promotes Contractile Phenotype of VSMCs

Similarly to miR-143 and miR-145, a number of additional microRNAs play a role in smooth muscle cell phenotypic switching and in vascular disease. In a recent study from our laboratory, we found that miR-133 has a potent inhibitory role on VSMC phenotypic switching. miR-133 specifically suppresses Sp-1 expression *in vitro* and *in vivo* and participates in a complex network with Serum response Factor (SRF) in regulating smooth muscle gene expression. Sp-1 is a key regulator of KLF4 expression in phenotypically modulated SMCs [76]. KLF-4 is up-regulated in VSMCs following vascular injury and inhibits myocardin-induced SMC marker gene expression. Following balloon injury of the rat carotid artery there is an increase of Sp1 expression in the neointima related to the acquisition of proliferative/synthetic phenotype of VSMCs. Accordingly, over-expression of miR-133 reduces neointima formation and SMCs proliferation after balloon injury of the rat carotid artery [77].

4.2.3. microRNAs that Promote De-Differentiated Phenotype of VSMCs

Other miRNAs have been implicated in the phenotypic modulation of SMCs: miR-21, miR-146a, miR-221, and miR-222. Unlike the miR-143/145 cluster and miR-133, these microRNAs were found to be significantly up-regulated after vascular injury [78–80] and their inhibition reduces neointimal formation following balloon injury of rat carotid arteries *in vivo*. Therefore, expression profiles of microRNAs in human atherosclerotic plaques in comparison to control, demonstrate that miR-21 and miR-146a were up-regulated in human atherosclerotic plaques, whereas several predicted targets of these miRNAs were down-regulated [81]. miR-221 and miR-222 contribute to SMC phenotype by repressing specific targets such as p57Kip2 and p27Kip1 [79]. They are cyclin-dependent kinase inhibitors and have an antiproliferative effect on VSMCs. Interestingly, PDGF induces the expression of miR-221, leading to down-regulation of multiple target genes and promoting the proliferation of SMCs [82]. Accordingly, inhibition of miR-221 prevents reduction of p27Kip1 in response to PDGF, as well as VSMC proliferation. miR-146a promotes SMCs proliferation *in vitro* and neointimal hyperplasia *in vivo* [80]. Notably, miR-146a inhibits KLF4 expression by targeting its 3'-UTR. Inhibition of miR-146a increases KLF4 expression, while its over-expression induces an opposite effect. It has been reported that KLF4 has a critical role in the regulation of the SMCs phenotype [83] and its expression is correlated to different microRNAs. Particularly, miR-143 and miR-145 directly target KLF4 in SMCs [75]. miR-146a promotes the proliferative phenotype of SMCs through a reduction of KLF4 expression; in contrast, KLF4 regulation of miR-145 and miR-143 promote the contractile phenotype of SMCs. Thus, further studies are necessary to define the regulatory mechanisms of KLF4 in SMCs. miR-21 promotes the cellular response that leads to proliferative thickening of the vessel by directly targeting Phosphatase and Tensin Homolog (PTEN), a critical regulator of SMCs function both *in vivo* and *in vitro* [78]. SRF is a transcription factor that plays a critical role in SMCs biology, influencing both proliferation and differentiation depending on the types of coactivators or repressors present at specific cellular stages. A recent report demonstrated that Serum Response Factor (SRF) regulates PTEN expression through a reduction of miR-21 levels [84]. Regulation SRF-mediated of miR-21 occurs through a miR-143-dependent signaling pathway. Recently, the role of miR-21 in the regulation of SMC phenotype was correlated with abdominal aortic aneurysm (AAA). The expression levels of miR-21 increase during development of AAA in two murine models [85]; moreover Lentivirus-mediated over-expression of miR-21 induced cell proliferation of SMCs, with protective effects on aneurysm expansion. In addition, miR-21 targets several signal molecules associated with SMCs phenotype such as Programmed Cell Death 4 (PDCD4) [86], B-cell leukemia/lymphoma 2 (BCL-2) [78], and Tropomyosin 1 (TPM1) [87]. These observations indicate that miR-21 could play important roles in diverse vascular diseases. Different microRNAs can be involved in the same process; for example recent studies [88] demonstrated that miR-424/322 also play a key role in modulation of SMCs phenotype in response to vascular injury. In particular, ectopic expression of miR-424/322 induces inhibition of proliferation and migration in SMCs and reduces restenosis in injured carotid arteries in rats. miR-424/322 regulates SMCs phenotype suppressing its direct targets cyclin D1 and CA+2 regulating proteins (calumenin). Interestingly, miR-424/322 is significantly up-regulated after vascular injury and this suggest that miR-424/322 is correlated to an adaptive response to counteract proliferation of SMCs. To summarize, these findings suggest a new therapeutic strategy for vascular diseases connected with phenotypic switching of SMCs. Interestingly, chemically modified antisense oligonucleotides, termed "antagomirs", have been used to decrease miRNA expression and function in different animal models [32,37,64]. Given that many microRNAs are increased after injury, it is possible that their specific inhibition by antagomirs could be considered as potential therapeutic targets for several vascular diseases. Nevertheless, further research is needed regarding the role of these aberrantly expressed microRNAs in SMCs.

4.3. Roles of microRNAs in Vascular Development

The formation of a vascular system requires the creation and remodeling of a continuous series of vessels. They are made mainly by endothelial cells, but smooth muscle cells ensure the correct tone and contractility of the vessels necessary for proper blood flow. After birth, VSMCs retain remarkable plasticity; they can switch between a contractile and proliferative phenotype, a characteristic fundamental in vascular development and remodeling. Both these processes are regulated by numerous factors, including microRNAs. miRNAs have been implicated in endothelial cell differentiation and are involved in the regulation of formation of blood vessels during vascular development. Knockout of DICER, the enzyme responsible for the maturation of microRNAs, reduces postnatal angiogenesis in response to several stimuli, such as exogenous VEGF, tumors, limb ischemia, and wound healing [89]. Specific silencing Dicer using siRNA, increases activation of the eNOS pathway but reduces proliferation and cord formation of human endothelial cell *in vitro* [41].

4.3.1. miRNAs Involved in Endothelial Development

The first miR shown to be essential for vessel formation and integrity is miR-126. It is a positive regulator of angiogenic signaling in endothelial cells and also of vascular integrity *in vivo*. miR-126-deficient endothelial cells failed to respond to various angiogenic factors, including VEGF, EGF and bFGF [43,90,91]. Studies on zebrafish show that down-regulation of this microRNA reduces vascular integrity and induces hemorrhages [43]; furthermore, studies in mice demonstrate that the deletion of miR-126 causes defects in endothelial cell proliferation, migration and angiogenesis [91]. However, both these studies demonstrate that miR-126 affects endothelial cell function but it is not essential for cell differentiation or embryonic vessel formation. Fish *et al.* show additionally that miR-146b, miR-625 and miR-197 appear up-regulated in mouse ESC-derived endothelial cells, however the role of these miRNAs in the vascular system is still unknown [43]. Kane *et al.*, in order to study the endothelial differentiation, carried out a Taqman low-density array (TLDA) analysis of miRNA levels in pluripotent hESCs and in hESCs differentiated for 10 days. They found that up-regulation of miR-126 causes an increase in levels of further microRNAs such as let-7 family, miR-210, miR-130a, miR-196, miR-133a [92]. However, despite the fact that all of these microRNAs are modulated during endothelial differentiation, not one of them has been shown to be directly correlated to endothelial cell control. Recently, Nicoli *et al.* investigated the role of miR-221 in endothelial cells in vascular development; they demonstrated that miR-221 promotes endothelial cell proliferation and migration through repression of two targets: cyclin dependent kinase inhibitor 1b (cdkn1b) and phosphoinositide-3-kinase regulatory subunit 1 (pik3r1). Also miR-221 expression is inhibited by Notch signaling [93].

4.3.2. microRNAs in Regulation of Vascular Smooth Muscle Cell Differentiation

Several studies have provided compelling evidence that microRNAs play a critical role in the initial specification of vascular smooth muscle cell lineage during development. Indeed, inactivation of Dicer in VSMCs results in late embryonic lethality due to decreased VSMC proliferation and differentiation and due to vascular abnormalities and extensive hemorrhage [94]. Thus, the function of

Dicer-generated miRNAs is essential during development of VSMC. A recent report demonstrated a crucial role for miR-143 and miR-145 in VSMC differentiation [65]. The down-regulation of miR-145 using cholesterol-modified antisense oligonucleotides inhibits myocardin-induced reprogramming of fibroblasts into SMC and represses expression of multiple SMC markers, such as ACTA2, MyH11 and Calponin. Also, miR-145 over-expression is sufficient to induce differentiation of multipotent neural crest stem cells into smooth muscle cells and to inhibit their proliferation. Accordingly, miR-143 and miR-145 target a network of factors, such as KLF4 and ELK-1, in order to promote VSMCs differentiation and repress proliferation. Other miRNAs that may promote VSMC differentiation are miR-10a, miR-1 and miR-21. Recently, it has been shown that miR-10a is involved in the differentiation to smooth muscle cell lineage from mouse ESCs in response to Retinoid acid [95]. Several studies indicate that Retinoid Signaling positively influences the SMC differentiation program from stem cells [96]. Notably, miR-10a is up-regulated during retinoid acid-induced SMC differentiation. Furthermore, miR-10a directly targets histone deacetylase 4 (HDAC4), which is a negative regulator of SMC differentiation [97]. miR-1 plays a critical role in the SMC lineage differentiation in embryonic stem cell-derived SMC cultures[98]. miR-1 expression is highly up-regulated during differentiation of mouse embryonic stem cell (ESC) to SMCs. miR-1 has been implicated in SMC differentiation by directly targeting the 3'UTR of KLF4 and enhancing expression of the smooth muscle-restricted markers gene. miR-21 was shown to promote differentiation of VSMCs in response to transforming growth factor and bone morphogenetic protein stimulation [99]. miR-21 directly targets PDCD4 (programmed cell death 4), which acts as a negative regulator of smooth muscle contractile genes.

4.4. Long Non Coding RNAs in Vascular Development and Disease

Although the heterogeneous group of lncRNAs play a wide range of roles in cellular function, their characterization pertaining to vascular development and disease is limited to only a few examples. Variation on chromosome 9p21 is associated with risk of coronary artery disease (CAD) [100,101]. This genomic region contains a long intergenic noncoding RNA, designated antisense noncoding RNA in the INK4 locus (ANRIL). ANRIL is a long non-coding RNA which is transcribed from the INK/ARF locus. ANRIL is expressed in tissues and cell types that are affected by atherosclerosis such as primary coronary smooth muscle cells, vascular endothelial cells, human monocyte-derived macrophage cells and RNA extracted from carotid and arterectomy [102]. Notably, increased expression of ANRIL transcripts was directly correlated with the severity of atherosclerosis [103]. However, despite the potential importance of ANRIL to vascular disease, the pathophysiology underlying the link between ANRIL and coronary artery disease remains currently unknown. ANRIL has been associated with epigenetic silencing of the INK4B-ARF-INK4A locus on chromosome 9p21.3 [104]. In fact, ANRIL binds the p15(INK4b) transcript and recruits the Polycomb Repressor Complex (PRC) to repress the transcription of genes at this locus. Therefore, it is possible that the increased expression of ANRIL is correlated with altered expression of p15INK4B leading to coronary artery disease. Future studies will be necessary to define the role of ANRIL in vascular disease. Using whole transcriptome sequencing, a recent publication revealed the expression profile of lncRNAs in VSMCs in response to Ang II [105]. In this paper, the authors showed that two miRNAs, miR-221 and miR-222, are co-transcribed with a specific lnc-RNA, Lnc-Ang362. These microRNAs have been found to play a critical role in smooth muscle cell proliferation and neointimal hyperplasia in response to vascular injury [79,82]. Interestingly, knockdown of Lnc-Ang362 reduces the expression of these miRNAs as well as cell growth. Correlations between the expression of lncRNAs and miRNAs raise the intriguing possibility of complex functional regulatory pathways in which several types of ncRNAs interact and influence the phenotype of VSMCs during vascular disease. Future studies are needed to dissect the exact roles of lncRNAs in phenotypic switching of VSMCs. A further example of a ncRNA correlated with vascular disease is a natural antisense transcript (NAT), termed sONE. A key function of this lncRNAs is the regulation of eNOS expression in a post-transcriptional manner under normoxic and hypoxic conditions [106,107]. Over-expression of sONE in endothelial cells reduces eNOS expression. Alterations of NO production by the vascular endothelium results in endothelial dysfunction, which occurs as a prelude to atherosclerosis. Thus it could be interesting to investigate the role of antisense lncRNA sONE in the post-transcriptional regulation of eNOS in order to define a potential therapeutic target in vascular disease. More recently, Keguo Li *et al.* showed that a new long noncoding antisense transcript, termed tie-1AS lncRNA, is required for the regulation of tyrosine kinase containing immunoglobulin and epidermal growth factor homology domain-1 gene (tie-1) levels *in vivo* and *in vitro* [108]. Analysis of tie-1AS lncRNA and tie-1 revealed that the ratio of tie-1 versus tie-1AS lncRNA is opposite in normal placenta tissue compared with vascular anomaly tissue. Also, the tie-1AS lncRNA selectively binds tie-1 mRNA, resulting in down-regulation of tie-1 protein and thus specific defects in endothelial cell contact junctions. Over-expression of tie-1 AS lncRNAs resulted in defects in endothelial cell junctions and tube formation. For the first time these results identified a lncRNA that plays a functional regulatory role with potential implications in the control of vascular development. In summary, the results described above indicate that lncRNAs are involved in different aspects of development and disease but their role in the cardiovascular system remains to be further investigated.

5. ncRNA in Heart Development and Pathophysiology

The heart is the first organ to form during embryo development, a complex process involving many classes of regulatory molecules. Heart uninterrupted contractility and correct function are essential for life; its alterations are associated with numerous diseases, including atherosclerosis and stroke [7]. Due to its importance, a complex system of transcription factors closely intertwined with a family of different molecules precisely controls multiple aspects of heart development, function and dysfunction. In these mechanisms, a central role is played by ncRNAs.

5.1. microRNA in Heart Development

5.1.1. miRNAs Encoded by MHC Genes

Two miRNAs highly expressed in the heart, miR-1 and miR-133, are implicated in the control of cardiac growth, regulating fundamental aspects of heart development *in vivo*. There are two isoforms of miR-1, miR-1-1 and miR-1-2, whereas miR-133 presents three isoforms, miR-133a-1, miR-133a-2 and miR-133b. These miRNAs are strictly related, in fact miR-1 and miR-133a are encoded by the same gene, in particular miR-1-1/miR-133a-2 and miR-1-2/miR133a-1. The expression of both these clusters is under the control of SRF and Mef2. SRF enhances the expression of these miRNAs in ventricular and atrial myocytes through a serum response element. Instead, Mef2 binds an intronic enhancer of these miRNAs to activate their expression in ventricular myocytes [109]. During differentiation from ES cells, miR-1 and miR-133 are expressed to promote mesoderm induction and to suppress the differentiation in other lineages, in mice [110,111]. Furthermore, the over-expression of miR-133 causes the inhibition of cardiomyocyte proliferation. Particularly, the up-regulation of miR-133 results in embryonic lethality due to the thinning of the ventricular walls and VSDs (Ventricular Septal Defects); whereas miR-133a-null mice present an ectopic expression of smooth muscle genes in the developing heart as well as aberrant cardiomyocyte proliferation [112]. The up-regulation of miR-1 causes embryonic lethality due to a deficiency in cardiomyocytes [109]. Furthermore deficient miR-1 mice exhibit an increased number of proliferating cardiomyocytes [113]. Other microRNAs encoded by MHC genes are implicated in heart development and stress responsiveness. α-MHC encodes for miR-208a, a heart-specific miRNA, while a closely related microRNA, miR-208b, is encoded by β-MHC. Both these miRs have the same seed sequence, but differ in their 3'UTR region. miR-208a is highly expressed in adult mouse heart, whereas mir-208b is very abundant in embryonic heart but it is present at low levels in an adult heart. Furthermore, while miR-208b is expressed in the heart and in other tissues like skeletal muscle, miR-208a is only expressed in the heart [114]. Interestingly the expression of MHC genes, essential for cardiac muscle contraction, is not equal during life; in fact α-MHC is the predominant myosin isoform in the adult heart, whereas β-MHC is highly expressed in the developing heart but is down-regulated after birth. Cardiac stress and diseases modulate MHC gene transcription, causing a switch in myosin content in the heart, which has a marked effect on cardiac contractility and function [115]. Clearly miR-208a and miR-208b follow the trend of their host genes, hinting that they have a role in the regulation of the α-MHC to β-MHC switch and consequently in cardiac conduction, in arrhythmias and in other aspects of the stress response. Nevertheless none of these microRNAs are essential for function of the adult heart, but they appear to function primarily to adapt adult cardiac gene expression to physiological and pathological signaling.

5.1.2. microRNAs in Heart Chamber Morphogenesis

The heart is composed of cells with similar origin that develop divergent patterns of gene expression with numerous transcriptional networks that establish chamber or domain-specific gene expression and function. A precise regulation in time and space of gene expression and protein activity is necessary for a correct cardiac patterning. The most important microRNA in this process is miR-138; it was studied in zebrafish which represents an excellent model to study heart development, despite it having a heart containing a single atrium and ventricle [116]. Morton *et al.* show that miR-138 is required for cardiac maturation, in fact knock-down of this miR in zebrafish embryo causes the failure of ventricular cardiomyocytes to fully mature. So, miR-138 is necessary to establish an appropriate chamber-specific gene expression pattern during embryo development [116]. This miRNA is expressed in specific domains of the heart and targets various members of different pathways, particularly of retinoic acid (RA). It establishes discrete temporal and spatial domains of gene expression during cardiac morphogenesis, ensuring a correct development of the heart. Another microRNA required for proper morphogenesis of heart chambers is miR-143. It directly targets adducin3 (add3), an F-actin capping protein. As reviewed by Taber [117], alterations in cytoskeletal dynamics could drive cardiac morphogenesis by promoting regional changes in cell size and shape. So, the regulation operated by miR-143 on add3-pathway modulates cytoskeletal protein and, consequently, it could influence heart development and particularly chamber formation through active adjustment of myocardial cell morphology [118].

Figure 2. miRNAs involved in Heart Development. Schematic representation of the relevant microRNAs involved in heart development with a subset of their principal targets: miRNAs which regulate heart-chambers morphogenesis (miR-143, miR-138); miRNAs involved in valves morphogenesis and elongation defects (miR-126, miR-23); miRNAs regulating heart differentiation and development.

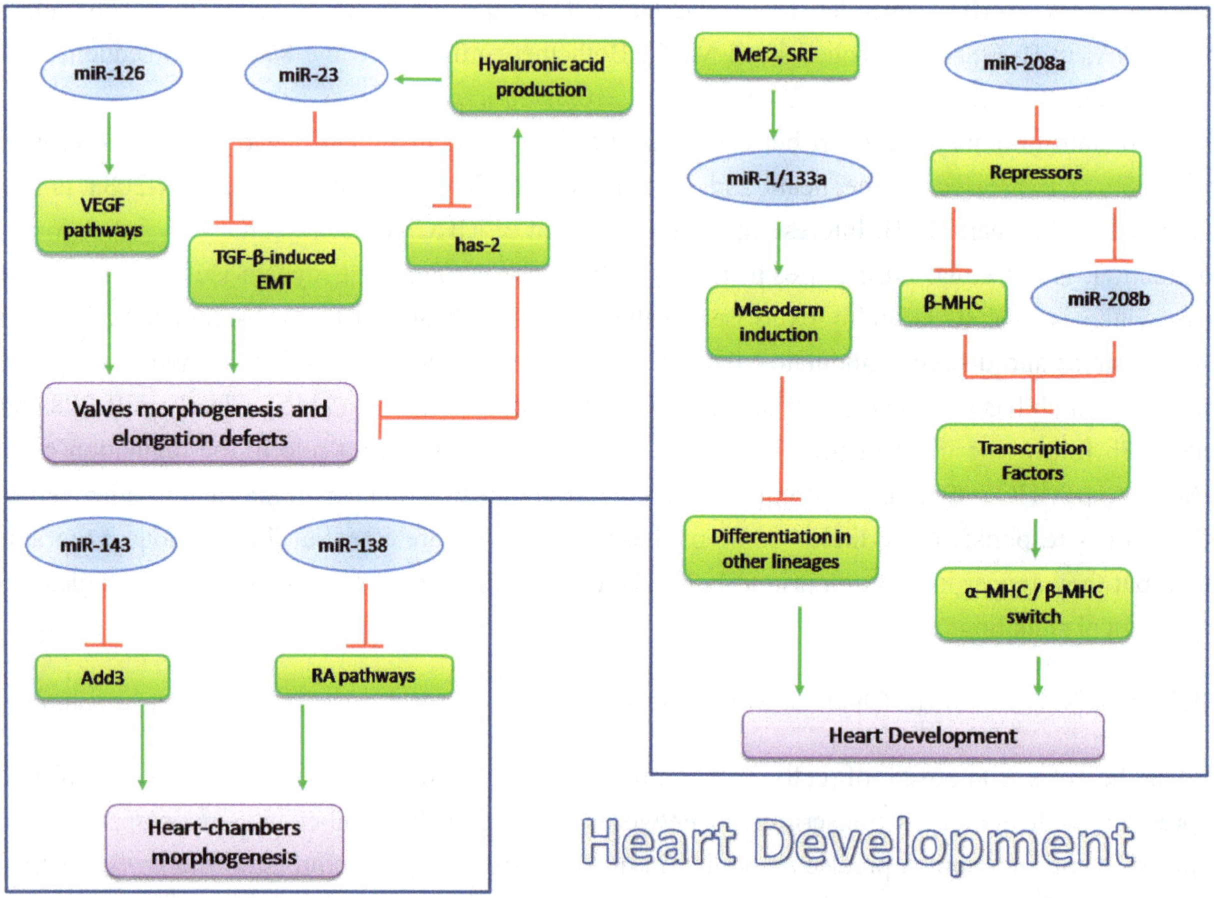

5.1.3. microRNAs in Valves Development

Defects in cardiac valves are the most common subtype of cardiovascular malformation and, in adults, are a major cause of morbidity and mortality. One gene was found which was involved in the regulation of this process: miR-23. Lagendijk *et al.* studied miR-23 in endothelial cells of mice and demonstrated that in this model miR-23 was able to inhibit a TGF-β-induced endothelial to mesenchymal transition (EMT), a process that normally occurs during heart valve development. They proposed that miR-23, has-2 (hyaluronic acid synthase 2) and hyaluronic acid (HA) create a regulatory feedback loop that could respond to various signals, including TGF-β [119]. Recently it has been shown in zebrafish that the loss of this microRNA causes endocardial defects, including cushion formation. miR-23 acts by the down-regulation of has-2, an extracellular remodeling enzyme required for endocardial cushion and valve formation. So this microRNA in the embryonic heart is required to restrict endocardial cushion formation by inhibiting has-2 expression and extracellular hyaluronic acid production [119,120]. A recent study shows that miR-126 is also implicated in valve elongation defects [121]. As shown above, this miR has a role in VEGF signaling in the development of endocardial cells. miR-126 targets a subunit of PI3K and Spred1, two negative regulators of VEGF pathway, so it positively regulates VEGF signaling in heart valve morphogenesis [122]. Despite current knowledge on microRNA functions in cardiovascular development, our complete understanding of their role is far from complete. Figure 2 summarizes the role of microRNAs in heart development.

5.2. microRNA in Heart Pathophysiology

The possibility that microRNAs might participate in heart disease was first suggested by the discovery of distinctive patterns of microRNA expression in the hearts of normal mice *vs.* mice that suffered from heart disease [123]. Recent studies on miRNA expression in heart have identified a subset of miRNAs highly expressed in the normal heart and modulated during cardiovascular disease [124].

5.2.1. miR-195 and miR-98/let-7b in Cardiac Hypertrophy

The first-characterized miRNA involved in inducing hypertrophic growth in the adult heart was miR-195. Adenoviral-mediated over-expression of this microRNA leads to dilated cardiomyopathy and heart dysfunction *in vivo*, it is also sufficient to induce hypertrophy in neonatal rat cardiomyocytes [123]. Chen and colleagues demonstrated the requirement of proper LKB1/STRAD/MO25 complex formation for full activation of AMPK signaling; miR-195 is sufficient to suppress MO25 expression and downstream targets of the LKB1/STRAD/MO25 pathway [125]. They hypothesized that miR-195 targets the LKB1/AMPK signaling axis in hypertrophic cardiomyopathy progression, implicating a functional role of this microRNA in this process.

Conversely, miR-98/let-7b has been demonstrated to mediate the anti-hypertrophic effect of thioredoxin (Trx1), an ubiquitously expressed antioxidant that inhibits NF-κB (nuclear factor kappa-light-chain enhancher of activated B cells), Ras and ASK1 (apoptosis signal-regulating kinase 1). Trx1 negatively regulates the protein kinase cascade known to stimulate hypertrophy. Particularly, Yang *et al.* studied the effects of miR-98 up-regulation or down-regulation on cardiac hypertrophy *in vivo*, at baseline and in response to Ang-II. These studies show that Trx1 negatively regulates Ang-II-induced cardiac hypertrophy through up-regulation of miR-98/let-7b but does not affect heart morphology at baseline [126]. A validated target of miR-98/let-7 is cyclin D2, a cyclin that plays a key role in hypertrophy mediated by this microRNA; the down-regulation of Trx1 causes the up-regulation of miR-98 and the inhibition of hypertrophy.

5.2.2. miR-1 in Heart Physiopathology

The most abundantly expressed microRNA in human heart is miR-1; as described before it is clear that it has a key role in the developing heart. After subjecting the heart of mice to increased pressure overload a down-regulation of miR-1 resulting in an increase in cardiac mass and contractile dysfunction was observed [113,127]. Furthermore, miR-1 is down-regulated in several models of cardiac hypertrophy and heart failure, conversely its over-expression attenuates cardiomyocyte hypertrophy indicating that miR-1 down-regulation has a causative role in the pathogenesis of this disease [113]. It is interesting to note that miR-1 was shown to regulate different pathways implicated in heart hypertrophy. First, it modulates calmodulin and Mef2a, two mediators of calcium signaling, and in addition the transcriptional effectors MEF2A and GATA4, suggesting that miR-1 controls calcium signaling by different modalities simultaneously [113,128]. Second, a dysregulation of insulin-like growth factor (IGF-1) has also been involved in pathological hypertrophy; it is a validated target of miR-1. In exercised trained rats and cardiac-specific Akt transgenic mice, which are models of physiological cardiac hypertrophy, miR-1, as well as miR-133, are down-regulated [110]. There is an inverse correlation between miR-1 and IGF-1: the microRNA controls the expression of IGF-1 and IGF-1 receptor and reciprocally it is down-regulated by IGF-1 stimulation depending on the activation of PI3K/AKT pathway and repression of Foxo3 transcription factor. Accordingly, acromegalic patients, in whom there is an atypical synthesis of IGF-1, display increased cardiac mass and wall thickness [127]. Finally, miR-1 targets twifilin 1 (Twf1), a cytoskeletal regulatory protein that binds to actin monomers preventing their assembly into filaments. The level of Twf1 is inversely correlated with expression of miR-1, so it is expressed at low levels in an adult heart. Moreover down-regulation of miR-1 induced by hypertrophic stimuli results in increased Twf1 expression; likewise Twf1 over-expression is sufficient to induce cardiac hypertrophy in neonatal rat cardiomyocytes, suggesting the therapeutic relevance of modulation of Twf1 expression in attenuating cardiac hypertrophy [129]. Moreover, miR-1 was studied in a model of acute ischemic heart disease (IHD): in cardiomyocytes from ischemia/reperfused (I/R) rats, this microRNA appears up-regulated and inversely correlated with the anti-apoptotic protein Bcl-2 [130], suggesting a potential role of this miRNA in cardiomyocyte apoptosis. In heart failure models levels of miR-1, like miR-133, appear decreased; the same effect is observed in the hearts of patients with hypertrophic cardiomyopathy or atrial dilation [110,131]. The down-regulation of miR-1, and likewise that of miR-133, is associated with the increased levels of two members of the HCN ion channel family, HCN2/HCN4, in hypertrophic hearts. Probably the up-regulation of these channels may contribute to enhanced automaticity and arrhythmias in heart failure [131]. Moreover, a recent study demonstrated that miR-1 directly targets connexin 43 (Cx43), the main cardiac connexin, which has an aberrant increased expression in hypertrophic cardiomyocytes *in vitro* and *in vivo* [132]. Given the numerous processes regulated by miR-1, it could be an important therapeutic target, however alterations of miR-1 levels could alter several mechanisms, so other studies are necessary before its application in the medical field.

5.2.3. miRNA Implicated in Calicineurin/NFATs Pathway

Cellular and *in vivo* models of cardiac hypertrophy induced by transverse aortic constriction and phenylephrine (PE) treatment involve increased activity and expression of calcineurin and decreased expression of miR-133 [133]. NATFc4, a member of calcineurin-activated NFAT family, has two functional binding sites for miR-133. Gain-of-function approaches show that miR-133 decreases NFAT mRNA levels as well as the hypertrophic response to PE-mediated stimulation in primary cardiomyocytes, and miR-133 loss-of-function increases NFATc4 expression and a hypertrophic response [123,134]. Moreover this microRNA decreases cardiac hypertrophy targeting RhoA, Cdc42 and Nelf-A/WHSC2 [110]. Various studies on this microRNA show that the over-expression of miR-133 attenuates agonist-induced hypertrophy [135]; conversely silencing of miR-133 makes the myocardium more sensitive to excessive cardiac growth [136]. Another member of NFAT family, NFATc3, is positively regulated by Myocardin, a transcriptional co-activator that promotes cardiac hypertrophy [137]. Under physiological conditions Myocardin is expressed at low levels but upon hypertrophic stimulation its expression is increased and consequently NFATc3 is up-regulated. Myocardin is a direct target of miR-9, so it was studied like a potential regulator of cardiac hypertrophy. Studies of miR-9 over-expression or inhibition, under hypertrophic stimulation, demonstrate that this microRNA negatively regulates cardiac hypertrophy, *in vivo* and *in vitro*, by targeting Myocardin [137]. Furthermore NFATC3 positively regulates miR-23a, which is up-regulated at transcriptional level by this factor. In fact, the expression of miR-23a is required to mediate hypertrophic growth in response to activation of the calcineurin/NFAT pathway; it directly targets an anti-hypertrophic protein: the muscle-specific ring finger protein 1 MuRF1 [134,138]. Another microRNA implicated in this process is miR-199b: it is a direct target of calcineurin-NFAT signaling, with an increased expression in heart failure. This microRNA modulates calcineurin-NFAT signaling-mediated hypertrophy in a positive feedback loop. Calcineurin induces miR-199b expression through a functional NFAT site upstream of the miR's gene; then the miR targets Dyrk1a, the dual-specificity tyrosine (Y) phosphorylation-regulated kinase 1a, in a process that constitutes a pathogenic feed-forward mechanism affecting calcineurin-responsive gene expression. Mice over-expressing miR-199b exhibit a strong hypertrophic phenotype induced by calcineurin/NFAT signaling, whereas inhibition of miR-199b normalizes Dyrk1a expression, reduces nuclear NFAT activity, and inhibits and even reverses the cardiac hypertrophy and fibrosis in mice models of heart failure [139].

5.2.4. miRNAs Regulated by Thyroid Hormone

miR-208a and miR-499 are implicated in the regulation of myosin gene expression and cardiac stress response. They play a redundant role: *in vivo* deletion of miR-208a resulting in viable animals with normal cardiac size at baseline, but these animals show a decline in cardiac function up to five months of age [140]. However miR-208a is under the thyroid hormone receptor TR, and its over-expression induces hypertrophic growth in mice by suppressing two negative regulators: thyroid hormone-associated protein-1 (Thap1) and myostatin [141]. Consistently elevated levels of miR-499 led to cardiomyopathy and cardiac hypertrophy in a dose-dependent manner [142]. miR-208a is up-regulated in response to a hemodynamic pressure overload and in heart failure [141]. It is a positive regulator of β-MHC, required for the development of cardiac hypertrophy and myocardial fibrosis [140].

5.2.5. miRNAs Regulated by TGF-β in Heart Physiopathology

Fibroblasts are the most abundant class of non-cardiomyocyte cells in the heart; they produce ECM proteins as well as paracrine factors that can regulate the function of cardiomyocytes. In particular, in response to TGF-β, fibroblasts produce ECM and reduce collagenase production, leading to an excessive matrix accumulation. In this way, a key role is played by miRNAs regulated by TGF-β, a known agonist in the production and deposition of collagens in the heart, which contribute to cardiac hypertrophy. Numerous miRNAs are dysregulated in excessive fibrosis; including miR-29 and miR-21 [143]. The miR-29 family is composed of three members, 29a-b and -c, which are preferentially expressed in fibroblasts as compared with cardiomyocytes. All miR-29 family members target mRNA encoding multiple collagens, fibrillins and elastins and another multitude of ECM-related proteins involved in fibrosis. Interestingly, these microRNAs are down-regulated after TGF-β stimulation in cardiac fibroblasts, suggesting that they could contribute to TGF-β-induced fibrosis [129]. Another microRNA induced by TGF-β and dysregulated in fibroblasts, including in cardiac fibroblasts, in multiple types of stress, is miR-21 [144,145]. Studies on miR-21 reveal that this microRNA contributes to myocardial remodeling through regulation of ERK-MAPK-signaling, which is crucial in fibroblast survival and activation. The over-expression of miR-21 indirectly enhances the activity of ERK-MAP kinase; in fact it targets directly Sprouty-1 (SPRY1), a negative regulator of this pathway. In this manner miR-21 positively regulates cardiac fibroblast survival and growth factor secretion that eventually controls interstitial fibrosis and cardiac hypertrophy [143]. However there is a disagreement in the literature on the role of this microRNA: miR-21, in fact, induces the expression of matrix metalloproteinase-2 by targeting the phosphatase and tension homolog (PTEN) in fibroblasts [146]. Furthermore miR-21-null mice display fibrosis levels comparable to wild-type littermates, suggesting that this microRNA is not essential for pathological cardiac remodeling [147]. miR-21 has a potential role in the regulation of different mechanism for the decrease in cardiac contractile function in heart failure. Studies in tumor cells demonstrate that it targets the tropomyosin I (TPM1) [148], but it has not yet been studied in human heart failure.

5.2.6. The Role of miR-378 in MAPK Signaling

MAPK signaling is also controlled by miR-378. This microRNA is sufficient to repress cardiomyocyte hypertrophy regulating this pathway by targeting four members: MAPK1 itself, IGF-1, GRB2 and KSR1. Ganesan *et al.* studied miR-378 *in vivo* in a mouse model for chronic pressure overload (TAC); they demonstrate that the restoration of this microRNA in mice in which it was down-regulated, partially prevents cardiomyocyte hypertrophy and also does not trigger apoptosis *in vivo* [149]. So the tissue-specific up-regulation of miR-378 may be the basis of future therapeutic approaches to counteract cardiomyocyte hypertrophy. miRNAs are interesting targets for therapeutic use because they are selective. In fact they act on diseased tissues but they seem to have minimal effects on healthy tissues. Moreover, unlike their use in cardiovascular development, new delivery methods have been found, catheter-based delivery systems which allow the injury site to be directly targeted, bypassing effects on other tissue [150]. Figure 3 summarizes the main microRNAs involved in heart pathophysiology.

Figure 3. miRNAs involved in Heart Pathophysiology. A schematic overview of the relevant microRNAs implicated in principal heart disease (*i.e.*, cardiac hypertrophy, heart failure and arrhythmias).

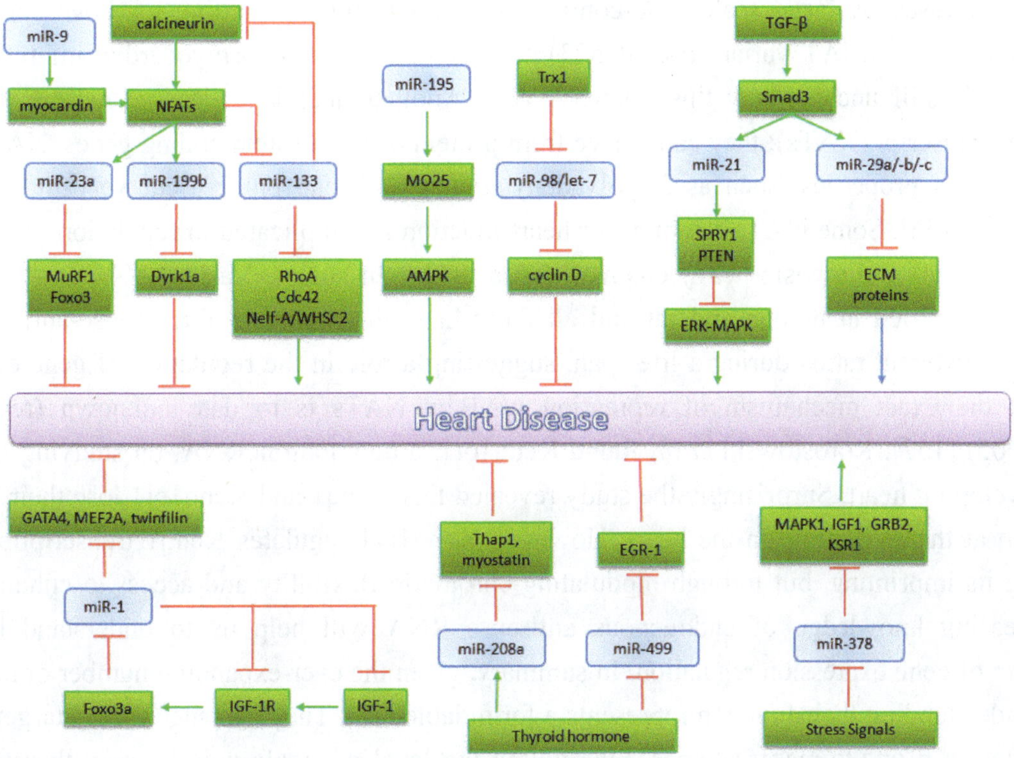

5.3. lncRNA in Heart Pathophysiology

miRNA are not the only ncRNAs implicated in regulatory processes. Recently another class of ncRNA, long non-coding RNA, so called because they are longer than 200 nt, has aroused much interest in cardiovascular function and disease (Table 2). lncRNAs are involved in cardiomyocyte differentiation, e.g., a new lncRNA (AK143260) was identified as a regulator of the cardiac lineage *in vitro*; it is required for mediating the transition from mesoderm to multipotent cardiac progenitors, regulating the activation of a network of cardiac differentiation specific genes [151]. Klattenhoff and colleagues named this lncRNA Braveheart because it is highly expressed in the heart, they identified Braveheart as a critical regulator of cardiovascular commitment from nascent mesoderm [15]. Recently, ANRIL, a multi-exonic lncRNA, has been shown to be implicated in epigenetic modulation in cardiac development and adult heart and also it has been associated with a locus implicated in cardiovascular disease [152,153]. Recently, it was shown that Fendrr, a lateral-mesoderm specific ncRNA, is fundamental in heart development in mouse. It mediates the epigenetic modification of target promoters thereby causing attenuation of the expression of transcription factors which are important in lateral mesoderm differentiation. Fendrr acts as chromatin modulator regulating PRC2 and TrxG/MLL, two histone-modifying complexes. PRC2 performs the methylation of histone H3 at lysine 27 whereas TrxG/MLL catalyzes the methylation at lysine 4; both are essential for embryonic development. Generally, the action of transcription factors is restricted in time and space; they have effects only in the cell in which they are expressed and for a limited time. Conversely, the epigenetic pattern can persist for different stages during differentiation. Fendrr follows this trend and it has a long-term effect [154]. Moreover the regulation operated by ncRNA might confer susceptibility to various diseases: e.g., a myocardial infarction-associated transcript (MIAT), also known as RNCR2 or Gomafu, is a long intergenic non-coding RNA that presents many genetic variants implicated in different processes. A large scale case-control association study regarding cardiovascular disease demonstrates that a MIAT variant (rs2301523) confers susceptibility to myocardial infarction [155]. However it is still unclear how this lincRNA acts. Another abundant class of ncRNA is Natural Antisense Transcript (NATs); they can derive from protein or non protein coding genes. NATs exhibit typical mRNAs properties, such as 3' polyadenylation and 5' cap but present very little sequence conservation [156]. Some NATs essential for heart function are implicated in regulation of the cardiac troponin I (cTNI) and myosin heavy chains (MHCs) and light chains. Various NATS of these genes have been identified in human and rat and all these have the ability to form sense-antisense RNA duplexes at different ratios during a life span, suggesting a role in the regulation of gene expression. However, the exact mechanism of repression used by NATs is to date unknown (reviewed in Luther 2005) [157]. Korostowski *et al.* found Kcnq1ot1, a new long ncRNA, on studying imprinting in the developing heart. Surprisingly the study revealed that Kcnq1 and Kcnq1ot1 lose their imprinted expression at the same time in the heart. However, Kcnq1ot1 regulates Kcnq1 transcription, not by regulating its imprinting, but through modulating chromatin flexibility and access to enhancers [16]. The increasing knowledge of endogenous antisense RNA will help us to understand better the mechanism of gene expression regulation. In summary, given the ever-expanding number of non-coding RNAs, understanding their function represents a formidable task. They can specifically target different genes, often in a one-to-many manner. Fine tuning the level of single ncRNA may therefore affect many pathways in a pleiotropic manner. New therapeutic strategies face the major challenge of developing standardized methods that combine high transfection efficiency with targeted delivery of miRNA to act on specific pathways. The new technologies might provide the ability to translate laboratory potential into clinical practice to prevent or treat cardiovascular diseases.

Table 2. Long-ncRNA in Heart Development and Pathophysiology.

Long-ncRNA	Function	Reference
BraveHeart (AK143260)	Cardiomyocytes differentiation	[15,151]
ANRIL	Epigenetic modulation in cardiac development.	[152,153]
Fendrr	Heart development	[154]
MIAT	Susceptibility to myocardial infarction.	[155]
NATs	Regulation of gene expression	[157]
Kcnq1ot1	Regulation of gene expression	[16]

6. Conclusions

Non coding-RNAs critically affect the main molecular mechanisms involved in cardiovascular development and disease. It is well-established that dysregulation of miRNAs leads to development of vascular as well as cardiac disorders. Moreover, the signaling pathways regulated by miRNAs and miRNAs themselves could be potential therapeutic targets in cardiovascular diseases. Nevertheless, further investigation is necessary to define how the cross-talk between microRNAs and their targets can affect different physiological and pathological pathways in the cardiovascular system. Recently, increasing interest has been aroused by lncRNAs in cardiovascular research. Despite rapid progress in our understanding of lncRNAs, data on their role in cardiovascular patophysiology is still poor. Given that long ncRNAs are associated with several cellular processes, an improved understanding of the functional roles of long non-coding RNA is needed to identify new therapeutic targets in cardiovascular diseases. For this purpose, it will be interesting to investigate the expression of lncRNA in cardiovascular tissues and subsequently to define the mechanisms in which these molecules are involved.

Acknowledgments

This study was partly supported by a grant from the Italian Ministry of Education, University and Research (MIUR): PON01_02833 "CARDIOTECH".

References

1. Birney, E.; Stamatoyannopoulos, J.A.; Dutta, A.; Guigó, R.; Gingeras, T.R.; Margulies, E.H.; Weng, Z.; Snyder, M.; Dermitzakis, E.T.; Thurman, R.E.; *et al.* Identification and analysis of functional elements in 1% of the human genome by the ENCODE pilot project. *Nature* **2007**, *447*, 799–816.
2. Alexander, R.P.; Fang, G.; Rozowsky, J.; Snyder, M.; Gerstein, M.B. Annotating non-coding regions of the genome. *Nat. Rev. Genet.* **2010**, *11*, 559–571.
3. He, L.; Hannon, G.J. MicroRNAs: Small RNAs with a big role in gene regulation. *Nat. Rev. Genet.* **2004**, *5*, 522–531.
4. Mercer, T.R.; Dinger, M.E.; Mattick, J.S. Long non-coding RNAs insight into functions. *Nat. Rev. Genet.* **2009**, *10*, 155–159.
5. Mercer, T.R.; Mattick, J.S. Structure and function of long noncoding RNAs in epigenetic regulation. *Nat. Struct. Mol. Biol.* **2013**, *20*, 300–307.
6. Esteller, M. Non-coding RNAs in human disease. *Nat. Rev. Genet.* **2011**, *12*, 861–874.
7. World Health Organization. *The Global Burden of Disease 2004 Update*. WHO: Geneva, Switzerland, 2008.
8. Small, E.M.; Olson, E.N. Pervasive roles of microRNAs in cardiovascular biology. *Nature* **2011**, *469*, 336–342.
9. Fiedler, J.; Thum, T. MicroRNAs in myocardial infarction. *Arterioscler. Thromb. Vasc. Biol.* **2013**, *33*, 201–205.
10. Holdt, L.M.; Teupser, D. Recent studies of the human chromosome 9p21 locus, which is associated with atherosclerosis in human populations. *Arterioscler. Thromb. Vasc. Biol.* **2012**, *32*, 196–206.
11. Ambros, V. The functions of animal microRNAs. *Nature* **2004**, *431*, 350–355.
12. Small, E.M.; Frost, R.J.; Olson, E.N. MicroRNAs add a new dimension to cardiovascular disease. *Circulation* **2010**, *121*, 1022–1032.
13. Batista, P.J.; Chang, H.Y. Long noncoding RNAs: Cellular address codes in development and disease. *Cell* **2013**, *152*, 1298–1307.

14. Hung, T.; Chang, H.Y. Long noncoding RNA in genome regulation: Prospects and mechanisms. *RNA Biol.* **2010**, *7*, 582–585.
15. Klattenhoff, C.A.; Scheuermann, J.C.; Surface, L.E.; Bradley, R.K.; Fields, P.A.; Steinhauser, M.L.; Ding, H.; Butty, V.L.; Torrey, L.; Haas, S.; *et al.* Braveheart, a long noncoding RNA required for cardiovascular lineage commitment. *Cell* **2013**, *152*, 570–583.
16. Korostowski, L.; Sedlak, N.; Engel, N. The Kcnq1ot1 long non-coding RNA affects chromatin conformation and expression of Kcnq1, but does not regulate its imprinting in the developing heart. *PLoS Genet.* **2012**, *8*, e1002956.
17. Chen, C.; Ridzon, D.A.; Broomer, A.J.; Zhou, Z.; Lee, D.H.; Nguyen, J.T.; Barbisin, M.; Xu, N.L.; Mahuvakar, V.R.; Andersen, M.R.; *et al.* Real-time quantification of microRNAs by stem-loop RT-PCR. *Nucleic Acids Res.* **2005**, *33*, doi:10.1093/nar/gni178.
18. Ule, J.; Jensen, K.; Mele, A.; Darnell, R.B. CLIP: A method for identifying protein-RNA interaction sites in living cells. *Methods* **2005**, *37*, 376–386.
19. Zhao, J.; Ohsumi, T.K.; Kung, J.T.; Ogawa ,Y.; Grau, D.J.; Sarma, K.; Song, J.J.; Kingston, R.E.; Borowsky, M.; Lee, J.T. Genome-wide identification of polycomb-associated RNAs by RIP-seq. *Mol. Cell* **2010**, *40*, 939–953.
20. Bittencourt, D.; Auboeuf, D. Analysis of co-transcriptional RNA processing by RNA-ChIP assay. *Methods Mol. Biol.* **2012**, *809*, 563–577.
21. Hu, W.; Yuan, B.; Flygare, J.; Lodish, H.F. Long noncoding RNA-mediated anti-apoptotic activity in murine erythroid terminal differentiation. *Genes Dev.* **2011**, *25*, 2573–2578.
22. Michelhaugh, S.K.; Lipovich, L.; Blythe, J.; Jia, H.; Kapatos, G.; Bannon, M.J. Mining Affymetrix microarray data for long non-coding RNAs: Altered expression in the nucleus accumbens of heroin abusers. *J. Neurochem.* **2011**, *116*, 459–466.
23. Bentwich, I.; Avniel, A.; Karov, Y.; Aharonov, R.; Gilad, S.; Barad, O.; Barzilai, A.; Einat, P.; Einav, U.; Meiri, E.; *et al.* Identification of hundreds of conserved and nonconserved human microRNAs. *Nat. Genet.* **2005**, *37*, 766–770.
24. Rinn, J.L.; Kertesz, M.; Wang, J.K.; Squazzo, S.L.; Xu, X.; Brugmann, S.A.; Goodnough, L.H.; Helms, J.A.; Farnham, P.J.; Segal, E.; *et al.* Functional demarcation of active and silent chromatin domains in human HOX loci by noncoding RNAs. *Cell* **2007**, *129*, 1311–1323.
25. Mortazavi, A.; Williams, B.A.; McCue, K.; Schaeffer, L.; Wold, B. Mapping and quantifying mammalian transcriptomes by RNA-Seq. *Nat. Methods.* **2008**, *5*, 621–628.
26. Tripathi, V.; Ellis, J.D.; Shen, Z.; Song, D.Y.; Pan, Q.; Watt, A.T.; Freier, S.M.; Bennett, C.F.; Sharma, A.; Bubulya, P.A.; *et al.* The nuclear-retained noncoding RNA MALAT1 regulates alternative splicing by modulating SR splicing factor phosphorylation. *Mol. Cell* **2010**, *39*, 925–938.
27. Yoon, J.H.; Abdelmohsen, K.; Srikantan, S.; Yang, X.; Martindale, J.L.; De, S.; Huarte, M.; Zhan, M.; Becker, K.G.; Gorospe, M. LincRNA-p21 suppresses target mRNA translation. *Mol. Cell* **2012**, *47*, 648–655.
28. Guttman, M.; Donaghey, J.; Carey, B.W.; Garber, M.; Grenier, J.K.; Munson, G.; Young, G.; Lucas, A.B.; Ach, R.; Bruhn, L.; *et al.* lincRNAs act in the circuitry controlling pluripotency and differentiation. *Nature* **2011**, *477*, 295–300.

29. Pandey, R.R.; Mondal, T.; Mohammad, F.; Enroth, S.; Redrup, L.; Komorowski, J.; Nagano, T.; Mancini-Dinardo, D.; Kanduri, C. Kcnq1ot1 antisense noncodingRNA mediates lineage-specific transcriptional silencing through chromatin-level regulation. *Mol. Cell* **2008**, *32*, 232–246.
30. Wamstad, J.A.; Alexander, J.M.; Truty, R.M.; Shrikumar, A.; Li, F.; Eilertson, K.E.; Ding, H.; Wylie, J.N.; Pico, A.R.; Capra, J.A.; *et al.* Dynamic and coordinated epigenetic regulation of developmental transitions in the cardiac lineage. *Cell* **2012**, *151*, 206–220.
31. Tsai, M.C.; Manor, O.; Wan, Y.; Mosammaparast, N.; Wang, J.K.; Lan, F.; Shi, Y.; Segal, E.; Chang, H.Y. Long noncoding RNA as modular scaffold of histone modification complexes. *Science* **2010**, *329*, 689–693.
32. Hung, T.; Wang, Y.; Lin, M.F.; Koegel, A.K.; Kotake, Y.; Grant, G.D.; Horlings, H.M.; Shah, N.; Umbricht, C.; Wang, P.; *et al.* Extensive and coordinated transcription of noncoding RNAs within cell-cycle promoters. *Nat. Genet.* **2011**, *43*, 621–629.
33. Cesana, M.; Cacchiarelli, D.; Legnini, I.; Santini, T.; Sthandier, O.; Chinappi, M.; Tramontano, A.; Bozzoni, I. A long noncoding RNA controls muscle differentiation by functioning as a competing endogenous RNA. *Cell* **2011**, *147*, 358–369.
34. Poliseno, L.; Salmena, L.; Zhang, J.; Carver, B.; Haveman, W.J.; Pandolfi, P.P. A coding-independent function of gene and pseudogene mRNAs regulates tumour biology. *Nature* **2010**, *465*, 1033–1038.
35. Memczak, S.; Jens, M.; Elefsinioti, A.; Torti, F.; Krueger, J.; Rybak, A.; Maier, L.; Mackowiak, S.D.; Gregersen, L.H.; Munschauer, M.; *et al.* Circular RNAs are a large class of animal RNAs with regulatory potency. *Nature* **2013**, *495*, 333–338.
36. Hansen, T.B.; Jensen, T.I.; Clausen, B.H.; Bramsen, J.B.; Finsen, B.; Damgaard, C.K.; Kjems, J. Natural RNA circles function as efficient microRNA sponges. *Nature* **2013**, *495*, 384–388.
37. Hansson, G.K. Inflammation, atherosclerosis, and coronary artery disease. *N. Engl. J. Med.* **2005**, *352*, 1685–1695.
38. Polimeni, A.; de Rosa, S.; Indolfi, C. Vascular miRNAs after balloon angioplasty. *Trends Cardiovasc. Med.* **2013**, *23*, 9–14.
39. Curcio, A.; Torella, D.; Indolfi, C. Mechanisms of smooth muscle cell proliferation and endothelial regeneration after vascular injury and stenting: Approach to therapy. *Circ. J.* **2011**, *75*, 1287–1296.
40. Kuehbacher, A.; Urbich, C.; Zeiher, A.M.; Dimmeler, S. Role of Dicer and Drosha for endothelial microRNA expression and angiogenesis. *Circ. Res.* **2007**, *101*, 59–68.
41. Suárez, Y.; Fernández-Hernando, C.; Pober, J.S.; Sessa, W.C. Dicer dependent microRNAs regulate gene expression and functions in human endothelial cells. *Circ. Res.* **2007**, *100*, 1164–1173.
42. Urbich, C.; Kuehbacher, A.; Dimmeler, S. Role of microRNAs in vascular diseases, inflammation, and angiogenesis. *Cardiovasc. Res.* **2008**, *79*, 581–588.
43. Fish, J.E.; Santoro, M.M.; Morton, S.U.; Yu, S.; Yeh, R.F.; Wythe, J.D.; Ivey, K.N.; Bruneau, B.G.; Stainier, D.Y.; Srivastava, D. miR-126 regulates angiogenic signaling and vascular integrity. *Dev. Cell.* **2008**, *15*, 272–284.
44. Harris, T.A.; Yamakuchi, M.; Ferlito, M.; Mendell, J.T.; Lowenstein, C.J. MicroRNA-126 regulates endothelial expression of vascular cell adhesion molecule 1. *Proc. Natl. Acad. Sci. USA* **2008**, *105*, 1516–1521.

45. Bonauer, A.; Dimmeler, S. The microRNA-17–92 cluster: Still a miRacle? *Cell Cycle* **2009**, *8*, 3866–3873.
46. Bonauer, A.; Carmona, G.; Iwasaki, M.; Mione, M.; Koyanagi, M.; Fischer, A.; Burchfield, J.; Fox, H.; Doebele, C.; Ohtani, K.; *et al.* MicroRNA-92a controls angiogenesis and functional recovery of ischemic tissues in Mice. *Science* **2009**, *324*, 1710–1713.
47. Fang, Y.; Davies, P.F. Site-specific microRNA-92a regulation of Kruppel-like factors 4 and 2 in atherosusceptible endothelium. *Arterioscler. Thromb. Vasc. Biol.* **2012**, *32*, 979–987.
48. Hamik, A.; Lin, Z.; Kumar, A.; Balcells, M.; Sinha, S.; Katz, J.; Feinberg, M.W.; Gerzsten, R.E.; Edelman, E.R.; Jain, M.K. Kruppel-like factor 4 regulates endothelial inflammation. *J. Biol. Chem.* **2007**, *282*, 13769–13779.
49. SenBanerjee, S.; Lin, Z.; Atkins, G.B.; Greif, D.M.; Rao, R.M.; Kumar, A.; Feinberg, M.W.; Chen, Z.; Simon, D.I.; Luscinskas, F.W.; *et al.* KLF2 is a novel transcriptional regulator of endothelial proinflammatory activation. *J. Exp. Med.* **2004**, *199*, 1305–1315.
50. Lin, Z.; Kumar, A.; SenBanerjee, S.; Staniszewski, K.; Parmar, K.; Vaughan, D.E.; Gimbrone, M.A., Jr; Balasubramanian, V.; García-Cardeña, G.; Jain, M.K. KLF2 Is a novel transcriptional regulator of endothelial proinflammatory activation. *J. Exp. Med.* **2004**, *199*, 1305–1315.
51. Iaconetti, C.; Polimeni, A.; Sorrentino, S.; Sabatino, J.; Pironti, G.; Esposito, G.; Curcio, A.; Indolfi, C. Inhibition of mir-92a increases endothelial proliferation and migration *in vitro* as well as reduces neointimal proliferation *in vivo* after vascular injury. *Basic Res. Cardiol.* **2012**, *107*, doi:10.1007/s00395-012-0296-y.
52. Indolfi, C.; Torella, D.; Coppola, C.; Curcio, A.; Rodriguez, F.; Bilancio, A.; Leccia, A.; Arcucci, O.; Falco, M.; Leosco, D.; *et al.* Physical training increases eNOS vascular expression and activity and reduces restenosis after balloon angioplasty or arterial stenting in rats. *Circ. Res.* **2002**, *91*, 1190–1197.
53. Poliseno, L.; Tuccoli, A.; Mariani, L.; Evangelista, M.; Citti, L.; Woods, K.; Mercatanti, A.; Hammond, S.; Rainaldi, G. MicroRNAs modulate the angiogenic properties of HUVECs. *Blood* **2006**, *108*, 3068–3071.
54. Zhu, N.; Zhang, D.; Chen, S.; Liu, X.; Lin, L.; Huang, X.; Guo, Z.; Liu, J.; Wang, Y.; Yuan, W.; *et al.* Endothelial enriched microRNAs regulate angiotensin II-induced endothelial inflammation and migration. *Atherosclerosis* **2011**, *215*, 286–293.
55. Zhan, Y.; Brown, C.; Maynard, E.; Anshelevich, A.; Ni, W.; Ho, I.C.; Oettgen, P. Ets-1 is a critical regulator of Ang IImediated vascular flammation and remodeling. *J. Clin. Investig.* **2005**, *115*, 2508–2516.
56. Sun, H.X.; Zeng, D.Y.; Li, R.T.; Pang, R.P.; Yang, H.; Hu, Y.L.; Zhang, Q.; Jiang, Y.; Huang, L.Y.; Tang, Y.B.; *et al.* Essential role of microRNA-155 in regulating endothelium-dependent vasorelaxation by targeting endothelial nitric oxide synthase. *Hypertension* **2012**, *60*, 1407–1414.
57. Hansson, G.K.; Libby, P. The immune response in atherosclerosis: A double-edged sword. *Nat. Rev. Immunol.* **2006**, *6*, 508–519.
58. Gareus, R.; Kotsaki, E.; Xanthoulea, S.; van der Made, I.; Gijbels, M.J.; Kardakaris, R.; Polykratis, A.; Kollias, G.; de Winther, M.P.; Pasparakis, M. Endothelial cell-specific NF-kappaB inhibition protects mice from atherosclerosis. *Cell Metab.* **2008**, *8*, 372–383.

59. Sun, X,; Icli, B.; Wara, A.K.; Belkin, N.; He, S.; Kobzik, L.; Hunninghake, G.M.; Vera, M.P.; MICU Registry; Blackwell, T.S.; *et al.* MicroRNA-181b regulates NF-κB-mediated vascular inflammation. *J. Clin. Investig.* **2012**, *122*, 1973–1990.
60. Fang, Y.; Shi, C.; Manduchi, E.; Civelek, M.; Davies, P.F. MicroRNA-10a regulation of proinflammatory phenotype in athero-susceptible endothelium *in vivo* and *in vitro*. *Proc. Natl. Acad. Sci. USA* **2010**, *107*, 13450–13455.
61. Li, D.; Yang, P., Xiong, Q.; Song, X.; Yang, X.; Liu. L.; Yuan, W.; Rui, Y.C. MicroRNA-125a/b-5p inhibits endothelin-1 expression in vascular endothelial cells. *J. Hypertens.* **2010**, *28*, 1646–1654.
62. Ohkita, M.; Tawa, M.; Kitada, K.; Matsumura, Y. Pathophysiological roles of endothelin receptors in cardiovascular diseases. *J. Pharmacol. Sci.* **2012**, *119*, 302–313.
63. Alexander, M.R.; Owens, G.K. Epigenetic control of smooth muscle cell differentiation and phenotypic switching in vascular development and disease. *Annu. Rev. Physiol.* **2012**, *74*, 13–40.
64. Gomez, D.; Owens, G.K. Smooth muscle cell phenotypic switching in atherosclerosis. *Cardiovasc. Res.* **2012**, *95*, 156–164.
65. Indolfi, C.; Stabile, E.; Coppola, C.; Gallo, A.; Perrino, C.; Allevato, G.; Cavuto, L.; Torella, D.; Di Lorenzo, E.; Troncone, G.; *et al.* Membrane-bound protein kinase A inhibits smooth muscle cell proliferation *in vitro* and *in vivo* by amplifying cAMP-protein kinase A signals. *Circ. Res.* **2001**, *88*, 319–324.
66. Indolfi, C.; Avvedimento, E.V.; Rapacciuolo, A.; Di Lorenzo, E.; Esposito, G.; Stabile, E.; Feliciello, A.; Mele, E.; Giuliano, P.; Condorelli, G.; *et al.* Inhibition of cellular ras prevents smooth muscle cell proliferation after vascular injury *in vivo*. *Nat. Med.* **1995**, *1*, 541–545.
67. Cordes, K.R.; Sheehy, N.T.; White, M.P.; Berry, E.C.; Morton, S.U.; Muth, A.N.; Lee, T.H.; Miano, J.M.; Ivey, K.N.; Srivastava, D. miR-145 and miR-143 regulate smooth muscle cell fate and plasticity. *Nature* **2009**, *460*, 705–710.
68. Cheng, Y.; Liu, X.; Yang, J.; Lin, Y.; Xu, D.Z.; Lu, Q.; Deitch, E.A.; Huo, Y.; Delphin, E.S.; Zhang, C. MicroRNA-145, a novel smooth muscle cell phenotypic marker and modulator, controls vascular neointimal lesion formation. *Circ. Res.* **2009**, *105*, 158–166.
69. Elia, L.; Quintavalle, M.; Zhang, J.; Contu, R.; Cossu, L.; Latronico, M.V.; Peterson, K.L.; Indolfi, C.; Catalucci, D.; Chen, J.; *et al.* The knockout of miR-143 and -145 alters smooth muscle cell maintenance and vascular homeostasis in mice: Correlates with human disease. *Cell Death Differ.* **2009**, *16*, 1590–1598.
70. Owens, G.K. Molecular control of vascular smooth muscle cell differentiation and phenotypic plasticity. *Novartis Found Symp.* **2007**, *283*, 174–191.
71. Quintavalle, M.; Elia, L.; Condorelli, G.; Courtneidge, S.A. MicroRNA control of podosome formation in vascular smooth muscle cells *in vivo* and *in vitro*. *J. Cell Biol.* **2010**, *189*, 13–22.
72. Lagna, G.; Ku, M.M.; Nguyen, P.H.; Neuman, N.A.; Davis, B.N.; Hata, A. Control of phenotypic plasticity of smooth muscle cells by bone morphogenetic protein signaling through the myocardin-related transcription factors. *J. Biol. Chem.* **2007**, *282*, 37244–37255.
73. Ten Dijke, P.; Arthur, H.M. Extracellular control of TGFβ signalling in vascular development and disease. *Nat. Rev. Mol. Cell Biol.* **2007**, *8*, 857–869.

74. King, K.E.; Iyemere, V.P.; Weissberg, P.L.; Shanahan, C.M. Krüppel-like factor 4 (KLF4/GKLF) is a target of bone morphogenetic proteins and transforming growth factor beta 1 in the regulation of vascular smooth muscle cell phenotype. *J. Biol. Chem.* **2003**, *278*, 11661–11669.
75. Davis-Dusenbery, B.N.; Chan, M.C.; Reno, K.E.; Weisman, A.S.; Layne, M.D.; Lagna, G.; Hata, A. Down-regulation of Kruppel-like factor-4 (KLF4) by microRNA-143/145 is critical for modulation of vascular smooth muscle cell phenotype by transforming growth factor-beta and bone morphogenetic protein 4. *J. Biol. Chem.* **2011**, *286*, 28097–28110.
76. Deaton, R.A.; Gan, Q.; Owens, G.K. Sp1-dependent activation of KLF4 is required for PDGF-BB-induced phenotypic modulation of smooth muscle. *Am. J. Physiol. Heart Circ. Physiol.* **2009**, *296*, 1027–1037.
77. Torella, D.; Iaconetti, C.; Catalucci, D.; Ellison, G.M.; Leone. A.; Waring, C.D.; Bochicchio, A.; Vicinanza, C.; Aquila, I.; Curcio, A.; *et al.* MicroRNA-133 controls vascular smooth muscle cell phenotypic switch *in vitro* and vascular remodeling *in vivo*. *Circ. Res.* **2011**, *109*, 880–893.
78. Ji, R.; Cheng, Y.; Yue, J.; Yang, J.; Liu, X.; Chen, H.; Dean, D.B.; Zhang, C. MicroRNA expression signa-ture and antisense-mediated depletion reveal an essential role of microRNA in vascular neointimal lesion formation. *Circ. Res.* **2007**, *100*, 1579–1588.
79. Liu, X.; Cheng, Y.; Zhang, S.; Lin, Y.; Yang, J.; Zhang, C. A necessary role of miR-221and miR-222 in vascular smooth muscle cell proliferation and neointimal hyperplasia. *Circ. Res.* **2009**, *104*, 476–487.
80. Sun, S.G.; Zheng, B.; Han, M.; Fang, X.M.; Li, H.X.; Miao, S.B.; Su, M.; Han, Y.; Shi, H.J.; Wen, J.K. miR-146a and Kruppel-like factor 4 form a feedback loop to participate in vascular smooth musclecell proliferation. *EMBO Reports* **2011**, *12*, 56–62.
81. Raitoharju, E.; Lyytikäinen, L.P.; Levula, M.; Oksala, N.; Mennander, A.; Tarkka, M.; Klopp, N.; Illig, T.; Kähönen, M.; Karhunen, P.J.; *et al.* miR-21, miR-210, miR-34a, and miR-146a/b are up-regulated in human atherosclerotic plaques in the Tampere Vascular Study. *Atherosclerosis* **2011**, *219*, 211–217.
82. Davis, B.N.; Hilyard, A.C.; Nguyen, P.H.; Lagna, G.; Hata, A. Induction of microRNA-221 by platelet-derived growth factor signaling is critical for modulation of vascular smooth muscle phenotype. *J. Biol. Chem.* **2009**, *284*, 3728–3738.
83. Zheng, B.; Han, M.; Wen, J.K. Role of Krüppel-like factor 4 in phenotypic switching and proliferation of vascular smooth muscle cells. *IUBMB Life* **2010**, *62*, 132–139.
84. Horita, H.N.; Simpson, P.A.; Ostriker, A.; Furgeson, S.; van Putten, V.; Weiser-Evans, M.C.; Nemenoff, R.A. Serum response factor regulates expression of phosphatase and tensin homolog through a microRNA network in vascular smooth muscle cells. *Arterioscler. Thromb. Vasc. Biol.* **2011**, *31*, 2909–2919.
85. Maegdefessel, L.; Azuma, J.; Toh, R.; Deng, A.; Merk, D.R.; Raiesdana, A.; Leeper, N.J.; Raaz, U.; Schoelmerich, A.M.; McConnell, M.V.; *et al.* MicroRNA-21 blocks abdominal aortic aneurysm development and nicotine-augmented expansion. *Sci. Transl. Med.* **2012**, *22*, doi:10.1126/scitranslmed.3003441.
86. Cheng, Y.; Liu, X.; Zhang, S.; Lin, Y.; Yang, J.; Zhang, C. MicroRNA-21 protects against the H(2)O(2)-induced injury on cardiac myocytes via its target gene PDCD4. *J. Mol. Cell. Cardiol.* **2009**, *47*, 5–14.

87. Wang, M.; Li, W.; Chang, G.Q.; Ye, C.S.; Ou, J.S.; Li, X.X.; Liu, Y.; Cheang, T.Y.; Huang, X.L.; Wang, S.M. MicroRNA-21 regulates vascular smooth muscle cell function via targeting tropomyosin 1 in arteriosclerosis obliterans of lower extremities. *Arterioscler. Thromb. Vasc. Biol.* **2011**, *31*, 2044–2053.
88. Merlet, E.; Atassi, F.; Motiani, R.K.; Mougenot, N.; Jacquet, A.; Nadaud, S.; Capiod, T.; Trebak, M.; Lompré, A.M.; Marchand, A. miR-424/322 regulates vascular smooth muscle cell phenotype and neointimal formation in the rat. *Cardiovasc. Res.* **2013**, *98*, 458–468.
89. Suárez, Y.; Fernández-Hernando, C.; Yu, J.; Gerber, S.A.; Harrison, K.D.; Pober, J.S.; Iruela-Arispe, M.L.; Merkenschlager, M.; Sessa, W.C. Dicer-dependent endothelial microRNAs are necessary for postnatal angiogenesis. *Proc. Natl. Acad. Sci. USA* **2008**, *105*, 14082–14087.
90. Fish, J.E.; Srivastava, D. MicroRNAs: Opening a new vein in angiogenesis research. *Sci. Signal.* **2009**, doi:10.1126/scisignal.252pe1.
91. Wang, S.; Aurora, A.B.; Johnson, B.A.; Qi, X.; McAnally, J.; Hill, J.A.; Richardson, J.A.; Bassel-Duby, R.; Olson, E.N. The endothelial-specific microRNA miR-126 governs vascular integrity and angiogenesis. *Dev. Cell* **2008**, *15*, 261–271.
92. Kane, N.M.; Meloni, M.; Spencer, H.L.; Craig, M.A.; Strehl, R.; Milligan, G.; Houslay, M.D.; Mountford, J.C.; Emanueli, C.; Baker, A.H. Derivation of endothelial cells from human embryonic stem cells by directed differentiation: Analysis of microRNA and angiogenesis *in vitro* and *in vivo*. *Arterioscler. Thromb. Vasc. Biol.* **2010**, *30*, 1389–1397.
93. Nicoli, S.; Knyphausen, C.P.; Zhu, L.J.; Lakshmanan, A.; Lawson, N.D. miR-221 is required for endothelial tip cell behaviors during vascular development. *Dev. Cell* **2012**, *22*, 418–429.
94. Albinsson, S.; Suarez, Y.; Skoura, A.; Offermanns, S.; Miano, J.M.; Sessa, W.C. MicroRNAs are necessary for vascular smooth muscle growth, differentiation, and function. *Arterioscler. Thromb. Vasc. Biol.* **2010**, *30*, 1118–1126.
95. Huang, H.; Xie, C.; Sun, X.; Ritchie, R.P.; Zhang, J.; Chen, Y.E. miR-10a contributes to retinoid acid-induced smooth muscle cell differentiation. *J. Biol. Chem.* **2010**, *285*, 9383–9389.
96. Xie, C.; Ritchie, R.P.; Huang, H.; Zhang, J.; Chen, Y.E. Smooth muscle cell differentiation: Models and underlying molecular mechanisms. *Arterioscler. Thromb. Vasc. Biol.* **2011**, *31*, 1485–1494.
97. Qiu, P.; Li, L. Histone acetylation and recruitment of serum responsive factor and CREB-binding protein onto SM22 promoter during SM22 gene expression. *Circ. Res.* **2002**, *90*, 858–865.
98. Xie, C.; Huang, H.; Sun, X.; Guo, Y.; Hamblin, M.; Ritchie, R.P.; Garcia-Barrio, M.T.; Zhang, J.; Chen, Y.E. MicroRNA-1 regulates smooth muscle cell differentiation by repressing Kruppel-like factor 4. *Stem Cells Dev.* **2011**, *20*, 205–210.
99. Davis, B.N.; Hilyard, A.C.; Lagna, G.; Hata, A. SMAD proteins control DROSHA-mediated microRNA maturation. *Nature* **2008**, *454*, 56–61.
100. Samani, N.J.; Erdmann, J.; Hall, A.S.; Hengstenberg, C.; Mangino, M.; Mayer, B.; Dixon, R.J.; Meitinger, T.; Braund, P.; Wichmann, H.E.; *et al.* WTCCC and the Cardiogenics Consortium. Genome wide association analysis of coronary artery disease. *N. Engl. J. Med.* **2007**, *357*, 443–453.
101. McPherson, R.; Pertsemlidis, A.; Kavaslar, N.; Stewart, A.; Roberts, R.; Cox, D.R.; Hinds, D.A.; Pennacchio, L.A.; Tybjaerg-Hansen, A.; Folsom, A.R.; *et al.* A common allele on chromosome 9 associated with coronary heart disease. *Science* **2007**, *316*, 1488–1491.

102. Broadbent, H.M.; Peden, J.F.; Lorkowski, S.; Goel, A.; Ongen, H.; Green, F.; Clarke, R.; Collins, R.; Franzosi, M.G.; Tognoni, G.; *et al.* PROCARDIS consortium. Susceptibility to coronary artery disease and diabetes is encoded by distinct, tightly linked SNPs in the ANRIL locus on chromosome 9p. *Hum. Mol. Genet.* **2008**, *17*, 806–814.
103. Holdt, L.M.; Beutner, F.; Scholz, M.; Gielen, S.; Gäbel, G.; Bergert, H.; Schuler, G.; Thiery, J.; Teupser, D. ANRIL expression is associated with atherosclerosis risk at chromosome 9p21. *Arterioscler. Thromb. Vasc. Biol.* **2010**, *30*, 620–627.
104. Yap, K.L.; Li, S.; Muñoz-Cabello, A.M.; Raguz, S.; Zeng, L.; Mujtaba, S.; Gil, J.; Walsh, M.J.; Zhou, M.M. Molecular interplay of the noncoding RNA ANRIL and methylated histone H3 lysine 27 by polycomb CBX7 in transcriptional silencing of INK4a. *Mol. Cell* **2010**, *38*, 662–674.
105. Leung, A.; Trac, C.; Jin, W.; Lanting, L.; Akbany, A.; Sætrom, P.; Schones, D.E.; Natarajan, R. Novel long non-coding RNAs are regulated by angiotensin II in vascular smooth muscle cells. *Circ. Res.* **2013**, *113*, 266–278.
106. Fish, J.E.; Matouk, C.C.; Yeboah, E.; Bevan, S.C.; Khan, M.; Patil, K.; Ohh, M.; Marsden, P.A. Hypoxia-inducible expression of a natural *cis*-antisense transcript inhibits endothelial nitric-oxide synthase. *J. Biol. Chem.* **2007**, *282*, 15652–15666.
107. Robb, G.B.; Carson, A.R.; Tai, S.C.; Fish, J.E.; Singh, S.; Yamada, T.; Scherer, S.W.; Nakabayashi, K.; Marsden, P.A. Post-transcriptional regulation of endothelial nitric-oxide synthase by an overlapping antisense mRNA transcript. *J. Biol. Chem.* **2004**, *279*, 37982–37996.
108. Li, K.; Blum, Y.; Verma, A.; Liu, Z.; Pramanik, K.; Leigh, N.R.; Chun, C.Z.; Samant, G.V.; Zhao, B.; Garnaas, M.K.; *et al.* A noncoding antisense RNA in tie-1 locus regulates tie-1 function *in vivo*. *Blood* **2010**, *115*, 133–139.
109. Zhao, Y.; Samal, E.; Srivastava, D. Serum response factor regulates a muscle-specific microRNA that targets Hand2 during cardiogenesis. *Nature* **2005**, *436*, 214–220.
110. Carè, A.; Catalucci, D.; Felicetti, F.; Bonci, D.; Addario, A.; Gallo, P.; Bang, M.L.; Segnalini, P.; Gu, Y.; Dalton, N.D.; *et al.* MicroRNA-133 controls cardiac hypertrophy. *Nat. Med.* **2007**, *13*, 613–618.
111. Chen, J.F.; Mandel E.M.; Thomson J.M.; Wu. Q.; Callis, T.E.; Hammond, S.M.; Conlon, F.L.; Wang, D.Z. The role of microRNA-1 and microRNA-133 in skeletal muscle proliferation and differentiation. *Nat. Genet.* **2006**, *38*, 228–233.
112. Liu, N.; Bezprozvannaya, S.; Williams, A.H.; Qi, X.; Richardson, J.A.; Bassel-Duby, R.; Olson, E.N. microRNA-133a regulates cardiomyocyte proliferation and suppresses smooth muscle gene expression in the heart. *Genes Dev.* **2008**, *22*, 3242–3254.
113. Ikeda, S.; He, A.; Kong, S.W.; Lu, J.; Bejar, R.; Bodyak, N.; Lee, K.H.; Ma, Q.; Kang, P.M.; Golub, T.R.; *et al.* MicroRNA-1 negatively regulates expression of the hypertrophy-associated calmodulin and Mef2a genes. *Mol. Cell Biol.* **2009**, *29*, 2193–2204.
114. Van Rooij, E.; Quiat, D.; Johnson, B.A.; Sutherland, L.B.; Qi, X.; Richardson, J.A.; Kelm, R.J., Jr; Olson, E.N. A family of microRNAs encoded by myosin genes governs myosin expression and muscle performance. *Dev. Cell* **2009**, *17*, 662–673.
115. Gupta, M.P. Factors controlling cardiac myosin-isoform shift during hypertrophy and heart failure. *J. Mol. Cell. Cardiol.* **2007**, *43*, 388–403.

116. Morton, S.U.; Scherz, P.J.; Cordes, K.R.; Ivey, K.N.; Stainier, D.Y.; Srivastava, D. microRNA-138 modulates cardiac patterning during embryonic development. *Proc. Natl. Acad. Sci. USA* **2008**, *105*, 17830–17835.

117. Taber, L.A. Biophysical mechanisms of cardiac looping. *Int. J. Dev. Biol.* **2006**, *50*, 323–332.

118. Deacon, D.C.; Nevis, K.R.; Cashman, T.J.; Zhou, Y.; Zhao, L.; Washko, D.; Guner-Ataman, B.; Burns, C.G.; Burns, C.E. The miR-143-adducin3 pathway is essential for cardiac chamber morphogenesis. *Development* **2010**, *137*, 1887–1896.

119. Lagendijk, A.K.; Goumans, M.J.; Burkhard, S.B.; Bakkers J. MicroRNA-23 restricts cardiac valve formation by inhibiting Has2 and extracellular hyaluronic acid production. *Circ. Res.* **2011**, *109*, 649–657.

120. Camenisch, T.D.; Spicer, A.P.; Brehm-Gibson, T.; Biesterfeldt, J.; Augustine, M.L.; Calabro, A., Jr.; Kubalak, S.; Klewer, S.E.; McDonald, J.A. Disruption of hyaluronan synthase-2 abrogates normal cardiac morphogenesis and hyaluronan-mediated transformation of epithelium to mesenchyme. *J. Clin. Investig.* **2000**, *106*, 349–360.

121. Stankunas, K.; Ma, G.K.; Kuhnert, F.J.; Kuo, C.J.; Chang, C.P. VEGF signaling has distinct spatiotemporal roles during heart valve development. *Dev. Biol.* **2010**, *347*, 325–336.

122. Kuhnert, F.; Mancuso, M.R.; Hampton, J.; Stankunas, K.; Asano, T.; Chen, C.Z.; Kuo, C.J. Attribution of vascular phenotypes of the murine Egfl7 locus to the microRNA miR-126. *Development* **2008**, *135*, 3989–3993.

123. Li, Q.; Lin, X.; Yang, X.; Chang, J. NFATc4 is negatively regulated in miR-133a-mediated cardiomyocyte hypertrophic repression. *Am. J. Physiol. Heart Circ. Physiol.* **2010**, *298*, H1340–H1347.

124. Rao, P.K.; Toyama, Y.; Chiang, H.R.; Gupta, S.; Bauer, M.; Medvid, R.; Reinhardt, F.; Liao, R.; Krieger, M.; Jaenisch, R.; *et al.* Loss of cardiac microRNA-mediated regulation leads to dilated cardiomyopathy and heart failure. *Circ. Res.* **2009**, *105*, 585–594.

125. Chen, H.; Untiveros, G.M.; McKee, L.A.; Perez, J.; Li, J.; Antin, P.B.; Konhilas, J.P. Micro-RNA-195 and -451 regulate the LKB1/AMPK signaling axis by targeting MO25. *PLoS One* **2012**, *7*, e41574.

126. Yang, Y.; Ago, T.; Zhai, P.; Abdellatif, M.; Sadoshima, J. Thioredoxin 1 negatively regulates angiotensin II-induced cardiac hypertrophy through upregulation of miR-98/let-7. *Circ. Res.* **2011**, *108*, 305–313.

127. Elia, L.; Contu, R.; Quintavalle, M.; Varrone, F.; Chimenti, C.; Russo, M.A.; Cimino, V.; de Marinis, L.; Frustaci, A.; Catalucci, D.; *et al.* Reciprocal regulation of microRNA-1 and insulin-like growth factor-1 signal transduction cascade in cardiac and skeletal muscle in physiological and pathological conditions. *Circulation* **2009**, *120*, 2377–2385.

128. Obata, K.; Nagata, K.; Iwase, M.; Odashima, M.; Nagasaka, T.; Izawa, H.; Murohara, T.; Yamada, Y.; Yokota, M. Overexpression of calmodulin induces cardiac hypertrophy by a calcineurin-dependent pathway. *Biochem. Biophys. Res. Commun.* **2005**, *338*, 1299–1305.

129. Li, Q.; Song, X.W.; Zou, J.; Wang, G.K.; Kremneva, E.; Li, X.Q.; Zhu, N.; Sun, T.; Lappalainen, P.; Yuan, W.J.; *et al.* Attenuation of microRNA-1 derepresses the cytoskeleton regulatory protein twinfilin-1 to provoke cardiac hypertrophy. *J. Cell Sci.* **2010**, *123*, 2444–2452.

130. Tang, Y.; Zheng, J.; Sun, Y.; Wu, Z.; Liu, Z.; Huang, G. MicroRNA-1 regulates cardiomyocyte apoptosis by targeting bcl-2. *Int. Heart J* **2009** *50*, 377–387.

131. Sayed, D.; Hong, C.; Chen, I.Y.; Lypowy, J.; Abdellatif, M. MicroRNA play an essential role in development of cardiac hypertrophy. *Circ. Res.* **2007**, *100*, 416–424.
132. Curcio, A.; Torella, D.; Iaconetti, C.; Pasceri, E.; Sabatino, J.; Sorrentino, S.; Giampà, S.; Micieli, M.; Polimeni, A.; Henning, B.J.; *et al.* MicroRNA-1 downregulation increases connexin 43 displacement and induces ventricular tachyarrhythmias in rodent hypertrophic hearts. *PLoS One* **2013**, *8*, e70158.
133. Dong, D.L.; Chen, C.; Huo, R.; Wang, N.; Li, Z.; Tu, Y.J.; Hu, J.T.; Chu, X.; Huang, W.; Yang, B.F. Reciprocal repression between microRNA-133 and calcineurin regulates cardiac hypertrophy: A novel mechanism for progressive cardiac hypertrophy. *Hypertension* **2010**, *55*, 946–952.
134. Van Rooij, E.; Sutherland, L.B.; Liu, N.; Williams, A.H.; McAnally, J.; Gerard, R.D.; Richardson, J.A.; Olson, E.N. A signature pattern of stress-responsive microRNAs that can evoke cardiac hypertrophy and heart failure. *Proc. Natl. Acad. Sci. USA* **2006**, *103*, 18255–18260.
135. Matkovich, S.J.; Wang, W.; Tu, Y.; Eschenbacher, W.H.; Dorn, L.E.; Condorelli, G.; Diwan, A.; Nerbonne, J.M.; Dorn, G.W., 2nd. MicroRNA-133a protects against myocardial fibrosis and modulates electrical repolarization without affecting hypertrophy in pressure-overloaded adult hearts. *Circ. Res.* **2010**, *106*, 166–175.
136. Krützfeldt, J.; Rajewsky, N.; Braich, R.; Rajeev, K.G.; Tuschl, T.; Manoharan, M.; Stoffel, M. Silencing of microRNAs *in vivo* with "antagomirs". *Nature***2005**, *438*, 685–689.
137. Wang, K.; Long, B.; Zhou, J.; Li, P.F. miR-9 and NFATc3 regulate myocardin in cardiac hypertrophy. *J. Biol. Chem.* **2010**, *285*, 11903–11912.
138. Lin, Z.; Murtaza, I.; Wang, K.; Jiao, J.; Gao, J.; Li, P.F. miR-23a functions downstream of NFATc3 to regulate cardiac hypertrophy. *Proc. Natl. Acad. Sci. USA* **2009**, *106*, 12103–12108.
139. Da Costa Martins, P.A.; Salic, K.; Gladka, M.M.; Armand, A.S.; Leptidis, S.; El Azzouzi, H.; Hansen, A.; Coenen-de Roo, C.J.; Bierhuizen, M.F.; van der Nagel, R.; *et al.* MicroRNA-199b targets the nuclear kinase Dyrk1a in an auto-amplification loop promoting calcineurin/NFAT signalling. *Nat. Cell Biol.* **2010**, *12*, 1220–1227.
140. Van Rooij, E.; Sutherland, L.B.; Qi, X.; Richardson, J.A.; Hill, J.; Olson, E.N. Control of stress-dependent cardiac growth and gene expression by a microRNA. *Science* **2007**, *316*, 575–579.
141. Callis, T.E.; Pandya, K.; Seok, H.Y.; Tang, R.H.; Tatsuguchi, M.; Huang, Z.P.; Chen, J.F.; Deng, Z.; Gunn, B.; Shumate, J.; *et al.* MicroRNA-208a is a regulator of cardiac hypertrophy and conduction in mice. *J. Clin. Investig.* **2009**, *119*, 2772–2786.
142. Shieh, J.T.; Huang, Y.; Gilmore, J.; Srivastava, D. Elevated miR-499 levels blunt the cardiac stress response. *PLoS One* **2011**, *6*, e19481.
143. Van Rooij, E.; Sutherland, L.B.; Thatcher, J.E.; DiMaio, J.M.; Naseem, R.H.; Marshall, W.S.; Hill, J.A.; Olson, E.N. Dysregulation of microRNAs after myocardial infarction reveals a role of miR-29 in cardiac fibrosis. *Proc. Natl. Acad. Sci. USA* **2008**, *105*, 13027–13032.
144. Thum, T.; Gross, C.; Fiedler, J.; Fischer, T.; Kissler, S.; Bussen, M.; Galuppo, P.; Just, S.; Rottbauer, W.; Frantz, S.; *et al.* MicroRNA-21 contributes to myocardial disease by stimulating MAP kinase signalling in fibroblasts. *Nature* **2008**, *456*, 980–984.
145. Zavadil, J.; Narasimhan, M.; Blumenberg, M.; Schneider, R.J. Transforming growth factor-beta and microRNA:mRNA regulatory networks in epithelial plasticity. *Cells Tissues Organs* **2007**, *185*, 157–161.

146. Roy, S.; Khanna, S.; Hussain, S.R.; Biswas, S.; Azad, A.; Rink, C.; Gnyawali, S.; Shilo, S.; Nuovo, G.J.; Sen, C.K. MicroRNA expression in response to murine myocardial infarction: miR-21 regulates fibroblast metalloprotease-2 via phosphatase and tensin homologue. *Cardiovasc. Res.* **2009**, *82*, 21–29.
147. Patrick, D.M.; Montgomery, R.L.; Qi, X.; Obad, S.; Kauppinen, S.; Hill, J.A.; van Rooij, E.; Olson, E.N. Stress-dependent cardiac remodeling occurs in the absence of microRNA-21 in mice. *J. Clin. Investig.* **2010**, *120*, 3912–3916.
148. Zhu, S.M.; Si, M.L.; Wu, H.L.; Mo, Y.Y. MicroRNA-21 targets the tumor suppressor gene tropomyosin I (TPM1). *J. Biol. Chem.* **2007**, *282*, 14328–14336.
149. Ganesan, J.; Ramanujam, D.; Sassi, Y.; Ahles, A.; Jentzsch, C.; Werfel, S.; Leierseder, S.; Loyer, X.; Giacca, M.; Zentilin, L.; *et al.* MiR-378 controls cardiac hypertrophy by combined repression of mitogen-activated protein kinase pathway factors. *Circulation* **2013**, *127*, 2097–2106.
150. Van Rooij, E.; Marshall, W.S.; Olson, E.N. Toward microRNA-based therapeutics for heart disease: The sense in antisense. *Circ. Res.* **2008**, *103*, 919–928.
151. Schonrock, N.; Harvey, R.P.; Mattick, J.S. Long noncoding RNAs in cardiac development and pathophysiology. *Circ. Res.* **2012**, *111*, 1349–1362.
152. Congrains, A.; Kamide, K.; Oguro, R.; Yasuda, O.; Miyata, K.; Yamamoto, E.; Kawai, T.; Kusunoki, H.; Yamamoto, H.; Takeya, Y.; *et al.* Genetic variants at the 9p21 locus contribute to atherosclerosis through modulation of ANRIL and CDKN2A/B. *Atherosclerosis* **2012**, *220*, 449–455.
153. He, A.; Ma, Q.; Cao, J.; von Gise, A.; Zhou, P.; Xie, H.; Zhang, B.; Hsing, M.; Christodoulou, D.C.; Cahan, P.; *et al.* Polycomb repressive complex 2 regulates normal development of the mouse heart. *Circ. Res.* **2012**, *110*, 406–415.
154. Grote, P.; Wittler, L.; Hendrix, D.; Koch, F.; Währisch, S.; Beisaw, A.; Macura, K.; Bläss, G.; Kellis, M.; Werber, M.; *et al.* The tissue-specific lncRNA Fendrr is an essential regulator of heart and body wall development in the mouse. *Dev. Cell.* **2013**, *24*, 206–214.
155. Ishii, N.; Ozaki, K.; Sato, H.; Mizuno, H.; Saito, S.; Takahashi, A.; Miyamoto, Y.; Ikegawa, S.; Kamatani, N.; Hori, M.; *et al.* Identification of a novel non-coding RNA, MIAT, that confers risk of myocardial infarction. *J. Hum. Genet.* **2006**, *51*, 1087–1099.
156. Guttman, M.; Garber, M.; Levin, J.Z.; Donaghey, J.; Robinson, J.; Adiconis, X.; Fan, L.; Koziol, M.J.; Gnirke, A.; Nusbaum, C.; *et al.* Ab initio reconstruction of cell type-specific transcriptomes in mouse reveals the conserved multi-exonic structure of lincRNAs. *Nat. Biotechnol.* **2010**, *28*, 503–510.
157. Luther, H.P. Role of endogenous antisense RNA in cardiac gene regulation. *J. Mol. Med.* **2005**, *83*, 26–32.

5

Involvement of miR-20a in Promoting Gastric Cancer Progression by Targeting Early Growth Response 2 (EGR2)

Xiangsheng Li, Zhichao Zhang, Ming Yu, Liqi Li, Guangsheng Du, Weidong Xiao * and Hua Yang *

Department of General Surgery, Xinqiao Hospital, Third Military Medical University, Xinqiao Road, Chongqing 400037, China; E-Mails: xiangsonlee@163.com (X.L.); zhichao_zhang@163.com (Z.Z.); yumimianbao@163.com (M.Y.); liliqi198610@163.com (L.L.); guangsheng_du@126.com (G.D.)

* Authors to whom correspondence should be addressed; E-Mails: weidong.xiao@126.com (W.X.); hwbyang@163.com (H.Y.);

Abstract: Gastric cancer (GC) is one of the most common cancers, with high incidences in East Asia. microRNAs (miRNAs) play essential roles in the carcinogenesis of GC. miR-20a was elevated in GC, while the potential function of miR-20a was poorly understood. miR-20a expression was examined in GC tissues and cell lines. The effects of miR-20a on the growth, migration, invasion, and chemoresistance of GC cells were examined. Luciferase reporter assay and Western blot were used to screen the target of miR-20a. miR-20a was increased in GC tissues and cell lines. miR-20a promoted the growth, migration and invasion of GC cells, enhanced the chemoresistance of GC cells to cisplatin and docetaxel. Luciferase activity and Western blot confirmed that miR-20a negatively regulated EGR2 expression. Overexpression of EGR2 significantly attenuated the oncogenic effect of miR-20a. miR-20a was involved in the carcinogenesis of GC through modulation of the EGR2 signaling pathway.

Keywords: gastric cancer; miR-20a; early growth response 2; carcinogenesis; chemoresistance

---

1. Introduction

Gastric cancer (GC) is one of the most common cancers and the second most common malignancy of cancer death worldwide, especially in East Asia [1]. Despite a steady decrease in GC incidence and mortality during the last decade, GC still ranked second in global cancer mortality. The carcinogenesis of GC was complicated involving dysregualtion of oncogenes and tumor suppressors [2]. Recently, emerging evidence found that a new group of RNAs, known as microRNAs (miRNAs), regulated a large variety of genes, including both oncogenes and tumor suppressor genes [3].

miRNAs were a family of endogenous, non-coding small RNAs (approximately 20–25 nucleotides), which negatively regulated gene expression by inhibiting translation or inducing mRNA degradation via binding to the 3' untranslated region (3' UTR) of target mRNAs. miRNAs played critical roles in the development, proliferation, differentiation, metabolism, and apoptosis. In addition, aberrant expression of miRNAs was related to carcinogenesis [4]. Abnormal expression profiles of miRNAs had been reported in numerous cancers, including breast, colon, lung, prostate, and GC [5–8]. miRNAs functioned as oncogenes or tumor suppressors by regulating different target gene expression in different cancers. miR-9, miR-22, and miR-146a had all been shown to act as tumor suppressors [9–11], whereas miR-19, miR-23a, and miR-301a had been shown to function as oncogenes [12–14]. These studies suggested that dysregulation of miRNAs might be involved in carcinogenesis and cancer progression.

miR-20a belonged to the miR-17-92 cluster, which were widely overexpressed oncogenes in diverse cancer subtypes [14]. Previous studies had shown that certain cancer suppressors, such as BH3-only protein (BIM), Phosphatase and tensin homolog (PTEN), were direct targets of the miR-17-92 cluster [12,15]. In human cervical cancer cells, miR-20a was reported to promote cancer cell migration and invasion by targeting Tankyrase 2 (TNKS2) [16]. Similarly, in osteosarcoma, miR-20a increased the metastatic potential of osteosarcoma cells by targeting Fas, indicating an oncogenic function of miR-20a [17]. However, the role of miR-20a in the progression of GC and its underlying mechanism remained unclear.

In this study, we explored the biological effects and the potential mechanisms of miR-20a in GC by detecting the expression of miR-20a in GC, validating previous finding that miR-20a was elevated in GC. Overexpression of miR-20a promoted the growth, migration, invasion, as well as chemoresistance of GC cells. By bioinformatics analysis, we predicted the tumor-suppressor, early growth response 2 (EGR2), as a putative target of miR-20a. Subsequent luciferase activity assay and Western blot confirmed that miR-20a repressed the expression of EGR2 by inducing EGR2 mRNA decay. Overexpression of EGR2 significantly attenuated the oncogenic effect of miR-20a.

2. Results

2.1. miR-20a Was Increased in GC Tissue Samples and Cell Lines

To validate the expression levels of miR-20a, we conducted quantitative real time PCR (qRT-PCR) in 28 GC tissues and the corresponding normal tissues and 3 GC cell lines and a gastric epithelial cell line. Expression of miR-20a was significantly increased in GC tissues and cell lines (Figure 1A,B). The clinicopathological features of 28 patients with GC were shown in Table 1, and miR-20a was correlated with the metastasis of GC patients, while the miR-20a expression had no relationship with other clinicopathological parameters. Thus, it was suggested that elevation of miR-20a in GC might have essential roles in GC carcinogenesis.

Figure 1. miR-20a was increased in gastric cancer (GC) tissue samples and cell lines. (**A**) qRT-PCR for miR-20a was performed using 28 GC tissue samples and matched with adjacent non-tumour normal tissues; (**B**) qRT-PCR for miR-20a was performed using 3 GC cell lines and human gastric mucosa cell line GES-1. The data represented triplicate measurements. * $p < 0.05$, ** $p < 0.01$ compared with control.

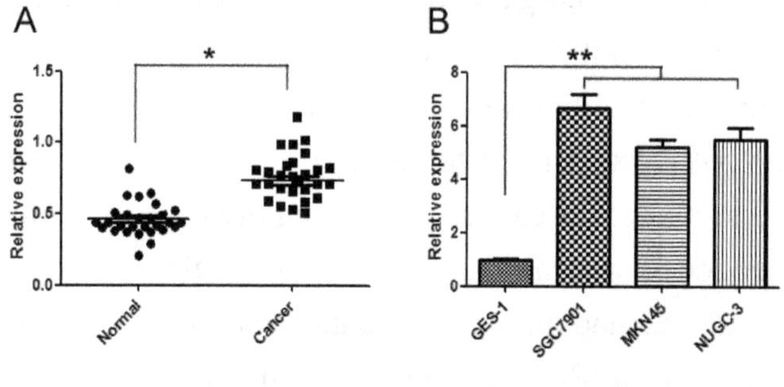

Table 1. Correlation of the expression of miR-20a with clinicopathologic features.

Clinicopathologic features	No.	Relative expression of miR-20a [a]	*p*-value [b]
Gender			0.727
Male	18	0.72 (0.51–1.33)	
Female	10	0.70 (0.52–1.02)	
Site of tumor			0.810
Upper stomach	8	0.68 (0.52–1.00)	
Middle stomach	6	0.70 (0.51–1.22)	
Lower stomach	14	0.71 (0.51–1.32)	
Differentiation			0.655
Poor	12	0.73 (0.59–1.33)	
Moderate	16	0.72 (0.52–1.12)	
Metastasis			0.003
N0	4	0.55 (0.51–0.61)	
N1	5	0.62 (0.51–0.64)	
N2	9	0.74 (0.52–1.09)	
N3	9	0.92 (0.51–1.33)	

[a] Median of relative expression; [b] Mann-Whitney U test between two groups and Kruskall-Wallis test for three groups.

2.2. miR-20a Promoted Growth of GC Cells

As miR-20a was markedly increased in GC, it might function as a cancer promoter or suppressor. Therefore, we tested the role of miR-20a by gain- and loss- function experiments in SGC7901 and MKN45 cells. In a 3-(4,5-dimethylthiazol-2-yl)-2,5-diphenyl-tetrazolium bromide (MTT) assay, cells transfected with miR-20a precursor grew more rapidly than the control group, while miR-20a inhibitor inhibited the growth (Figure 2A,B). To further study the effect of miR-20a on the growth of GC cells, colony formation assay was performed. GC cells transfected with miR-20a precursor showed higher colony formation than cells transfected with control. However, cells transfected with miR-20a inhibitor showed lower colony formation than cells transfected with control (Figure 2C). We further used Flow cytometric analysis to determine the effect of miR-20a on apoptosis of GC cells; no significant difference was detected between miR-20a precursor and the control in these cells (Figure 2D). Results from cell cycle assay indicated that overexpression of miR-20a precursor had less cells in G0/G1 phase. Furthermore, overexpression group had more cells in S and G2M phases (Figure 2E). Proliferation index (PI) = (S + G2M)/(G0 + GS + G2M). The PI was higher in the overexpression group than that in the control group. These data suggested that miR-20a might promote SGC7901 and MKN45 cell proliferation *in vitro*. The transfection efficiency was detected by qRT-PCR (Figure 2F).

Figure 2. MiR-20a promoted growth of GC cell lines. SGC7901 or MKN45 cells were transfected with miR-20a precursor/inhibitor or the corresponding control, respectively. (**A**) At 24, 48, 36, 72, or 96 h after transfection, MTT assay was performed to examine SGC7901 proliferation; (**B**) MTT assay of MKN45 cells; (**C**) Representative results of colony formation assay in SGC7901 and MKN45 cells transfected with miR-20a precursor/inhibitor or the corresponding control, respectively; (**D**) The apoptosis of SGC7901 and MKN45 cells after miR-20a precursor transfection; (**E**) The proliferation index of SGC7901 and MKN45 cells transfected with miR-20a precursor; (**F**) qRT-PCR for miR-20a was performed in SGC7901 and MKN45 cells transfected with miR-20a precursor/inhibitor or the corresponding control, respectively. The data represented at least four measurements. * $p < 0.05$, ** $p < 0.01$ compared with control.

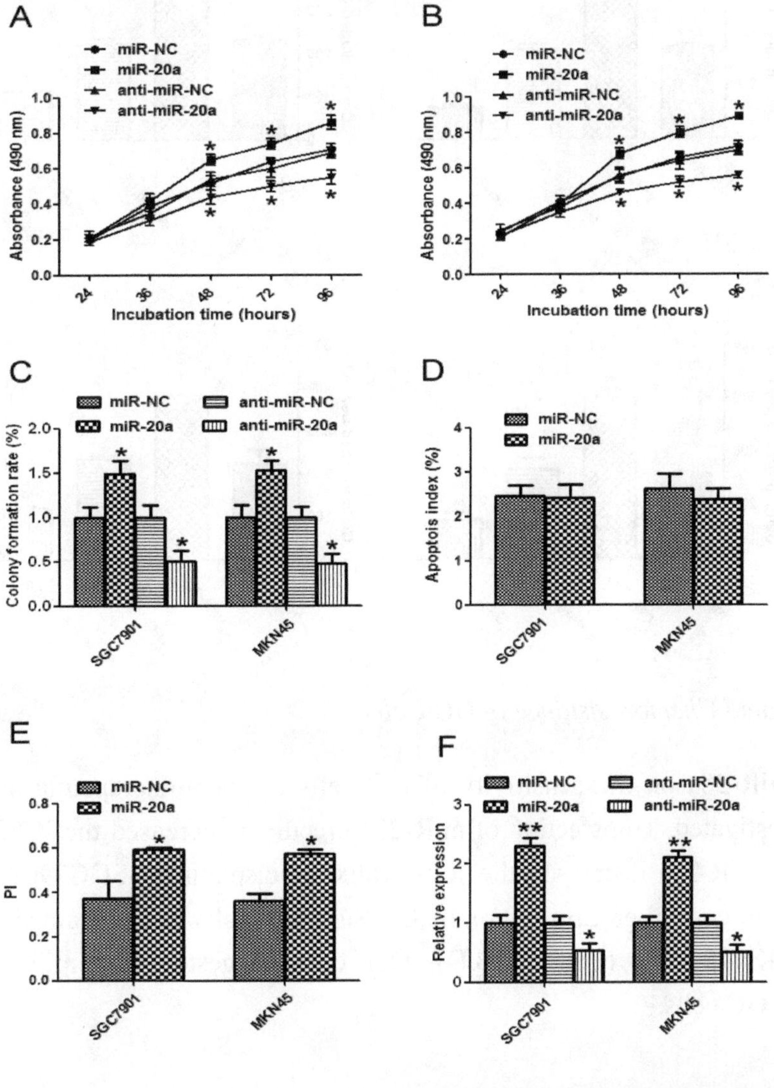

2.3. *miR-20a Promoted Migration and Invasion of GC Cells*

To investigate the role of miR-20a in GC metastasis, miR-20a precursor/inhibitor was transfected into SGC-7901 and MKN45 cells and *in vitro* migration and invasion assays were performed. Results showed that miR-20a significantly increased the *in vitro* migration ability of GC cells, while miR-20a inhibitor remarkably decreased the *in vitro* migration ability of GC cells (Figure 3A,C). Similar results were observed in *in vitro* invasion ability of GC cells (Figure 3B,D). Collectively, these data suggested that miR-20a promoted the migration and invasion abilities of GC cells.

Figure 3. miR-20a promoted migration and invasion of GC cells. (**A**) *In vitro* migration assay of SGC7901 cells transfected with miR-20a precursor/inhibitor or the corresponding control, respectively; (**B**) *In vitro* invasion assay of SGC7901 cells transfected with miR-20a precursor/inhibitor or the corresponding control, respectively; (**C**) *In vitro* migration assay of MKN4 cells; (**D**) *In vitro* invasion assay of MKN4 cells. The data represented at least four measurements. * $p < 0.05$ compared with control.

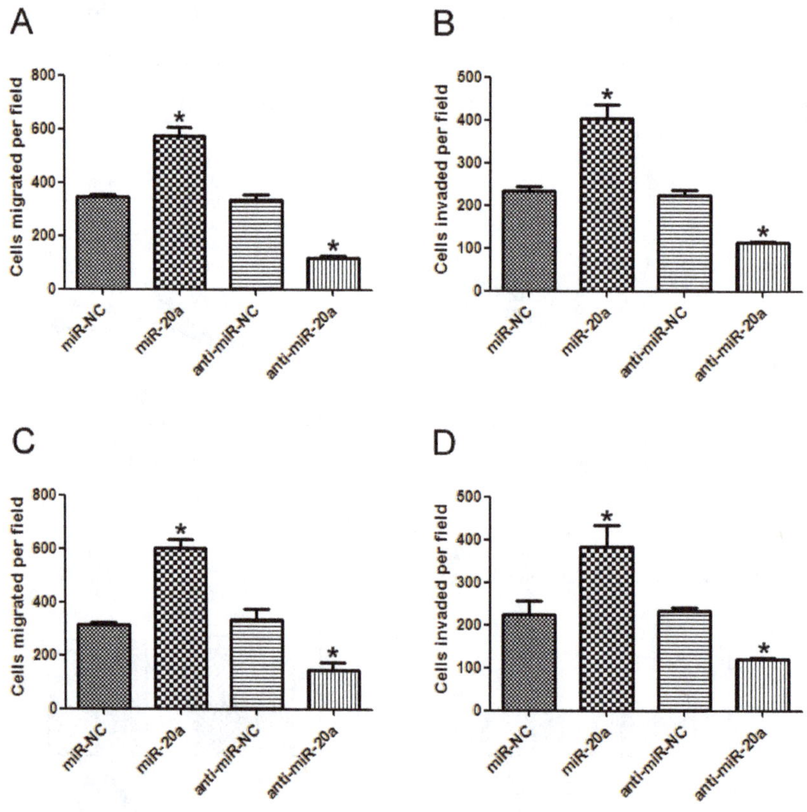

2.4. *miR-20a Promoted Chemoresistance of GC Cells*

The effect of miR-20a on the sensitivity of GC cells to chemotherapeutic agents, cisplatin and docetaxel, was investigated. Transfection of miR-20a precursor increased the IC50 value of cisplatin, while inhibition of miR-20a decreased the IC50 value of cisplatin in SGC7901 and MKN45 cells compared with that in control group (Figure 4A,C). Similar results were obtained in docetaxel treated SGC7901 and MKN45 cells (Figure 4B,D). Our data suggested that miR-20a might promote chemoresistance of GC cells.

Figure 4. miR-20a promoted chemoresistance of GC cells. (**A**) Alteration of IC50 values (cisplatin) in SGC7901 cells transfected with miR-20a precursor/inhibitor, and the corresponding control was analyzed by MTT; (**B**) Alteration of IC50 values (docetaxel) in SGC7901 cells; (**C**) Alteration of IC50 values (cisplatin) in MKN45 cells; (**D**) Alteration of IC50 values (docetaxel) in MKN45 cells. The data represented at least four measurements. * $p < 0.05$, ** $p < 0.01$ compared with control.

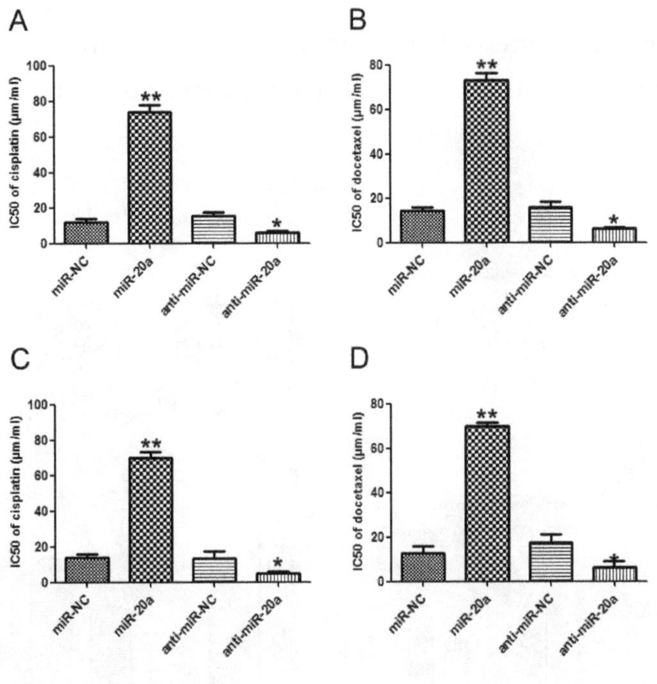

2.5. EGR2 Was a Direct Target of miR-20a

To investigate the downstream target of miR-20a, Targetscan 6.2 was used to screen its target. Early growth response 2 (EGR2) was predicted to be a target of miR-20a. To confirm that, we amplified the EGR2 3' UTR containing the target sequences, or the mutants, into a luciferase reporter vector (Figure 5A). As shown in Figure 5B, miR-20a suppressed the luciferase activity of the wild type EGR2 3' UTR (WT), while mutation of the miR-20a binding sites (Mut) blocked this suppression in SGC7901 cells. Western blot demonstrated that transfection of miR-20a precursor in SGC7901 cells inhibited EGR2 expression while miR-20a inhibitor elevated EGR2 protein level (Figure 5C). qRT-PCR showed that miR-20a precursor decreased EGR2 mRNA level, while miR-20a inhibitor elevated EGR2 mRNA level (Figure 5D), indicating that miR-20a suppressed EGR2 expression post-transcriptionally.

2.6. miR-20a Was inversely Correlated with EGR2 Expression

qRT-PCR was performed to detect the mRNA levels of EGR2 in 28 GC and adjacent non-tumor normal tissue samples. EGR2 mRNA was significantly decreased in GC group (Figure 6A). Furthermore, EGR2 protein levels were also down-regulated in GC tissues compared with the non-tumor normal tissue samples (Figure 6B). Moreover, we correlated EGR2 mRNA level with miR-20a expression in the same GC tissues. EGR2 mRNA level was inversely correlated with miR-20a expression in GC tissues (Figure 6C).

Figure 5. Early growth response 2 (EGR2) was a direct target of miR-20a. (**A**) Wild type of the mutated sequences of EGR2 3' UTR (nucleotides 458-465); (**B**) SGC7901 cells were co-transfected with miR-20a precursor or negative control (miR-NC) with EGR2 3' UTR fragment with either the miR-20a target sequence (WT), or a mutant (Mut). Luciferase activity was detected; (**C**) EGR2 protein level was detected by Western blot in cells transfected with miR-20a precursor/inhibitor or the corresponding control; (**D**) expression of EGR2 mRNA was detected by qRT-PCR in cells transfected with miR-20a precursor/inhibitor or the corresponding control. GAPDH was used as an internal control. The data represented at least four measurements. * $p < 0.05$ compared with control.

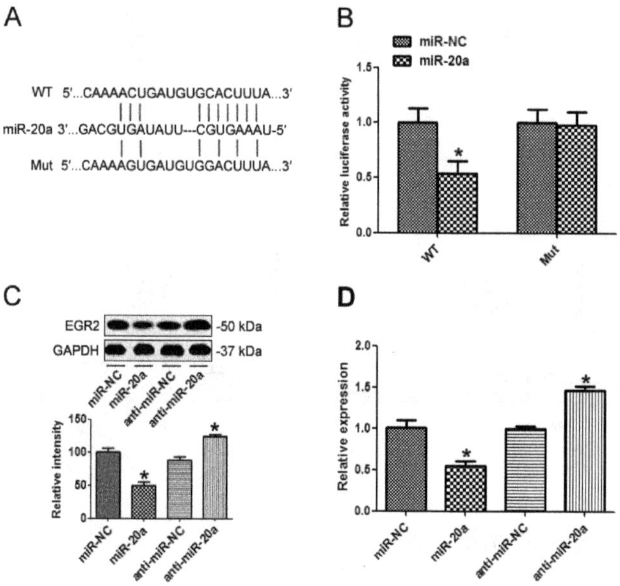

Figure 6. miR-20a was inversely correlated with EGR2 expression. (**A**) EGR2 mRNA levels in GC tissues were analyzed by qPCR; (**B**) EGR2 protein levels in GC tissues were analyzed by Western blot; (**C**) Correlation of miR-20a expression and EGR2 mRNA level was analyzed. The data represented triplicate measurements. * $p < 0.05$, ** $p < 0.01$ compared with control.

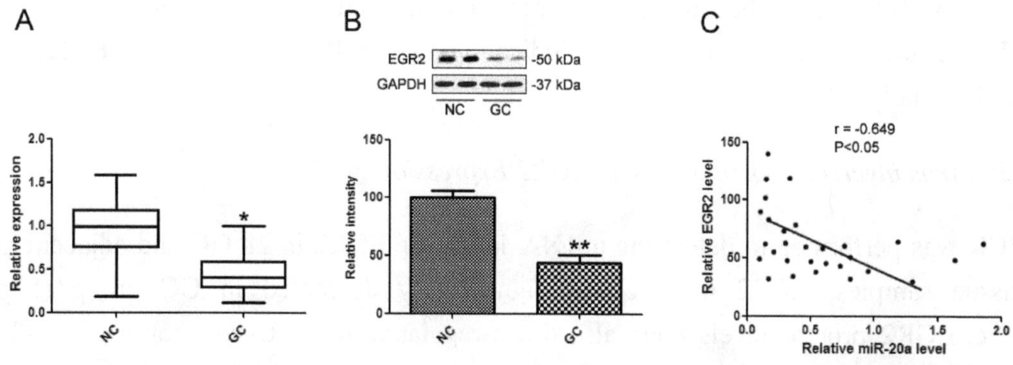

2.7. miR-20a Promoted GC Progression by Targeting EGR2

Since EGR2 was found to be a target of miR-20a, we further investigated whether overexpression of EGR2 could attenuate the oncogenic effect of miR-20a. MTT assay (Figure 7A), colony formation assay (Figure 7B), cell migration (Figure 7C) and invasion (Figure 7D) all showed that supplement of EGR2 by an EGR2 overexpression plasmid could significantly attenuate the oncogenic effect of miR-20a. The effect of EGR2 plasmid was validated by qRT-PCR (Figure 7E) and Western blot (Figure 7F). These data suggested that miR-20a promoted GC progression partially by targeting EGR2.

Figure 7. miR-20a promoted GC progression by targeting EGR2. (**A**) SGC7901 cells were co-transfected with miR-20a precursor and EGR2 overexpression plasmid or the control, at 24, 48, 36, 72, or 96 h after transfection, MTT assay was performed to examine SGC7901 proliferation; (**B**) Representative results of colony formation assay in SGC7901 cells co-transfected with miR-20a precursor and EGR2 overexpression plasmid or the control; (**C**) *In vitro* migration assay of SGC7901 cells transfected with miR-20a precursor and EGR2 overexpression plasmid or the control; (**D**) *In vitro* invasion assay of SGC7901 cells transfected with miR-20a precursor and EGR2 overexpression plasmid or the control; (**E**) qRT-PCR was used to detect the mRNA level of EGR2 in in cells transfected with EGR2 overexpression plasmid or the control; (**F**) EGR2 protein level was detected by Western blot in cells transfected with EGR2 overexpression plasmid or the control. GAPDH was used as an internal control. The data represented at least four measurements. * $p < 0.05$, ** $p < 0.01$ compared with control; $^{\#}$ $p < 0.05$, compared with miR-20a precursor transfected group.

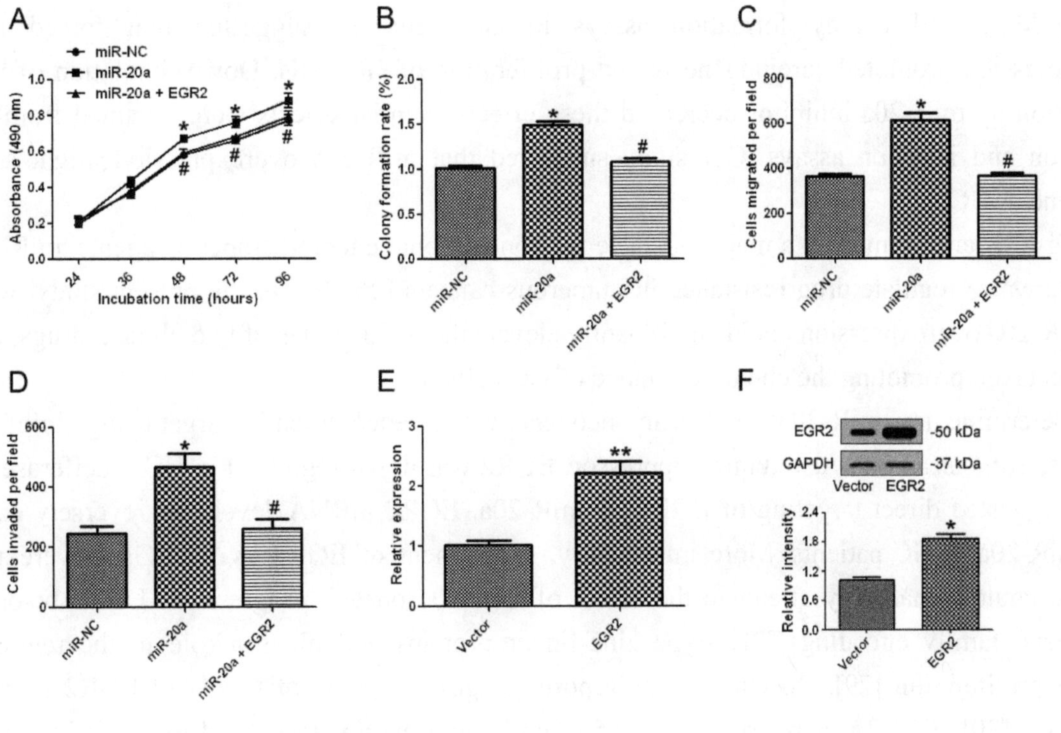

3. Discussion

Although numerous miRNAs had been identified in GC carcinogenesis, their underlying molecular mechanisms in GC development still remained poorly understood. Hence, exploring the function of miRNAs specifically involved in GC carcinogenesis would greatly help expand our understanding of GC and screening new targets for its diagnosis and therapy [18]. The aberrant expression of miR-20a in cervical cancer, colorectal cancer, GC, and prostate cancer had been found [16,19–21]. A recent study suggested that miR-20a overexpression in rheumatoid fibroblast-like synoviocytes decreased apoptosis signal-regulating kinase 1 (ASK1) activity, indicating an anti-apoptotic effect [22]. miR-20a was up-regulated in gallbladder carcinoma [23], but down-regulated in hepatocellular carcinoma [24]. The difference indicated that dysregulation of miR-20a in different cancers depended on the cellular microenvironment.

In this study, we validated that miR-20a was increased in a number of GC tissue samples and cell lines SGC7901, MKN45, and NUGC-3 cells compared to GES-1 cells. Ectopic expression of miR-20a promoted proliferation of GC cells, while suppression of miR-20a with inhibitor had the opposite effect. miR-20a was found to inhibit EGR2 expression partially by inducing mRNA decay. Similarly Wang M and colleagues reported that the levels of circulating miR-17-5p/20a might be a molecular marker for GC [20]. They found that overexpression of miR-17-5p/20a promoted GC cell cycle progression and inhibited apoptosis, whereas knockdown of miR-17-5p/20a resulted in cell cycle arrest and increased apoptosis in AGS cells and *in vivo* as well by targeting p21 and tumor protein p53-induced nuclear protein 1 (TP53INP1) [25]. In our study, overexpression of miR-20a promoted proliferation without effect on apoptosis in SGC7901 and MKN45 cell lines *in vitro*. The contradictory results might due to different cell lines, Wang M *et al.* used AGS cell line, and we used SGC7901 and MKN45 cells. Secondly, different molecules were involved in the two studies.

The MTT and colony formation assays in GC cells all suggested that forced miR-20a overexpression promoted carcinogenesis and proliferation of GC cells. Down-regulation of miR-20a expression by miR-20a inhibitor decreased these effects. Similar effects were obtained from *in vitro* migration and invasion assays. Our study suggested that miR-20a overexpression might act as an oncogene in GC.

Drug resistance remained a major obstacle for conventional chemotherapeutic agents. miRNAs had been shown to regulate drug resistance in numerous cancers [26–28]. In the present study, we found that miR-20a overexpression could significantly elevate the IC50 values of two clinical drugs, cisplatin and docetaxel, promoting the chemoresistance of GC cells.

To determine how miR-20a acted as an oncogene, we screened potential target using bioinformatics analysis, Targetscan 6.2. As a tumor suppressor, EGR2 was down-regulated in GC. Luciferase activity assay suggested direct targeting of EGR2 by miR-20a. EGR2 mRNA level was reversely correlated with miR-20a in GC patients. More importantly, supplement of EGR2 by an EGR2 overexpression plasmid could remarkably attenuate the effect of miR-20a on GC progression. EGR2 belongs to a multi-gene family encoding C2H2-type zinc-finger proteins and plays a role in the regulation of cellular proliferation [29]. Recently, some reports suggested an essential role of EGR2 in apoptosis regulation [30]. EGR2 was a target of the p53 family, and overexpression of EGR2 led to apoptosis, while down-regulation of EGR2 attenuated p53-mediated apoptosis [31]. EGR2 could be also regulated by other miRNAs. Wu Q *et al.* has reported that miR-150 could promote GC proliferation by negatively regulating EGR2 expression [32]. Our study shed new light on the regulation of EGR2.

In summary, our study validated that miR-20a was dramatically increased in GC tissues and cell lines and that ectopic expression of miR-20a promoted proliferation, migration, invasion and chemoresistance of GC cells. Moreover, down-regulation of miR-20a had the opposite effect on GC cells by targeting EGR2. Further studies for the functional and clinical implications of miR-20a and its target EGR2 might contribute to the early diagnosis and treatment of GC.

4. Materials and Methods

4.1. Cell lines and Tissue Samples

Human GC cell lines SGC7901, MKN45, NUGC-3, and human gastric mucosa cell line GES-1 were from the Chinese Academy of Sciences (Shanghai, China). Cells were grown in Dulbecco's modified Eagle medium supplemented with 10% fetal bovine serum at 37 °C with 5% CO_2. Twenty eight paired tissues of GC and matched normal tissues (located >5 cm away from the tumor) were collected from our department. Informed consent was obtained from each patient and this work was approved by the Ethics Committee of Third Military Medical University.

4.2. Plasmids and Transfection

miR-20a precursor (Catalog# HmiR0202-MR01, shown as miR-20a in figures) and inhibitor (Catalog# HmiR-AN0312-AM01, shown as anti-miR-20a) constructs were purchased from GeneCopoeia (Rockville, MD, USA). pEGFP-N1-EGR2 plasmid was generated by using the following primers, sense 5'-CCCTCGAGATCCCAGGCTCAGTCCAACC-3', antisense 5'-CCAAGCTTAG GTGTCCGGGTCCGAGA-3'. The amplified sequences were inserted into pEGFP-N1 within XhoI/HindIII sites. Transfection was performed using Lipofectamine 2000 reagent (Invitrogen, Carlsbad, CA, USA). Transfected cells were harvested after 24 h for migration and invasion assays and after 48 h for RNA isolation and Western blot.

4.3. Quantitative Real Time PCR

Total RNA was extracted using TRIzol (Invitrogen, Carlsbad, CA, USA) then reverse-transcribed into cDNA with a TaqMan MicroRNA Reverse-Transcription Kit (Applied Biosystems, Foster City, CA, USA). PCR reactions were performed using the ABI Stepone plus Detection System (Applied Biosystems, Foster City, CA, USA). The relative expression of miRNA was normalized with U6. Samples were compared by using the relative CT method. The fold change was determined relative to a control after normalizing to a housekeeping gene by using $2^{-\Delta\Delta CT}$, where ACT is (gene of interest CT) minus (GAPDH CT), and ΔΔCT is (ΔCT cancer) minus (ΔCT control). The relative expression of EGR2 was normalized with GAPDH. All experiments were carried out at least in triplicate.

4.4. In Vitro *Cell Proliferation Assay*

Forty eight hours after transfection, cells were seeded into 96-well plates at 6×10^3 cells/well. The MTT assay was used to measure cell viability. Optical densities at 490 nm were measured.

4.5. Colony Formation Assay

Cells transfected with miR-20a precursor/inhibitor or the corresponding control were seeded in a 10 cm dish and maintained in complete culture medium. After 21 days, SGC7901 cells were fixed with methanol and then stained with 0.1% crystal violet. Colonies were manually counted.

4.6. In Vitro *Migration and Invasion Assay*

In vitro migration assay was performed using 8 μm pore size Transwell plates (Millipore, Billerica, MA, USA). After transfection, the cells (1×10^5 cells/100 μL serum-free medium) were added to the upper chamber. RPMI 1640 containing 10% FCS was added to the bottom chamber as a chemoattractant. After 24 h, cells on the upper surface were removed, while cells attached on the bottom were fixed and stained with 0.1% crystal violet. The images of invaded cells were counted with a photomicroscope (Olympus, Tokyo, Japan). For the *in vitro* invasion assay, the plates were coated with matrigel (BD Biosciences, San Jose, CA, USA) diluted in serum-free medium and performed the same as migration assay.

4.7. Luciferase Reporter Assay

The 3' UTR of human EGR2 was PCR-amplified and cloned into psiCHECK-2 vector. These constructs (1 μg) were co-transfected with 1 μg of control or miR-20a precursor into SGC7901 cell. Luciferase activity was assayed 48 h after transfection by using the Dual-luciferase activity assay system (Promega, Madison, WI, USA). All the experiments were performed at least in four times.

4.8. Western Blot

Cells were harvested 48 h after transfection, and then proteins were extracted and separated by 10% SDS-PAGE. Gels were transferred to nitrocellulose filter membrane and probed with anti-EGR2 or anti-GAPDH. Horseradish peroxidase-conjugated secondary antibodies were applied and detected by enhanced chemiluminescence (Thermo Scientific, Rockford, IL, USA).

4.9. Flow Cytometric Analysis of Apoptosis

Cells were harvested at the above indicated time points, at least 5×10^5 cells were recovered by centrifugation for evaluation of apoptotic cells with the use of double staining with annexin V-fluoresein isothiocyanate (annexin V-FITC) and propidium iodide (PI) (BioVision, St. Pete Beach, FL, USA) according to the manufacturer's instructions, followed by flow cytometric analysis with the use of Cell Quest software (Version 5.1, Becton, Rutherford, NJ, USA).

4.10. Cell Cycle Analysis by Flow Cytometry

Cells were harvested 48 h after transfection and fixed in 75% ethanol. Then, cells were treated with RNase A and PI (50 μg/mL) and incubated for 30 min. Cell cycle analysis was performed by Flow cytometry.

4.11. Statistical Analysis

All experiments were repeated independently at least in triplicate, and the results were expressed as the mean ± SD. The results were assessed by a one-way ANOVA, or Student *t* test. A value of $p < 0.05$ was accepted to indicate statistical significance.

5. Conclusions

Our work demonstrated that miR-20a was involved in the carcinogenesis of GC through modulation of the EGR2 signaling pathway.

References

1. Jemal, A.; Bray, F.; Center, M.M.; Ferlay, J.; Ward, E.; Forman, D. Global cancer statistics. *CA Cancer J. Clin.* **2011**, *61*, 69–90.
2. Ali, Z.; Deng, Y.; Ma, C. Progress of research in gastric cancer. *J. Nanosci. Nanotechnol.* **2012**, *12*, 8241–8248.
3. Link, A.; Kupcinskas, J.; Wex, T.; Malfertheiner, P. Macro-role of microRNA in gastric cancer. *Dig. Dis.* **2012**, *30*, 255–267.
4. Song, J.H.; Meltzer, S.J. microRNAs in pathogenesis, diagnosis, and treatment of gastroesophageal cancers. *Gastroenterology* **2012**, *143*, 35–47.
5. Liu, H. microRNAs in breast cancer initiation and progression. *Cell. Mol. Life Sci.* **2012**, *69*, 3587–3599.
6. Rossi, S.; di Narzo, A.F.; Mestdagh, P.; Jacobs, B.; Bosman, F.T.; Gustavsson, B.; Majoie, B.; Roth, A.; Vandesompele, J.; Rigoutsos, I.; *et al.* microRNAs in colon cancer: A roadmap for discovery. *FEBS Lett.* **2012**, *586*, 3000–3007.
7. Du, L.; Pertsemlidis, A. microRNA regulation of cell viability and drug sensitivity in lung cancer. *Expert Opin. Biol. Ther.* **2012**, *12*, 1221–1239.
8. O'Kelly, F.; Marignol, L.; Meunier, A.; Lynch, T.H.; Perry, A.S.; Hollywood, D. microRNAs as putative mediators of treatment response in prostate cancer. *Nat. Rev. Urol.* **2012**, *9*, 397–407.
9. Guo, M.M.; Hu, L.H.; Wang, Y.Q.; Chen, P.; Huang, J.G.; Lu, N.; He, J.H.; Liao, C.G. miR-22 is down-regulated in gastric cancer, and its overexpression inhibits cell migration and invasion via targeting transcription factor Sp1. *Med. Oncol.* **2013**, *30*, 542.
10. Yao, Q.; Cao, Z.; Tu, C.; Zhao, Y.; Liu, H.; Zhang, S. microRNA-146a acts as a metastasis suppressor in gastric cancer by targeting WASF2. *Cancer Lett.* **2013**, *335*, 219–224.
11. Zheng, L.; Qi, T.; Yang, D.; Qi, M.; Li, D.; Xiang, X.; Huang, K.; Tong, Q. microRNA-9 suppresses the proliferation, invasion and metastasis of gastric cancer cells through targeting cyclin D1 and Ets1. *PLoS One* **2013**, *8*, e55719.
12. Wang, F.; Li, T.; Zhang, B.; Li, H.; Wu, Q.; Yang, L.; Nie, Y.; Wu, K.; Shi, Y.; Fan, D. microRNA-19a/b regulates multidrug resistance in human gastric cancer cells by targeting PTEN. *Biochem. Biophys. Res. Commun.* **2013**, *434*, 688–694.
13. Wang, M.; Li, C.; Yu, B.; Su, L.; Li, J.; Ju, J.; Yu, Y.; Gu, Q.; Zhu, Z.; Liu, B. Overexpressed miR-301a promotes cell proliferation and invasion by targeting RUNX3 in gastric cancer. *J. Gastroenterol.* **2013**, [Epub ahead of print].

14. An, J.; Pan, Y.; Yan, Z.; Li, W.; Cui, J.; Jiao, Y.; Tian, L.; Xing, R.; Lu, Y. miR-23a in Amplified 19p13.13 loci targets metallothionein 2A and promotes growth in gastric cancer cells. *J. Cell. Biochem.***2013**, *114*, 2160–2169.
15. Yan, H.J.; Liu, W.S.; Sun, W.H.; Wu, J.; Ji, M.; Wang, Q.; Zheng, X.; Jiang, J.T.; Wu, C.P. miR-17-5p inhibitor enhances chemosensitivity to gemcitabine via upregulating Bim expression in pancreatic cancer cells. *Dig. Dis. Sci.* **2012**, *57*, 3160–3167.
16. Kang, H.W.; Wang, F.; Wei, Q.; Zhao, Y.F.; Liu, M.; Li, X.; Tang, H. miR-20a promotes migration and invasion by regulating TNKS2 in human cervical cancer cells. *FEBS Lett.* **2012**, *586*, 897–904.
17. Huang, G.; Nishimoto, K.; Zhou, Z.; Hughes, D.; Kleinerman, E.S. miR-20a encoded by the miR-17-92 cluster increases the metastatic potential of osteosarcoma cells by regulating Fas expression. *Cancer Res.* **2012**, *72*, 908–916.
18. Gigek, C.O.; Chen, E.S.; Calcagno, D.Q.; Wisnieski, F.; Burbano, R.R.; Smith, M.A. Epigenetic mechanisms in gastric cancer. *Epigenomics* **2012**, *4*, 279–294.
19. Li, X.; Pan, J.H.; Song, B.; Xiong, E.Q.; Chen, Z.W.; Zhou, Z.S.; Su, Y.P. Suppression of CX43 expression by miR-20a in the progression of human prostate cancer. *Cancer Biol. Ther.* **2012**, *13*, 890–898.
20. Wang, M.; Gu, H.; Wang, S.; Qian, H.; Zhu, W.; Zhang, L.; Zhao, C.; Tao, Y.; Xu, W. Circulating miR-17-5p and miR-20a: Molecular markers for gastric cancer. *Mol. Med. Rep.* **2012**, *5*, 1514–1520.
21. Chai, H.; Liu, M.; Tian, R.; Li, X.; Tang, H. miR-20a targets BNIP2 and contributes chemotherapeutic resistance in colorectal adenocarcinoma SW480 and SW620 cell lines. *Acta Biochim. Biophys. Sin.* **2011**, *43*, 217–225.
22. Philippe, L.; Alsaleh, G.; Pichot, A.; Ostermann, E.; Zuber, G.; Frisch, B.; Sibilia, J.; Pfeffer, S.; Bahram, S.; Wachsmann, D.; *et al.* miR-20a regulates ASK1 expression and TLR4-dependent cytokine release in rheumatoid fibroblast-like synoviocytes. *Annu. Rheum. Dis.* **2013**, *72*, 1071–1079.
23. Chang, Y.; Liu, C.; Yang, J.; Liu, G.; Feng, F.; Tang, J.; Hu, L.; Li, L.; Jiang, F.; Chen, C.; *et al.* miR-20a triggers metastasis of gallbladder carcinoma. *J. Hepatol.* **2013**, doi:10.1016/j.jhep.2013.04.034.
24. Fan, M.Q.; Huang, C.B.; Gu, Y.; Xiao, Y.; Sheng, J.X.; Zhong, L. Decrease expression of microRNA-20a promotes cancer cell proliferation and predicts poor survival of hepatocellular carcinoma. *J. Exp. Clin. Cancer Res.* **2013**, *32*, 21.
25. Wang, M.; Gu, H.; Qian, H.; Zhu, W.; Zhao, C.; Zhang, X.; Tao, Y.; Zhang, L.; Xu, W. miR-17-5p/20a are important markers for gastric cancer and murine double minute 2 participates in their functional regulation. *Eur. J. Cancer* **2013**, *49*, 2010–2021.
26. Wang, F.; Song, X.; Li, X.; Xin, J.; Wang, S.; Yang, W.; Wang, J.; Wu, K.; Chen, X.; Liang, J.; *et al.* Noninvasive visualization of microRNA-16 in the chemoresistance of gastric cancer using a dual reporter gene imaging system. *PLoS One* **2013**, *8*, e61792.
27. Tekiner, T.A.; Basaga, H. Role of microRNA deregulation in breast cancer cell chemoresistance and stemness. *Curr. Med. Chem.* **2013**, [Epub ahead of print].

28. Wang, Y.Q.; Guo, R.D.; Guo, R.M.; Sheng, W.; Yin, L.R. microRNA-182 promotes cell growth, invasion, and chemoresistance by targeting programmed cell death 4 (PDCD4) in human ovarian carcinomas. *J. Cell. Biochem.* **2013**, *114*, 1464–1473.
29. Joseph, L.J.; le Beau, M.M.; Jamieson, G.A., Jr.; Acharya, S.; Shows, T.B.; Rowley, J.D.; Sukhatme, V.P. Molecular cloning, sequencing, and mapping of EGR2, a human early growth response gene encoding a protein with "zinc-binding finger" structure. *Proc. Natl. Acad. Sci. USA* **1988**, *85*, 7164–7168.
30. Dzialo-Hatton, R.; Milbrandt, J.; Hockett, R.D., Jr.; Weaver, C.T. Differential expression of Fas ligand in Th1 and Th2 cells is regulated by early growth response gene and NF-AT family members. *J. Immunol.* **2001**, *166*, 4534–4542.
31. Yokota, I.; Sasaki, Y.; Kashima, L.; Idogawa, M.; Tokino, T. Identification and characterization of early growth response 2, a zinc-finger transcription factor, as a p53-regulated proapoptotic gene. *Int. J. Oncol.* **2010**, *37*, 1407–1416.
32. Wu, Q.; Jin, H.; Yang, Z.; Luo, G.; Lu, Y.; Li, K.; Ren, G.; Su, T.; Pan, Y.; Feng, B.; *et al.* miR-150 promotes gastric cancer proliferation by negatively regulating the pro-apoptotic gene EGR2. *Biochem. Biophys. Res. Commun.* **2010**, *392*, 340–345.

6

MicroRNA Expression Changes during Interferon-Beta Treatment in the Peripheral Blood of Multiple Sclerosis Patients

Michael Hecker [1,2,†,*], Madhan Thamilarasan [2,†], Dirk Koczan [3], Ina Schröder [2], Kristin Flechtner [3], Sherry Freiesleben [4], Georg Füllen [4], Hans-Jürgen Thiesen [3] and Uwe Klaus Zettl [2]

[1] Steinbeis Transfer Center for Proteome Analysis, Schillingallee 68, 18057 Rostock, Germany

[2] Department of Neurology, Division of Neuroimmunology, University of Rostock, Gehlsheimer Str. 20, 18147 Rostock, Germany; E-Mails: madhangp@gmail.com (M.T.); ina.schroeder@med.uni-rostock.de (I.S.); uwe.zettl@med.uni-rostock.de (U.K.Z.)

[3] Institute of Immunology, University of Rostock, Schillingallee 70, 18055 Rostock, Germany; E-Mails: dirk.koczan@med.uni-rostock.de (D.K.); kristin.flechtner@googlemail.com (K.F.); hj.thiesen@gmx.de (H.-J.T.)

[4] Institute for Biostatistics and Informatics in Medicine and Ageing Research, University of Rostock, Ernst-Heydemann-Str. 8, 18057 Rostock, Germany; E-Mails: sherry.freiesleben@uni-rostock.de (S.F.); fuellen@uni-rostock.de (G.F.)

† These authors contributed equally to this work.

* Author to whom correspondence should be addressed; E-Mail: michael.hecker@rocketmail.com;

Abstract: MicroRNAs (miRNAs) are small non-coding RNA molecules acting as post-transcriptional regulators of gene expression. They are involved in many biological processes, and their dysregulation is implicated in various diseases, including multiple sclerosis (MS). Interferon-beta (IFN-beta) is widely used as a first-line immunomodulatory treatment of MS patients. Here, we present the first longitudinal study on the miRNA expression changes in response to IFN-beta therapy. Peripheral blood mononuclear cells (PBMC) were obtained before treatment initiation as well as after two days, four days, and one month, from patients with clinically isolated syndrome (CIS) and patients with relapsing-remitting MS (RRMS). We measured the expression of 651 mature miRNAs and about 19,000 mRNAs in parallel using real-time PCR arrays and Affymetrix microarrays.

We observed that the up-regulation of IFN-beta-responsive genes is accompanied by a down-regulation of several miRNAs, including members of the mir-29 family. These differentially expressed miRNAs were found to be associated with apoptotic processes and IFN feedback loops. A network of miRNA-mRNA target interactions was constructed by integrating the information from different databases. Our results suggest that miRNA-mediated regulation plays an important role in the mechanisms of action of IFN-beta, not only in the treatment of MS but also in normal immune responses. miRNA expression levels in the blood may serve as a biomarker of the biological effects of IFN-beta therapy that may predict individual disease activity and progression.

Keywords: interferon-beta; multiple sclerosis; peripheral blood; microRNA; gene expression

---

1. Introduction

Multiple sclerosis (MS) is a chronic disease of the central nervous system (CNS), which is characterized by multiple discrete areas of inflammatory demyelination, axonal degeneration, and glial scarring. The resulting loss of neurons and axons leads to diverse neurological symptoms, progressive disability, and a significant decrease in quality of life. The disease usually begins in early adulthood, and is more common in females. Different types of MS are distinguished: In about 85% of patients, the disease starts with a single demyelinating episode (clinically isolated syndrome, CIS) and progresses to a relapsing-remitting course (RRMS) with acute exacerbations and periods of remission [1–4].

A number of disease-modifying therapies for MS are available, and they are especially effective when applied in the early stages of the disease [4,5]. Injections of recombinant interferon-beta (IFN-beta) are considered a first-line option in the treatment of RRMS. IFN-beta has been shown to reduce the number of relapses and to suppress the accumulation of new inflammatory lesions in the brain. Three different preparations of IFN-beta are in clinical use. They differ in dose, route, and frequency of IFN-beta administration, but they are comparable regarding clinical efficacy [6].

IFN-beta has broad effects on the gene regulation of blood cells [7–11]. This has been shown by several studies that used microarray technology to analyze the gene expression dynamics in the peripheral blood of MS patients in response to IFN-beta therapy. In this way, more than a hundred genes have consistently been found differentially expressed during treatment [10]. The transcript levels of most of these genes are up-regulated within a few hours after IFN-beta injection, and they return to pre-treatment levels after a few days [12,13]. These IFN-beta-responsive genes are believed to mediate the beneficial effects of the treatment through immunomodulatory, antiproliferative, and antipathogenic processes [7–9].

While the therapeutic effects on the regulation of mRNAs have been extensively investigated, studies on the regulation of microRNAs (miRNAs) are lacking. miRNAs are a distinct class of small (~22 nt) non-coding RNA molecules [14]. They originate from precursor RNAs (pre-miRNAs) found in longer primary transcripts (pri-miRNAs), which often also contain the exons of an mRNA. Mature miRNAs act as post-transcriptional regulators. They repress gene expression via base-pairing with complementary sequences within the 3' untranslated regions (UTRs) of target mRNAs. This interaction results in gene silencing by translational repression or target degradation. A miRNA can have hundreds of different mRNA targets, and a target might be regulated by a combination of multiple miRNAs [15]. The human genome encodes over 1000 miRNAs [16,17]. miRNAs are thus likely involved in most biological processes, and they play essential roles in the immune system and in the correct function of the CNS [18,19].

Dysregulated expression of miRNAs is associated with pathological conditions, including neurological diseases. Human MS studies showed altered miRNA expression in peripheral blood samples, lymphocyte subpopulations, and active CNS lesions from MS patients [20–22]. Studies with the animal model of MS, experimental autoimmune encephalomyelitis (EAE), also support the involvement of miRNAs in this disease [23,24]. These findings provided important insights into the pathophysiology of MS and opened a new avenue in biomarker research. If miRNA levels in the blood or brain of MS patients correlate with disease stage and progression of disability, they may also support early diagnosis and effective treatment in future [25].

In order to better understand the molecular mechanisms of action of IFN-beta therapy, it is important to investigate the miRNA expression dynamics during therapy. The regulation of miRNAs may contribute to the immunomodulatory and clinical effects of the treatment. Moreover, miRNAs might be markers for characterizing the biological response to IFN-beta. miRNA biomarkers for treatment monitoring could be useful in the individual management of disease activity. However, so far, there is only one study on the expression of miRNAs in MS patients during IFN-beta therapy: Waschbisch *et al.* obtained peripheral blood mononuclear cells (PBMC) from patients with RRMS, and analyzed the expression of five selected miRNAs by real-time PCR [26]. They compared the miRNA levels between treatment-naive patients (n = 36), IFN-beta-treated patients (n = 18), and patients treated with glatiramer acetate (GA, n = 20). As a result, none of the five miRNAs was differentially expressed in IFN-beta-treated patients, but *hsa-miR-146a-5p* and *hsa-miR-142-3p* were expressed at significantly lower levels in GA-treated patients [26]. Other researchers used microarrays to study the expression of hundreds of miRNAs in IFN-stimulated cells. In this way, O'Connell *et al.* observed that *hsa-miR-155-5p* is induced in primary murine macrophages after exposure to IFN-beta for 6 h [27]. Pedersen *et al.* studied the regulation of miRNAs in Huh7 cells and primary hepatocytes, which were stimulated with different concentrations of IFN-beta for up to 48 h [28]. They observed increased and reduced miRNA expression in response to IFN-beta, and showed that some of the IFN-beta-induced miRNAs mediate antiviral effects against hepatitis C virus. This provides an example of miRNAs as components of the innate immune response.

In this study, we used microarrays to investigate in parallel the expression dynamics of mRNAs and miRNAs in PBMC of patients with CIS or RRMS in response to therapy with subcutaneous (sc.) IFN-beta. The blood samples were obtained longitudinally from six patients at four time points in the early phase of therapy, namely before the first (baseline), second, and third IFN-beta injection as well as after one month of treatment. We then screened for significant changes in miRNA and mRNA expression, and identified several miRNAs as differentially expressed during therapy. Information of different databases was then integrated [22,29] to examine whether the expression of these miRNAs is cell type-specific and correlates with the levels of their target mRNAs. Predicted and experimentally verified miRNA-mRNA interactions were compiled to construct a network of IFN-beta-responsive genes and miRNAs. To our knowledge, this is the first genome-wide miRNA profiling study on the *in vivo* effects of IFN-beta treatment in MS.

2. Results and Discussion

2.1. Study Population

Six female patients of Western European descent, and diagnosed with CIS (n = 2) or RRMS (n = 4), were recruited for this study (Pat1-6, mean age 37.5 years, Table 1). The patients were treatment-naive and started an immunomodulatory therapy with IFN-beta-1b (Betaferon, Bayer HealthCare) administered subcutaneously every other day. In the first weeks, the Betaferon titration pack was used, hence the patients started with a low dose (62.5 µg for the first three injections) that was gradually increased to the full dose (250 µg) after three weeks. All patients were continuously treated with IFN-beta-1b for at least one year. During follow-up, they were monitored for relapses and rated using the Expanded Disability Status Scale (EDSS). The individual disease activity during therapy was relatively low: Four of the patients (Pat1-4) were relapse-free and showed no disability progression within the first year of treatment (Table 1). The two patients with CIS (Pat1 and Pat5) did not convert to clinically definite MS in this period of time.

Table 1. Clinical data and demographic data of the patients.

Patient	Type	Age	Disease duration	EDSS (baseline)	EDSS (1 year)	Relapses (1 year)
Pat1	CIS	28	1	1.0	1.0	0
Pat2	RRMS	38	2	1.5	1.5	0
Pat3	RRMS	22	1	1.5	1.0	0
Pat4	RRMS	50	12	2.5	2.5	0
Pat5	CIS	60	2	1.5	2.5	0
Pat6	RRMS	27	2	2.0	1.0	2

Six female patients were recruited for the main cohort of this study. They were diagnosed with relapsing-remitting MS (RRMS) or clinically isolated syndrome (CIS) suggestive of MS. The age at study onset (in years) and the duration from the diagnosis to the start of IFN-beta-1b sc. therapy (in months) are shown. Additionally, the EDSS scores before treatment initiation (baseline) and after one year, as well as the number of relapses during the first year of clinical follow-up are given in the table.

Note that the patient group included only women. A differential hormonal regulation of immune system genes in blood cells has been described for different phases of the menstrual cycle [30]. Such differences in gene expression may have led to increased variance in the data. However, prior mRNA profiling studies observed no significant gender-specific differences in the gene expression signature in response to IFN-beta therapy [8,31], and this seems to be the case regarding the expression of miRNAs as well (see Section 2.5).

2.2. Parallel Measurement of mRNAs and MicroRNAs in Blood Cells

Patient blood samples were drawn immediately before first IFN-beta injection as well as two days, four days, and one month post therapy initiation. Total RNA of Ficoll-isolated PBMC from each sample was extracted to measure the levels of mRNAs and miRNAs with different platforms. We used

TaqMan Array Human MicroRNA cards (Applied Biosystems, Foster City, CA, USA) to quantify the expression of 651 mature miRNAs and Affymetrix HG-U133 Plus 2.0 microarrays (Affymetrix, Santa Clara, CA, USA) to quantify the expression of about 19,000 mRNAs. In this way, we obtained in parallel the mRNA and miRNA expression profiles from six patients (Pat1-6) within the first month of IFN-beta treatment.

The data were preprocessed as described in Section 3.6. Relatively low variation in the transcriptome profiles indicated high data quality, comparable to our previous microarray time course data sets [8,31,32]. In the miRNA data, systematic and stochastic variation was higher. For the TaqMan miRNA B-cards of Pat5, the raw threshold cycle (Ct) values were generally higher (Supplementary File 1) due to an unknown measurement bias. In the PBMC samples of the other five patients, approximately 400 miRNAs could be detected (Ct < 38) with the TaqMan miRNA arrays (Table 2). The raw TaqMan data were transformed to the linear scale, and coefficients of variation (*CV*) were calculated to assess the effects of data normalization. The normalization decreased the average *CV* over all 768 measured assays from 0.953 to 0.894. The *CV* for the housekeeping miRNA *hsa-miR-191-5p* [33] was 0.183. In comparison, assays for the non-coding RNAs *U6*, *U44* and *U48* had *CV*s of >0.35. The *CV* for the housekeeping mRNA GAPDH was 0.071.

Table 2. Numbers of microRNAs detected in the samples with the TaqMan cards set.

Patient	Baseline	~48 h	~96 h	1 month
Pat1	380	374	364	373
Pat2	375	407	392	362
Pat3	427	400	408	407
Pat4	390	387	388	389
Pat5	285	258	266	262
Pat6	431	451	446	391

Approximately 400 miRNAs were found to be expressed (raw Ct value < 38) per sample. The numbers are lower for the PBMC samples of patient Pat5, where generally lower miRNA amounts were measured with the B-cards for all four time points (Supplementary File 1).

2.3. Analysis of mRNA Expression Dynamics

We filtered for mRNAs showing strong expression changes in response to IFN-beta therapy by comparing the baseline expression levels with the expression levels at the three time points during treatment. The MAID filtering method [34] was used to analyze the mRNA dynamics. As a result, 14, 34, and 66 genes were found to be expressed at higher or lower levels after two days, four days, and one month, respectively. In total, 95 genes were identified as up-regulated ($n = 75$) or down-regulated ($n = 20$) in the early course of the therapy (Supplementary File 2). The gene expression changes in the first month of subcutaneous IFN-beta-1b treatment are visualized in the heat map in Figure 1.

A permutation test (see Section 3.7) revealed that the number of 95 differentially expressed genes is significantly higher than would be expected by chance. In randomly permuted data sets, 29.6 genes on average were filtered. The number of filtered genes was below 95 in 98.8% of the permutations, which implies an empirical p-value of <0.05, demonstrating that most of the mRNA expression changes that we found are indeed due to the therapy.

Figure 1. Heat map visualization of the mRNA expression changes in response to IFN-beta. Shown are the baseline and one month transcript levels of the 95 genes that were identified as differentially expressed during IFN-beta therapy. The patient samples are represented in the columns, the genes are represented in the rows, and the gene expression levels were centered and scaled in row direction (z-scores). The clustering analysis clearly separated the PBMC samples obtained at baseline and after one month of therapy. The row labels of the heat map (*i.e.*, the respective genes) are given in Supplementary File 2. The upper half of the heat map contains most of the IFN-beta-induced genes.

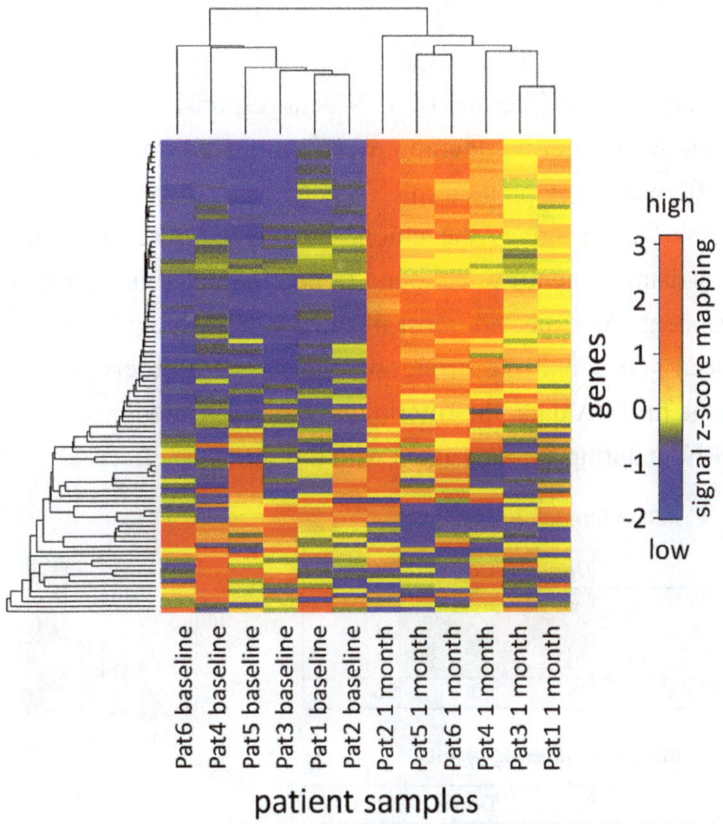

Despite the relatively small patient cohort ($n = 6$), the mRNA results were quite consistent with the literature. In another study, we already analyzed the PBMC gene expression profiles of a larger group of MS patients ($n = 25$) treated with IFN-beta-1b sc. [31]. In total, 63 of the 95 differentially expressed genes were also filtered in the previous study. Recently, we completed a similar microarray study on the effects of IFN-beta-1a sc. [8], in which 49 of the 95 genes were already identified as transcriptionally modulated (Supplementary File 2). Moreover, more up-regulated than down-regulated genes were filtered, which also confirms previous findings [10]. After the first IFN-beta injections, fewer genes were altered in expression than after one month (*cf.* Goertsches *et al.*, 2010 [31]). This can be explained by the fact that the patients started the first week of the therapy with a quarter of the full dose (Betaferon titration pack). Most of the filtered genes are part of an up-regulated type I IFN signature. For instance, *IFI6*, *IFI44L*, and *SIGLEC1* are known type I IFN-induced genes, which were up-regulated at all time points during therapy in comparison to baseline. In contrast, FCER1A was consistently down-regulated in response to therapy as has been described previously as well [8,31,32]. Regarding the functions of these genes, the reader is referred to the literature [7-9].

2.4. Analysis of MicroRNA Expression Dynamics

As for the mRNA data, the MAID filtering method [34] was used to identify miRNAs that are differentially expressed in PBMC within the first month of IFN-beta treatment. When we compared the expression levels at the three time points during therapy with the expression levels at baseline, 20 different miRNAs were filtered. Of these, seven miRNAs appeared as up-regulated and 13 miRNAs appeared as down-regulated in response to the therapy (Figure 2, Table 3 and Supplementary File 3). According to the permutation test, this is significantly more than expected by chance: In only 0.6% of the randomly permuted data sets, 20 (or more) miRNAs were filtered, and only 7.7 miRNAs were filtered on average.

Figure 2. Summary of the filtering of IFN-beta-responsive mRNAs and microRNAs. PBMC expression levels during therapy were compared to pre-treatment levels. The number (#) of differentially expressed mRNAs and miRNAs is depicted in the tables and bar plots. The row "Total" gives the union set over all three time point comparisons. (**A**) In the mRNA data, 95 genes were found to be modulated in expression in response to IFN-beta-1b treatment. As expected, most of them were up-regulated (n = 75) and known type I IFN-induced genes, and the strongest changes were observed at one month versus baseline; (**B**) In the miRNA data, the filtering method identified more down-regulated than up-regulated miRNAs during therapy, again with the strongest effects seen after one month.

A mRNA filtering result

	Up	Down	Total
48 h	10	4	14
96 h	22	12	34
1 month	60	6	66
Total	75	20	95

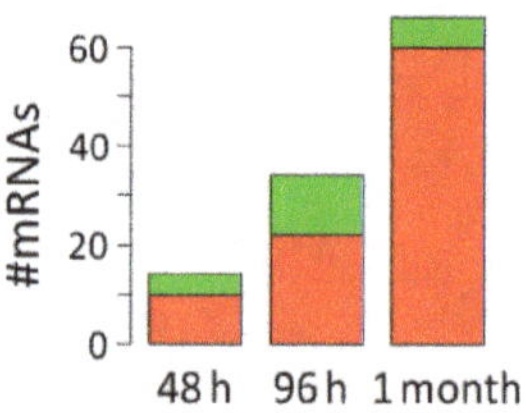

B microRNA filtering result

	Up	Down	Total
48 h	1	1	2
96 h	0	6	6
1 month	6	8	14
Total	7	13	20

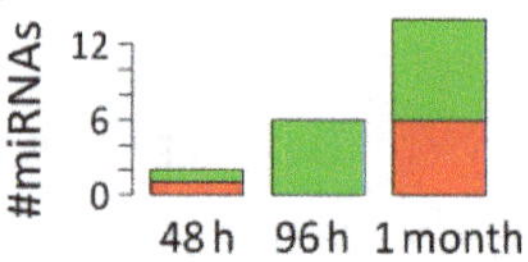

up-regulated
down-regulated

Two of the 20 miRNAs (*hsa-miR-149-5p* and *hsa-miR-708-5p*) were filtered at two different time points. For the remaining miRNAs, the expression changes were not very stable in the course of therapy. This may be due to the small number of patients and the fact that the accuracy of miRNA measurements is in general limited. Therefore, our list of 20 miRNAs represents candidates that have to be validated in a larger patient cohort using, e.g., single real-time PCR assays.

Most of the miRNAs (n = 14) were filtered as up-regulated or down-regulated one month after IFN-beta-1b sc. treatment initiation. This corresponds to the results of the gene expression profiling, where the strongest changes in mRNA levels were also observed after one month (Figure 2). Previous studies demonstrated that the majority of IFN-beta-responsive genes can be seen at this time point [8,31]. Therefore, we hypothesize that, similarly, the number of miRNAs that are modulated in expression is not much higher after long-term treatment. Instead, the development of neutralizing antibodies (NAb) to IFN-beta might impair the biological response to the drug in some patients [35]. However, further studies are needed to investigate the long-term regulation of miRNAs and the potential effects of NAb.

Table 3. Details of microRNAs differentially expressed during IFN-beta therapy.

Mature miRNA	Sequence	Expression change	Family	pre-miRNA	pri-miRNA	Location
hsa-let-7a-5p	UGAGGUAGUAGGUUGUAUAGUU	up-regulated	let-7	MIRLET7A1		chr9 (q22.32)
				MIRLET7A2	MIR100HG	chr11 (q24.1)
				MIRLET7A3	MIRLET7BHG	chr22 (q13.31)
hsa-let-7b-5p	UGAGGUAGUAGGUUGUGUGGUU	up-regulated	let-7	MIRLET7B	MIRLET7BHG	chr22 (q13.31)
hsa-miR-16-5p	UAGCAGCACGUAAAUAUUGGCG	up-regulated	mir-15	MIR16-1	DLEU2	chr13 (q14.2)
				MIR16-2	SMC4	chr3 (q25.33)
hsa-miR-27a-5p	AGGGCUUAGCUGCUUGUGAGCA	down-regulated	mir-27	MIR27A		chr19 (p13.13)
hsa-miR-29a-3p	UAGCACCAUCUGAAAUCGGUUA	down-regulated	mir-29	MIR29A		chr7 (q32.3)
hsa-miR-29b-1-5p	GCUGGUUUCAUAUGGUGGUUUAGA	down-regulated	mir-29	MIR29B1		chr7 (q32.3)
hsa-miR-29c-3p	UAGCACCAUUUGAAAUCGGUUA	down-regulated	mir-29	MIR29C		chr1 (q32.2)
hsa-miR-95	UUCAACGGGUAUUUAUUGAGCA	down-regulated	mir-95	MIR95	ABLIM2	chr4 (p16.1)
hsa-miR-149-5p	UCUGGCUCCGUGUCUUCACUCCC	down-regulated	mir-149	MIR149	GPC1	chr2 (q37.3)
hsa-miR-181c-3p	AACCAUCGACCGUUGAGUGGAC	down-regulated	mir-181	MIR181C		chr19 (p13.13)
hsa-miR-193a-3p	AACUGGCCUACAAAGUCCCAGU	down-regulated	mir-193	MIR193A		chr17 (q11.2)
hsa-miR-193a-5p	UGGGUCUUUGCGGGCGAGAUGA	down-regulated	mir-193	MIR193A		chr17 (q11.2)
hsa-miR-342-5p	AGGGGUGCUAUCUGUGAUUGA	up-regulated	mir-342	MIR342	EVL	chr14 (q32.2)
hsa-miR-346	UGUCUGCCCGCAUGCCUGCCUCU	up-regulated	mir-346	MIR346	GRID1	chr10 (q23.2)
hsa-miR-423-5p	UGAGGGGCAGAGAGCGAGACUUU	down-regulated	mir-423	MIR423	NSRP1	chr17 (q11.2)
hsa-miR-518b	CAAAGCGCUCCCCUUUAGAGGU	up-regulated	mir-515	MIR518B		chr19 (q13.42)
hsa-miR-532-5p	CAUGCCUUGAGUGUAGGACCGU	down-regulated	mir-188	MIR532	CLCN5	chrX (p11.23)
hsa-miR-708-5p	AAGGAGCUUACAAUCUAGCUGGG	down-regulated	mir-708	MIR708	TENM4	chr11 (q14.1)
hsa-miR-760	CGGCUCUGGGUCUGUGGGGA	up-regulated	mir-760	MIR760		chr1 (p22.1)
hsa-miR-874	CUGCCCUGGCCCGAGGGACCGA	down-regulated	mir-874	MIR874	KLHL3	chr5 (q31.2)

The table lists the 20 miRNAs found to be expressed at higher or lower levels in the PBMC of patients with CIS or MS in response to IFN-beta therapy. The base sequence, the gene regulatory effect of the treatment ("Expression change"), the miRNA family, the HGNC symbols of the precursor and primary miRNAs, as well as the genomic location are shown. Two of the mature miRNAs (*hsa-let-7a-5p* and *hsa-miR-16-5p*) are processed from more than one precursor miRNA. For 11 miRNAs the pri-miRNA transcript has been annotated. None of these pri-miRNAs appeared in the mRNA filtering result. Precursor miRNAs of *hsa-let-7a-5p* and *hsa-let-7b-5p*, and of *hsa-miR-29a-3p* and *hsa-miR-29b-1-5p* are clustered, *i.e.*, they share their transcription locus.

Apparently, there are more down-regulated than up-regulated miRNAs during therapy, which is the opposite of the mRNA results. This suggests that the induction of IFN-beta-responsive genes is paralleled by a preferential down-regulation of miRNAs, which is plausible given that miRNAs act as gene silencers. Therefore, we analyzed whether the miRNAs indeed participate in the regulation of the mRNA transcripts (see Section 2.7).

2.5. Validation of IFN-beta-Induced MicroRNA Expression Changes

We used Affymetrix miRNA microarrays to replicate the miRNA measurements of the PBMC samples from three patients (Pat1-3) before the start of IFN-beta therapy as well as after one month (see Section 3.4). These microarrays had a lower measurement range than the TaqMan miRNA arrays, though the comparability of the data was acceptable (Spearman's rho = 0.823). The statistical analysis of this additional data set is limited by the small number of patients. However, when we compared the mean miRNA expression levels before and one month after treatment initiation, 13 of the 20 filtered miRNAs showed the same trend of up-regulation or down-regulation (Supplementary File 3). At the significance threshold alpha = 0.10, four miRNAs (*hsa-miR-29a-3p*, *hsa-miR-29c-3p*, *hsa-miR-193a-3p*, and *hsa-miR-532-5p*) were confirmed to be down-regulated during treatment.

For further validation, we selected five of the 20 filtered miRNAs to quantify their expression in PBMC of an independent cohort of 12 patients using TaqMan single-tube assays (Supplementary File 4 and Supplementary File 5). These 12 patients (8 RRMS/4 CIS, 7 females/5 males, mean age 36.2 years) also started a therapy with IFN-beta-1b sc. The PBMC were obtained again in a longitudinal manner before the first drug injection and after one month of treatment. In this data set, *hsa-miR-29a-3p* and *hsa-miR-29c-3p* could be confirmed as differentially expressed in response to IFN-beta therapy (*t*-test *p*-values < 0.001). *hsa-miR-29c-3p* was expressed at lower levels during therapy in comparison to pre-treatment levels in all 12 patients, hence independent of disease stage (CIS or RRMS), age, and gender (Figure 3). Additionally, *hsa-miR-532-5p* was confirmed to be down-regulated (*p*-value = 0.048). *hsa-miR-16-5p* and *hsa-miR-149-5p* showed the same trend of expression change as in the other data sets (TaqMan miRNA arrays and Affymetrix miRNA arrays), but this was not statistically significant (*p*-values > 0.10, Supplementary File 3).

2.6. Functions of the MicroRNAs in the Context of Multiple Sclerosis

The filtered miRNAs (Table 3) affect diverse cellular functions and pathways, and some of them have been implicated in MS. In our data set, *hsa-let-7a-5p* and *hsa-let-7b-5p*, which belong to the let-7 family, were expressed at higher levels during therapy. Lehmann *et al.* showed that let-7 family members can activate TLR7 signaling in macrophages and microglia, thereby inducing neurodegeneration [36]. In CD4+ T cells, let-7 miRNAs reduce the expression of IL10, a cytokine with anti-inflammatory properties [37]. The up-regulation of *hsa-let-7b-5p* by IFN-beta has already been demonstrated *in vitro* in primary macrophages [38]. Interestingly, *hsa-let-7b-5p* is in turn capable of binding the endogenous IFN-beta transcript, forming a negative feedback loop for the regulation of IFN-beta protein [38]. Recently, Gandhi *et al.* found altered *hsa-let-7a-5p* levels in the blood plasma of patients with the secondary-progressive form of MS, but not of patients with RRMS [39].

The observed up-regulation of *hsa-miR-16-5p* in response to IFN-beta therapy may restore the aberrant expression of this miRNA in the disease. Two studies observed reduced levels of *hsa-miR-16-5p* in the blood of RRMS patients. One study compared the expression in PBMC and CD4+ T cells from untreated patients and healthy donors [40]. The other study showed that *hsa-miR-16-5p* is down-regulated in B cells as well [41].

Figure 3. Down-regulation of *hsa-miR-29c-3p* in response to IFN-beta therapy. The *hsa-miR-29c-3p* expression dynamics within the first month of IFN-beta treatment are presented. (**A**) TaqMan miRNA cards revealed reduced levels of this miRNA in the PBMC of 6 patients (Pat1-6, the main cohort); (**B**) Affymetrix miRNA arrays were then used to replicate the measurement for three of these patients (Pat1-3); (**C**) Finally, the down-regulation of *hsa-miR-29c-3p* was confirmed in an independent group of 12 patients (the validation cohort) using TaqMan single-tube assays. The Affymetrix analysis was based on hybridization of miRNA molecules to probes (probe set "*hsa-miR-29c_st*"), whereas the TaqMan analyses were based on real-time PCR. The TaqMan data are in linear scale, and the Affymetrix data are in log2 scale due to a different data preprocessing.

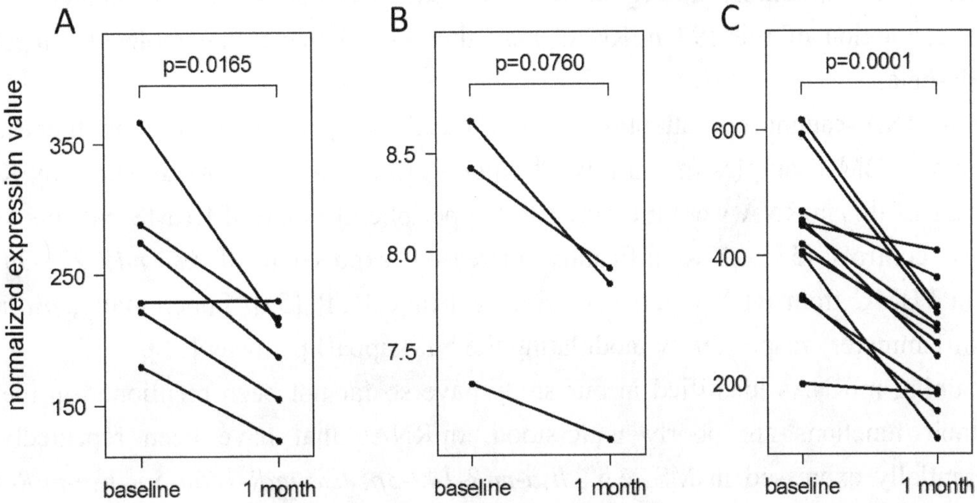

Three miRNAs of the mir-29 family were down-regulated one month after the start of therapy, and this was confirmed for two of them in an independent validation cohort of 12 patients (Figure 3). *hsa-miR-29a-3p* and *hsa-miR-29c-3p* were both expressed at relatively high levels (Supplementary File 3), and their mature sequences differ only in one base (Table 3). Therefore, it is likely that they play similar roles as post-transcriptional regulators. *hsa-miR-29a-3p* and *hsa-miR-29b-1-5p* belong to the same genomic cluster. Smith *et al.* demonstrated that IFN-gamma enhances the transcription of this *miR-29ab1* cluster, which acts in a negative feedback loop by regulating TBET and IFN-gamma [42]. Additionally, they showed decreased *hsa-miR-29b-3p* levels upon T cell activation in MS patients. This suggests a dysregulation of the feedback loop, which may bias T helper type 1 cell differentiation and may contribute to chronic inflammation [42]. Other studies also provided evidence that the members of the mir-29 family control innate and adaptive immune responses by targeting IFN-gamma [43,44]. The therapeutic down-modulation of mir-29 miRNAs might be mediated by NF-kappaB. The activation of NF-kappaB signaling, via ligation of Toll-like receptors, was shown to inhibit *miR-29ab1* expression [45]. Functionally, mir-29 promotes apoptosis, whereas repression of mir-29 levels is protective [45]. *hsa-miR-29a-3p* has been further shown to regulate myelin gene expression by Schwann cells [46].

hsa-miR-181c-3p was filtered as down-regulated during IFN-beta therapy. Several studies described the other strand of its pre-miRNA to be dysregulated in MS. Lower levels of *hsa-miR-181c-5p* were measured in PBMC [47] and in MS lesions [48] of patients in comparison to controls. On the other hand, *hsa-miR-181c-5p* seems to be overabundant in the cerebrospinal fluid of patients with MS [49]. However, the biological processes that are influenced by the *hsa-mir-181c* miRNAs and their role in MS therapy remain largely unknown.

The expression of the mir-193 family members *hsa-miR-193a-3p* and *hsa-miR-193a-5p* was repressed during the therapy (Table 3). A study by Lindberg *et al.* demonstrated increased expression of *hsa-miR-193a-5p* in CD4+ T cells of RRMS patients compared to healthy subjects [50]. Otaegui *et al.* confirmed the potential relevance of this miRNA duplex in MS. Based on a co-expression network analysis, they postulated that *hsa-miR-193a-3p* is related to the remission stage of MS [51]. Moreover, the precursor molecule *hsa-mir-193a* was found to modulate apoptotic processes by promoting CASP3 activation induced by TNFSF10 signaling [52]. TNFSF10 (=TRAIL), in turn, is a known IFN-beta-induced gene and was transcriptionally up-regulated in the patients' PBMC (Supplementary File 2). The concomitant regulation of mir-193 miRNAs may thus contribute to the molecular mechanisms of action of IFN-beta.

Another miRNA caught our attention: *hsa-miR-223-3p* appeared to be the highest expressed miRNA in most PBMC samples (mean raw Ct value = 13.6). In a microarray study by Keller *et al.*, elevated levels of this miRNA were measured in the peripheral blood of RRMS patients as compared with healthy controls [53]. A significantly increased expression of *hsa-miR-223-3p* was later confirmed in PBMC from RRMS patients using real-time PCR [54]. Functionally, *hsa-miR-223-3p* modulates inflammatory responses by modulating the NF-kappaB pathway [55].

The remaining miRNAs identified in our study have so far not been mentioned in the context of MS, and their functions are poorly understood. miRNAs that have been repeatedly described to be differentially expressed in MS, e.g., *hsa-miR-142-3p*, *hsa-miR-146a-5p*, *hsa-miR-155-5p* and *hsa-miR-326* [22], were not contained in the filtering result, thus the therapy did not normalize their abnormal expression (*cf.* Waschbisch *et al.* [26]). Further studies are needed to decipher the immunological pathways involved, and to better understand the role of miRNA-dependent regulatory mechanisms in the immunopathogenesis of MS.

2.7. Interactions between Filtered MicroRNAs and mRNAs

Interactions between the 20 filtered miRNAs and the 95 filtered mRNAs were derived from two databases providing potential target genes of miRNAs. The miRWalk database [56] was used to obtain miRNA-mRNA interactions consistently predicted by multiple computational algorithms. This resulted in 74 potential interactions for 15 of the 20 filtered miRNAs. The miRTarBase database [57] was used to extract interactions with experimental evidence in the literature. This resulted in two verified interactions: *hsa-miR-16-5p* was identified as a post-transcriptional regulator of HERC6 in both databases. Moreover, there is an experimentally determined interaction between *hsa-let-7b* and IFIT5. IFIT5 is a known IFN-beta-induced gene. A recent study by Abbas *et al.* characterized IFIT5 as an innate immune effector molecule acting as a sensor of viral single-stranded RNAs [58]. This confers antiviral defense by inhibiting viral replication. However, IFIT5 also recognizes cellular RNAs, including tRNAs [59]. The full network of miRNA-mRNA target interactions is visualized in Figure 4. The most connected regulators in the network were *hsa-miR-16-5p* and *hsa-miR-532-5p* with nine IFN-beta-responsive target genes each. *hsa-miR-29a-3p* and *hsa-miR-29c-3p*, which are closely related to each other, had eight predicted gene targets in common due to their similar mature sequences.

Figure 4. Verified and predicted interactions between IFN-beta-responsive microRNAs and mRNAs. The network of miRNA-target interactions between differentially expressed miRNAs and mRNAs was built using the databases miRWalk and miRTarBase. miRWalk reported results of 10 different prediction algorithms, and we only considered mRNA targets being computationally predicted by at least five of the 10 algorithms. miRTarBase provided experimentally validated miRNA-target interactions. In total, 15 of the filtered miRNAs were linked to 34 of the filtered genes by 74 predicted and two validated interactions. *hsa-miR-29a-3p* and *hsa-miR-29c-3p* (in the center-right) belong to the same miRNA family and are predicted to regulate eight target genes in common. The network is available as a Cytoscape session file (Supplementary File 6).

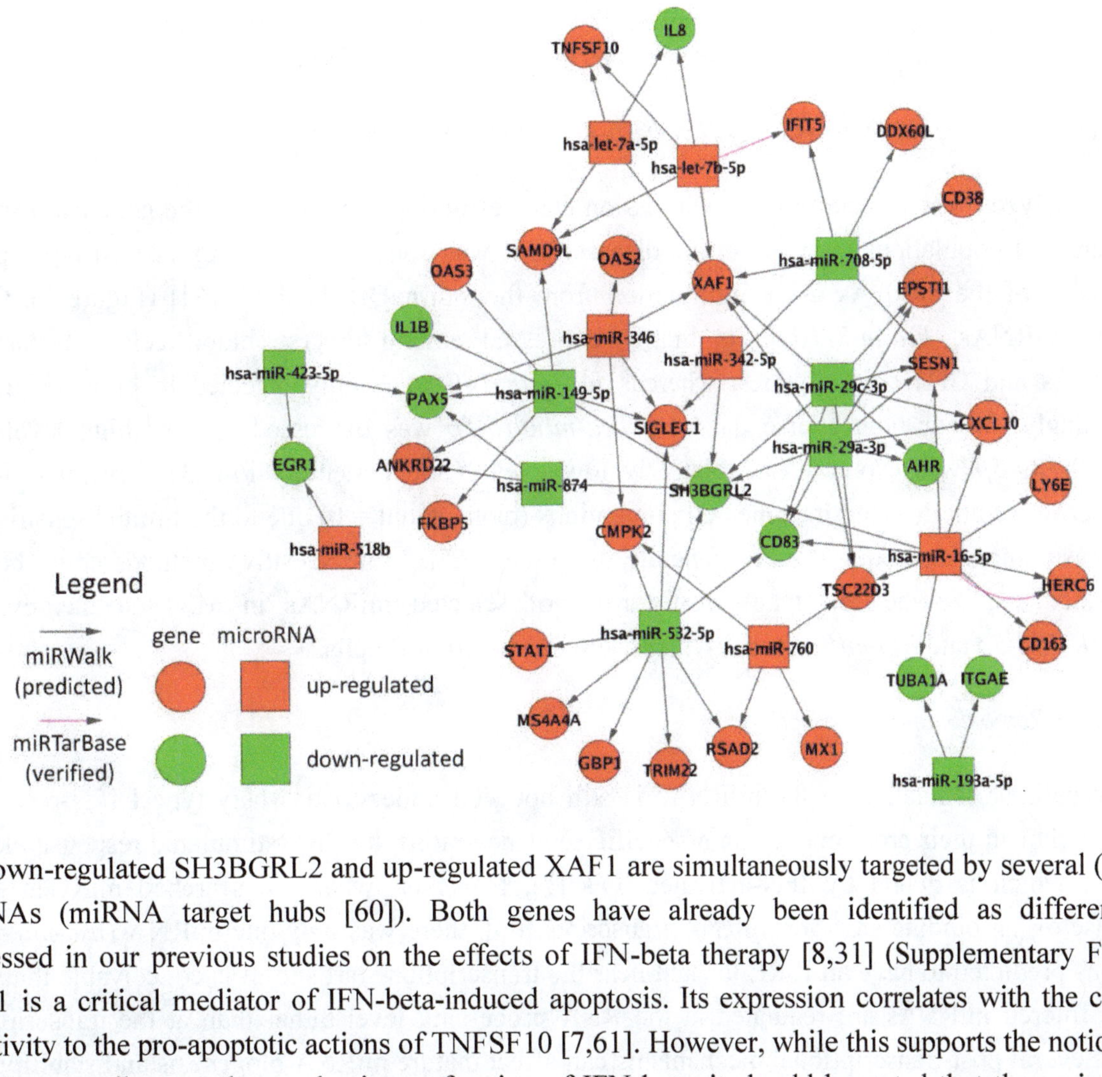

Down-regulated SH3BGRL2 and up-regulated XAF1 are simultaneously targeted by several ($n > 5$) miRNAs (miRNA target hubs [60]). Both genes have already been identified as differentially expressed in our previous studies on the effects of IFN-beta therapy [8,31] (Supplementary File 2). XAF1 is a critical mediator of IFN-beta-induced apoptosis. Its expression correlates with the cellular sensitivity to the pro-apoptotic actions of TNFSF10 [7,61]. However, while this supports the notion that miRNAs contribute to the mechanisms of action of IFN-beta, it should be noted that the majority of interactions in the network is predicted. For instance, the miRNA-mRNA interaction between *hsa-miR-16-5p* and SESN1 is predicted by nine out of 10 algorithms implemented in the miRWalk database, but has not yet been experimentally demonstrated. Another limitation is that we did not analyze whether the miRNAs bind their target mRNAs at multiple sites. Such a cooperative regulation through repetitive elements in the 3' UTR can increase repression efficacy [62]. Additional studies are needed to validate the miRNA targets, e.g., by luciferase assays.

Opposing effects exist in the network since several mRNAs are targeted by down-regulated and up-regulated miRNAs (Figure 4). Moreover, the network does not include the effects of the transcription factors (TF), which are activated through the IFN-beta signaling pathway and which are known to regulate the expression of most of the genes [31,63]. Therefore, it is difficult to disentangle the effects of miRNA expression changes on the mRNA levels measured during therapy. Further studies are thus required, e.g., specific miRNA transfection experiments examining the impact on potential target genes at both the mRNA and protein level.

Our network analysis was limited to the set of filtered genes, and the many other potential target genes of the filtered miRNAs were out of scope. Moreover, we did not investigate whether the miRNAs target viral RNAs, which is also worth to be explored in detail [28]. Despite these limitations, we conclude that miRNA-mediated regulation plays an important role in the pleiotropic effects of IFN-beta in normal immune responses as well as in the treatment of MS.

2.8. Cell Type-Specific Expression of IFN-beta-Responsive MicroRNAs

We analyzed whether the miRNA expression changes during therapy affect the gene expression in different cell populations involved in the disease. This was done by comparing the cell type-specific expression of the miRNAs using information from the smirnaDB database [64] (Figure 5). Of the filtered miRNAs, *hsa-miR-16-5p* is highly expressed among diverse blood cells, in particular monocytes and CD4+ T cell lines, whereas *hsa-miR-149-5p* is only detected in brain (Figure 5). Accordingly, in our time course data set, *hsa-miR-16-5p* was expressed at very high levels, and *hsa-miR-149-5p* was expressed at relatively low levels (Supplementary File 3). Several miRNAs, however, were not detected in some cell populations (clone count = 0) due to the limited sensitivity of the measurement technique used to generate these data [65]. More sensitive methods could be used to further analyze the cell type-specific roles of selected miRNAs in MS and therapy, e.g., *hsa-miR-29a-3p* and *hsa-miR-29c-3p*, which were expressed in B cells.

2.9. Final Remarks and Perspectives

The cellular regulation of the miRNAs is still not well understood. Many type I IFN-responsive genes harbor in their promoter region a specific sequence motif, the IFN-stimulated response element (ISRE), which is bound by IFN-activated TFs [7]. However, when we searched miRGen 2.0, a database of TF binding sites for miRNA transcripts [66], there was only one miRNA (*hsa-mir-203a*) that was predicted to have an ISRE located near the transcription start site. It is conceivable that some of the filtered miRNAs are regulated at the RNA processing level rather than at the transcriptional level. Several post-transcriptional mechanisms can affect mature miRNA biogenesis and stability [67]. Recent studies provided evidence that miRNAs can be suppressed by circular RNAs, which act as natural miRNA sponges [68]. On the other hand, the localization of the miRNAs might be altered, e.g., by microparticle shedding [69]. Additional research is needed to elucidate how the activity of miRNAs might be modulated during therapy.

Figure 5. Expression of 17 filtered miRNAs in different cell populations. The heat map visualizes the levels of IFN-beta-responsive miRNAs in 19 blood cell populations and four brain tissues. The data were downloaded from the smirnaDB database [64], which did not contain three of the 20 filtered miRNAs (*hsa-miR-27a-5p*, *hsa-miR-29b-1-5p* and *hsa-miR-181c-3p*). *hsa-let-7a-5p* and *hsa-miR-16-5p* are highly expressed in peripheral blood and brain. *hsa-let-7b-5p* and *hsa-miR-149-5p* are preferentially expressed in brain tissues. Several of the miRNAs (e.g., *hsa-miR-346*) were not detected in certain cell types [65].

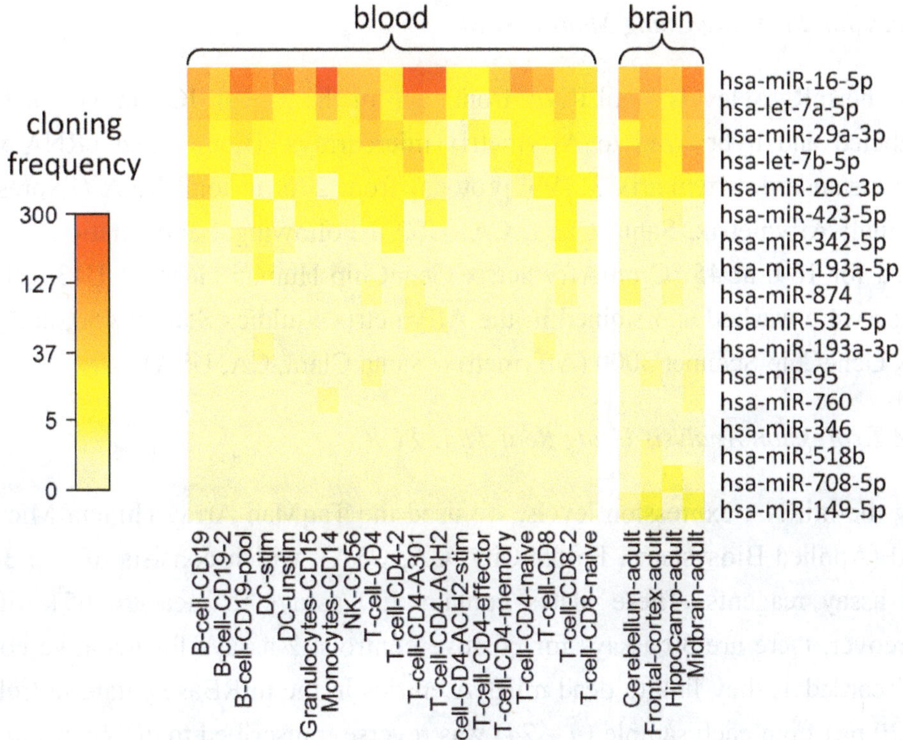

Currently, there is a clear lack of studies investigating the changes in miRNA expression during MS therapy. Apart from the study by Waschbisch *et al.* [26] (see Section 1), there is only another study by Sievers *et al.*, who found differentially expressed miRNAs in B cells of patients treated with natalizumab [41]. Future miRNA profiling analyses should use a longitudinal design and address both the short-term and the long-term effects of the available treatments. Such studies may also help to understand why some patients continue to have clinical relapses, disability progression or active lesions despite therapy. Defining the individual response to treatment is difficult, but miRNAs may have the potential to be used as prognostic biomarkers, thereby facilitating improved patient care. The identification of miRNA biomarkers should be supported by functional studies on how miRNAs affect complex biological processes by targeting multiple genes in different cell types.

3. Experimental Section

3.1. Samples

Fifteen milliliters peripheral venous EDTA blood samples were taken from six patients (Table 1) immediately before first (baseline), second, and third IFN-beta injection as well as after one month.

For validation, further blood samples were taken from an independent cohort of 12 patients (Supplementary File 4) before and one month after the start of IFN-beta-1b sc. therapy. The samples were always processed within one hour. PBMC were separated by isopycnic centrifugation in Ficoll density gradients, and total RNA enriched with small RNAs was isolated using the mirVana miRNA isolation kit (Invitrogen, Carlsbad, CA, USA) according to the manufacturers' protocols. The study was approved by the University of Rostock's ethics committee and carried out according to the Declaration of Helsinki. Written informed consent was obtained from all patients before study onset.

3.2. Gene Expression Profiling Using Microarrays

To quantify the mRNA levels, total RNA from each of the 24 PBMC samples of the main study cohort was labeled and hybridized to Affymetrix microarrays. Biotinylated cRNA were prepared according to the standard Affymetrix 3' IVT protocol from 200 ng total RNA (Expression Analysis Technical Manual; Affymetrix, Santa Clara, CA, USA). Following fragmentation, 15 µg of cRNA were hybridized for 16 h at 45 °C on Affymetrix GeneChip Human Genome U133 Plus 2.0 Arrays. The microarrays were washed and stained in the Affymetrix Fluidics Station 450, and scanned using the Affymetrix GeneChip Scanner 3000 (Affymetrix, Santa Clara, CA, USA).

3.3. MicroRNA Expression Analysis Using Real-Time PCR

To quantify the miRNA expression levels, we used the TaqMan Array Human MicroRNA A + B Cards Set v2.0 (Applied Biosystems, Foster City, CA, USA), which consists of two 384-well plates with TaqMan assay reagents. These plates contain 720 assays to measure 651 different human miRNAs. Moreover, there are 30 assays for positive controls, 2 assays for negative controls, and 16 assays were discarded as they link to dead miRNA entries in the miRBase database (release 17) [17]. Total RNA (120 ng) from each sample ($n = 24$) was reverse transcribed to cDNA using Megaplex RT Primers in combination with Megaplex PreAmp Primers (Life Technologies, Carlsbad, CA, USA). The real-time PCR measurements were then performed with predesigned primers and TaqMan probes with 45 cycles according to the manufacturer's instructions in a 7900HT Sequence Detection System (Applied Biosystems, Foster City, CA, USA). Raw Ct values were computed using the SDS 2.3 and RQ Manager 1.2 software (Applied Biosystems, Foster City, CA, USA), and undetermined data were set to Ct = 45.

3.4. Validation MicroRNA Analysis Using Microarrays

For validation, we replicated the miRNA expression measurements of the PBMC from three patients (Pat1-3) at two time points (before the first IFN-beta injection as well as after one month). Total RNA of these six samples was labeled and hybridized to Affymetrix GeneChip miRNA 2.0 arrays. Biotinylated RNA was prepared using the FlashTag Biotin HSR RNA labeling kit according to the standard Affymetrix protocol from 600 ng total RNA (Expression Analysis Technical Manual; Affymetrix). Following fragmentation, the biotin-labeled RNA was hybridized for 16 h at 45 °C on Affymetrix miRNA 2.0 arrays. The microarrays were washed and stained in the Affymetrix Fluidics Station 450, and scanned using the Affymetrix GeneChip Scanner 3000.

3.5. Validation MicroRNA Analysis Using Real-Time PCR

To verify the results in an independent cohort of patients, we measured 5 of the 20 filtered miRNAs in PBMC samples from 12 additional patients (Supplementary File 5). The blood samples were obtained before and after one month of IFN-beta-1b treatment. The miRNAs were selected based on a combination of different criteria, e.g., change of expression after one month according to both the TaqMan miRNA array data and the Affymetrix miRNA microarray data. The validation experiment was performed using TaqMan single-tube assays for *hsa-miR-16-5p* (Assay ID 000391), *hsa-miR-29a-3p* (Assay ID 002112), *hsa-miR-29c-3p* (Assay ID 000587), *hsa-miR-149-5p* (Assay ID 002255), and *hsa-miR-532-5p* (Assay ID 001518). Additionally, the housekeeping miRNA *hsa-miR-191-5p* (Assay ID 002299) was measured for normalization [33]. For each assay, 10 ng of total RNA from each sample ($n = 24$) were used to convert an individual miRNA to cDNA using an RT primer specific for the miRNA of interest (Applied Biosystems). The real-time PCR quantitation was performed in triplicates with predesigned primers and TaqMan probes according to the TaqMan MicroRNA Assay protocol with 45 cycles in a 7900HT Sequence Detection System (Applied Biosystems). An equivalent of 0.5 ng total RNA was used to obtain a single data point. Raw Ct values were computed using the SDS 2.3 and RQ Manager 1.2 software (Applied Biosystems).

3.6. Expression Data Preprocessing

In the case of the Affymetrix U133 Plus 2.0 gene expression microarrays, the raw probe-level signals were converted to expression values (signal intensities) using the MAS5.0 algorithm and custom GeneAnnot-based chip definition files (version 2.2.0) [70]. Data normalization was performed by loess normalization using the R package “affy”. Each Affymetrix GeneChip yielded mRNA levels of 19,204 human genes.

In case of the TaqMan miRNA cards, we first set the detection limit at Ct = 38 [71], and converted the raw Ct values to the linear scale using the equation $2^{-Ct} \times 10^{-9}$. After this step, a Ct value of 38 corresponds to an expression signal of 0.004, and a Ct value of 20 corresponds to an expression signal of 953. Systematic differences in the time course data were then corrected by loess normalization. This was done separately for each patient ($n = 6$) and each card (A and B). Finally, to reduce variation in the expression signals between the patients, we scaled the data of card A and of card B so that the respective 95% quantile was the same for each patient.

In case of the Affymetrix miRNA microarrays, the raw signals were converted to expression values using the RMA algorithm with quantile normalization. For each chip, this resulted in log-transformed expression levels for 20,706 probe sets interrogating small non-coding RNA transcripts. For each of the 20 filtered miRNAs (Table 3), there was one designated probe set.

In case of the TaqMan single-tube real-time PCR miRNA assays, we set the detection limit at Ct = 38 [71], calculated the mean Ct value of each triplicate, converted the raw Ct values to the linear scale using the equation 2^{-Ct}, normalized the results to the expression values of the housekeeping miRNA *hsa-miR-191-5p* [33], and scaled these ratios by a factor of 1000 for convenience.

The non-normalized and the normalized expression data of the six patients receiving IFN-beta therapy are available in the Gene Expression Omnibus (GEO) database via the SuperSeries record

GSE46293. This GEO entry links to all data from the Affymetrix U133 Plus 2.0 microarrays, the TaqMan MicroRNA Cards Sets v2.0, and the Affymetrix miRNA 2.0 microarrays.

3.7. Filtering of Differentially Expressed mRNAs and MicroRNAs

We filtered the Affymetrix U133 Plus 2.0 and TaqMan miRNA cards data sets for IFN-beta-responsive genes and miRNAs by comparing the PBMC expression levels immediately before treatment initiation (baseline) with the expression levels two days, four days, and one month after the start of IFN-beta therapy. Up-regulation and down-regulation of genes and miRNAs were quantified using signal intensity-dependent fold-changes (MAID-scores) as described in our previous studies [8,34] (see also http://www.hki-jena.de/index.php/0/2/490). MAID-scores represent adjusted fold-changes, where a higher fold-change (*i.e.*, relative change in expression) is required for genes expressed at low levels than for genes expressed at high levels. This is realized by an exponential function (MAID regression curve) that is fitted to the signal intensity-dependent variation in the data. For each time point comparison and each type of array, we computed the MAID-score for all patients and for all measured genes and miRNAs. We then selected the genes and miRNAs being up-regulated (MAID-score above the cutoff C) or being down-regulated (MAID-score $< -C$) in at least four of the six patients. We chose $C = 2$ for the gene expression data set, and $C = 1$ for the miRNA expression data set. For the latter, a lower MAID-score cutoff was chosen because the larger variation in the miRNA data already leads to a higher MAID regression curve.

To provide an estimate of the number of genes and miRNAs passing the MAID filtering by chance, a permutation test was performed. The data sets were permuted 1000 times by randomly rearranging the temporal sequence of the data of each patient. The same filtering criteria as described above were then applied to each permutation.

As an alternative filtering criterion, we statistically compared the PBMC expression levels of mRNAs and miRNAs before and during treatment using paired *t*-tests (Supplementary File 2 and Supplementary File 3). However, considering the relatively small number of patients in our study, the MAID filtering method is thought to be more robust to the variation in the data.

3.8. Visualization of the mRNA Expression Data

A heat map was used to visualize the expression changes of the 95 filtered mRNAs (Figure 1). The heat map displays the respective data of all 6 patients before the start of therapy and after one month. To reorder the rows and columns of the data matrix, hierarchical clustering was performed with the single linkage method and Pearson's correlation coefficient as a measure of similarity. For visualization purposes, the expression values (signal intensities) were centered and scaled row-wise (resulting in z-scores) with the standard R heat map function.

3.9. Evaluation of the MicroRNA Validation Data

Spearman's rank correlation coefficient (rho) was calculated to evaluate whether the TaqMan miRNA array data correlate with the Affymetrix miRNA microarray data (120 data pairs: 20 miRNAs, three patients, and two time points). Paired *t*-tests were computed for assessing the difference in expression after one month of IFN-beta-1b sc. therapy versus baseline in the preprocessed and normalized data of the Affymetrix microarrays (Pat1-3) and of the TaqMan single-tube assays (*n* = 12 additional patients).

3.10. Interaction Network Analysis

We studied the regulatory interactions between the miRNAs and their target mRNAs by integrating the information from two different databases. Experimentally verified and computationally predicted target genes of the 20 IFN-beta-responsive miRNAs were extracted from the databases miRTarBase (version 3.5) [57] and miRWalk (April 2013) [56], respectively. miRWalk contained predictions for all 20 filtered miRNAs, and miRTarBase contained interactions for 9 of the 20 miRNAs. For the prediction of targets with miRWalk, we applied the option of the web server to run the calculations with 10 different prediction algorithms on 3' UTRs of all human genes, and then gathered only the miRNA-mRNA interactions that were predicted by at least 5 of the 10 algorithms. Finally, interactions being associated with the 95 filtered genes were visualized as a network using the Cytoscape software (version 2.8.0) [72].

3.11. MicroRNA Expression in Different Cell Populations

To investigate the expression of the filtered miRNAs in different peripheral blood cell types and brain regions, we used the smirnaDB database, which provides expression levels of 692 human miRNAs for 170 cell populations and tissues [64]. This miRNA expression atlas is based on sequence analysis of small RNA clone libraries [65]. The relative cloning frequencies of miRNAs represent a measure of miRNA expression. However, in this data set, many miRNAs were identified at very low clone counts (*cf.* Landgraf *et al.* [65]). The data for 19 blood cell types (including three CD4+ T cell lines) and four brain tissues were downloaded from smirnaDB and visualized as a heat map in the R software environment.

4. Conclusions

To our knowledge, this is the first longitudinal genome-wide study examining the *in vivo* effects of IFN-beta treatment on miRNA expression in blood cells of patients with CIS or RRMS. The strongest changes in mRNA and miRNA expression were detected one month after the start of IFN-beta-1b sc. treatment. We observed that the induction of IFN-beta-responsive genes is paralleled by a preferential down-regulation of miRNAs. This suggests that the regulation of miRNAs contributes to the molecular mechanisms of action of IFN-beta in protective immune responses as well as in MS therapy. We confirmed the down-regulation of *hsa-miR-29a-3p*, *hsa-miR-29c-3p*, and *hsa-miR-532-5p* in an independent cohort of patients. We further analyzed the interactions between differentially expressed miRNAs and mRNAs. The largest number of predicted interactions to IFN-responsive genes was found for *hsa-miR-532-5p* and *hsa-miR-16-5p*. Up-regulated *hsa-miR-16-5p* was expressed at very high levels in different cell types of the blood, in particular monocytes. However, unraveling the complex gene regulatory interactions between TFs, miRNAs and genes remains a big challenge for the future. Functionally, some of the 20 filtered miRNAs (e.g., members of the mir-29 family) are associated with apoptosis and are involved in IFN signaling feedback loops. miRNA expression profiles in blood cells may provide biomarkers for monitoring the biological response to therapy to predict individual disease activity and progression. They may also help to better understand the pathogenetic mechanisms and to optimize the treatment of MS. Our results provide a rationale for subsequent studies in larger MS cohorts.

Acknowledgments

We thank our colleagues Ildikó Tóth, Brigitte Paap, Gabriele Gillwaldt and Caroline Siebrecht for laboratory assistance and clinical documentation. We thank Charlotte Angerstein for helpful discussions and comments on microRNA databases. We are grateful to study nurse Christa Tiffert for her excellent coordination of patient care and sample collection. We thank the United Europeans for the development of PHArmacogenomics in Multiple Sclerosis (UEPHA*MS) consortium for their funding.

The experiments were partially funded by Bayer HealthCare. The funder had no role in the study design, data collection and analysis, decision to publish, or preparation of the manuscript.

Supplementary Information

Supplementary File 1 (TIFF image): Distribution of the raw Ct values measured with the TaqMan miRNA arrays. The data of the A-cards (A) and the B-cards (B) are shown for all six patients and all four time points. The detection limit was set at Ct = 38.

Supplementary File 2 (XLS Excel spreadsheet): mRNA filtering result. This table provides the 95 genes that were up-regulated or down-regulated in the PBMC of the patients two days, four days or one month after the initiation of subcutaneous IFN-beta-1b therapy. Different types of information are given for each gene, e.g., gene symbol, official full name, Entrez Gene identifier, and the MAID filtering results.

Supplementary File 3 (XLS Excel spreadsheet): microRNA filtering result and validation. In comparison to pre-treatment levels, 20 miRNAs were found to be expressed at higher or lower levels in PBMC during IFN-beta treatment. The table provides, e.g., the TaqMan Detector identifiers, the MAID filtering outputs as well as the results from the validation analyses.

Supplementary File 4 (XLS Excel spreadsheet): Clinical data and demographic data of the patients in the validation cohort. Twelve patients were recruited to confirm the observed expression changes of five selected miRNAs within the first month of IFN-beta-1b sc. treatment.

Supplementary File 5 (XLS Excel spreadsheet): Validation real-time PCR data set. TaqMan single-tube assays were used to quantify the expression of five selected miRNAs before the start of IFN-beta therapy and after one month in PBMC of an independent cohort of 12 patients (Supplementary File 4). This table contains the raw Ct values of these five miRNAs and the housekeeping miRNA *hsa-miR-191-5p* [33]. The measurements were done in triplicates.

Supplementary File 6 (CYS Cytoscape session file): miRNA-mRNA interaction network. This Cytoscape file (http://www.cytoscape.org) [72] contains computationally predicted and experimentally determined interactions between IFN-beta-responsive miRNAs and mRNAs. The interactions were obtained from the databases miRWalk [56] and miRTarBase [57]. A visualization of the network is shown in Figure 4.

References

1. Compston, A.; Coles, A. Multiple sclerosis. *Lancet* **2008**, *372*, 1502–1517.
2. Miller, D.H.; Chard, D.T.; Ciccarelli, O. Clinically isolated syndromes. *Lancet Neurol.* **2012**, *11*, 157–169.
3. Polman, C.H.; Reingold, S.C.; Banwell, B.; Clanet, M.; Cohen, J.A.; Filippi, M.; Fujihara, K.; Havrdova, E.; Hutchinson, M.; Kappos, L.; *et al.* Diagnostic criteria for multiple sclerosis: 2010 revisions to the McDonald criteria. *Ann. Neurol.* **2011**, *69*, 292–302.
4. Vosoughi, R.; Freedman, M.S. Therapy of MS. *Clin. Neurol. Neurosurg.* **2010**, *112*, 365–385.
5. Buck, D.; Hemmer, B. Treatment of multiple sclerosis: Current concepts and future perspectives. *J. Neurol.* **2011**, *258*, 1747–1762.
6. Limmroth, V.; Putzki, N.; Kachuck, N.J. The interferon beta therapies for treatment of relapsing-remitting multiple sclerosis: Are they equally efficacious? A comparative review of open-label studies evaluating the efficacy, safety, or dosing of different interferon beta formulations alone or in combination. *Ther. Adv. Neurol. Disord.* **2011**, *4*, 281–296.
7. Borden, E.C.; Sen, G.C.; Uze, G.; Silverman, R.H.; Ransohoff, R.M.; Foster, G.R.; Stark, G.R. Interferons at age 50: Past, current and future impact on biomedicine. *Nat. Rev. Drug Discov.* **2007**, *6*, 975–990.
8. Hecker, M.; Hartmann, C.; Kandulski, O.; Paap, B.K.; Koczan, D.; Thiesen, H.J.; Zettl, U.K. Interferon-beta therapy in multiple sclerosis: The short-term and long-term effects on the patients' individual gene expression in peripheral blood. *Mol. Neurobiol.* **2013**, doi:10.1007/s12035-013-8463-1.
9. Dhib-Jalbut, S.; Marks, S. Interferon-beta mechanisms of action in multiple sclerosis. *Neurology* **2010**, *74*, S17–S24.
10. Goertsches, R.H.; Hecker, M.; Zettl, U.K. Monitoring of multiple sclerosis immunotherapy: From single candidates to biomarker networks. *J. Neurol.* **2008**, *255*, 48–57.
11. Goertsches, R.H.; Zettl, U.K.; Hecker, M. Sieving treatment biomarkers from blood gene-expression profiles: A pharmacogenomic update on two types of multiple sclerosis therapy. *Pharmacogenomics* **2011**, *12*, 423–432.
12. Reder, A.T.; Velichko, S.; Yamaguchi, K.D.; Hamamcioglu, K.; Ku, K.; Beekman, J.; Wagner, T.C.; Perez, H.D.; Salamon, H.; Croze, E. IFN-beta1b induces transient and variable gene expression in relapsing-remitting multiple sclerosis patients independent of neutralizing antibodies or changes in IFN receptor RNA expression. *J. Interferon Cytokine Res.* **2008**, *28*, 317–331.
13. Gilli, F.; Marnetto, F.; Caldano, M.; Sala, A.; Malucchi, S.; di Sapio, A.; Capobianco, M.; Bertolotto, A. Biological responsiveness to first injections of interferon-beta in patients with multiple sclerosis. *J. Neuroimmunol.* **2005**, *158*, 195–203.
14. Bartel, D.P. MicroRNAs: Target recognition and regulatory functions. *Cell* **2009**, *136*, 215–233.
15. Friedman, R.C.; Farh, K.K.; Burge, C.B.; Bartel, D.P. Most mammalian mRNAs are conserved targets of microRNAs. *Genome Res.* **2009**, *19*, 92–105.
16. Bentwich, I.; Avniel, A.; Karov, Y.; Aharonov, R.; Gilad, S.; Barad, O.; Barzilai, A.; Einat, P.; Einav, U.; Meiri, E.; *et al.* Identification of hundreds of conserved and nonconserved human microRNAs. *Nat. Genet.* **2005**, *37*, 766–770.

17. Kozomara, A.; Griffiths-Jones, S. miRBase: Integrating microRNA annotation and deep-sequencing data. *Nucleic Acids Res.* **2011**, *39*, D152–D157.
18. Xiao, C.; Rajewsky, K. MicroRNA control in the immune system: Basic principles. *Cell* **2009**, *136*, 26–36.
19. Im, H.I.; Kenny, P.J. MicroRNAs in neuronal function and dysfunction. *Trends Neurosci.* **2012**, *35*, 325–334.
20. Tufekci, K.U.; Oner, M.G.; Genc, S.; Genc, K. MicroRNAs and multiple sclerosis. *Autoimmune Dis.* **2010**, *2011*, 807426.
21. Guerau-de-Arellano, M.; Alder, H.; Ozer, H.G.; Lovett-Racke, A.; Racke, M.K. miRNA profiling for biomarker discovery in multiple sclerosis: From microarray to deep sequencing. *J. Neuroimmunol.* **2012**, *248*, 32–39.
22. Angerstein, C.; Hecker, M.; Paap, B.K.; Koczan, D.; Thamilarasan, M.; Thiesen, H.J.; Zettl, U.K. Integration of MicroRNA databases to study microRNAs associated with multiple sclerosis. *Mol. Neurobiol.* **2012**, *45*, 520–535.
23. Thamilarasan, M.; Koczan, D.; Hecker, M.; Paap, B.; Zettl, U.K. MicroRNAs in multiple sclerosis and experimental autoimmune encephalomyelitis. *Autoimmun. Rev.* **2012**, *11*, 174–179.
24. Mycko, M.P.; Cichalewska, M.; Machlanska, A.; Cwiklinska, H.; Mariasiewicz, M.; Selmaj, K.W. MicroRNA-301a regulation of a T-helper 17 immune response controls autoimmune demyelination. *Proc. Natl. Acad. Sci. USA* **2012**, *109*, E1248–E1257.
25. Paap, B.K.; Hecker, M.; Koczan, D.; Zettl, U.K. Molecular biomarkers in multiple sclerosis. *J. Clin. Cell. Immunol.* **2013**, *S10*, doi:10.4172/2155-9899.S10-009.
26. Waschbisch, A.; Atiya, M.; Linker, R.A.; Potapov, S.; Schwab, S.; Derfuss, T. Glatiramer acetate treatment normalizes deregulated microRNA expression in relapsing remitting multiple sclerosis. *PLoS One* **2011**, *6*, e24604.
27. O'Connell, R.M.; Taganov, K.D.; Boldin, M.P.; Cheng, G.; Baltimore, D. MicroRNA-155 is induced during the macrophage inflammatory response. *Proc. Natl. Acad. Sci. USA* **2007**, *104*, 1604–1609.
28. Pedersen, I.M.; Cheng, G.; Wieland, S.; Volinia, S.; Croce, C.M.; Chisari, F.V.; David, M. Interferon modulation of cellular microRNAs as an antiviral mechanism. *Nature* **2007**, *449*, 919–922.
29. Schmitz, U.; Wolkenhauer, O. Web resources for microRNA research. *Adv. Exp. Med. Biol.* **2013**, *774*, 225–250.
30. Dosiou, C.; Lathi, R.B.; Tulac, S.; Huang, S.T.; Giudice, L.C. Interferon-related and other immune genes are downregulated in peripheral blood leukocytes in the luteal phase of the menstrual cycle. *J. Clin. Endocrinol. Metab.* **2004**, *89*, 2501–2504.
31. Goertsches, R.H.; Hecker, M.; Koczan, D.; Serrano-Fernandez, P.; Moeller, S.; Thiesen, H.J.; Zettl, U.K. Long-term genome-wide blood RNA expression profiles yield novel molecular response candidates for IFN-beta-1b treatment in relapsing remitting MS. *Pharmacogenomics* **2010**, *11*, 147–161.
32. Hecker, M.; Goertsches, R.H.; Fatum, C.; Koczan, D.; Thiesen, H.J.; Guthke, R.; Zettl, U.K. Network analysis of transcriptional regulation in response to intramuscular interferon-beta-1a multiple sclerosis treatment. *Pharmacogenomics J.* **2012**, *12*, 134–146.

33. Peltier, H.J.; Latham, G.J. Normalization of microRNA expression levels in quantitative RT-PCR assays: Identification of suitable reference RNA targets in normal and cancerous human solid tissues. *RNA* **2008**, *14*, 844–852.

34. Hecker, M.; Goertsches, R.H.; Engelmann, R.; Thiesen, H.J.; Guthke, R. Integrative modeling of transcriptional regulation in response to antirheumatic therapy. *BMC Bioinforma.* **2009**, *10*, 262.

35. Killestein, J.; Polman, C.H. Determinants of interferon beta efficacy in patients with multiple sclerosis. *Nat. Rev. Neurol.* **2011**, *7*, 221–228.

36. Lehmann, S.M.; Krüger, C.; Park, B.; Derkow, K.; Rosenberger, K.; Baumgart, J.; Trimbuch, T.; Eom, G.; Hinz, M.; Kaul, D.; *et al.* An unconventional role for miRNA: Let-7 activates Toll-like receptor 7 and causes neurodegeneration. *Nat. Neurosci.* **2012**, *15*, 827–835.

37. Swaminathan, S.; Suzuki, K.; Seddiki, N.; Kaplan, W.; Cowley, M.J.; Hood, C.L.; Clancy, J.L.; Murray, D.D.; Méndez, C.; Gelgor, L.; *et al.* Differential regulation of the Let-7 family of microRNAs in CD4+ T cells alters IL-10 expression. *J. Immunol.* **2012**, *188*, 6238–6246.

38. Witwer, K.W.; Sisk, J.M.; Gama, L.; Clements, J.E. MicroRNA regulation of IFN-beta protein expression: Rapid and sensitive modulation of the innate immune response. *J. Immunol.* **2010**, *184*, 2369–2376.

39. Gandhi, R.; Healy, B.; Gholipour, T.; Egorova, S.; Musallam, A.; Shuja, M.; Nejad, P.; Patel, B.; Hei, H.; Khoury, S.; *et al.* Circulating microRNAs as biomarkers for disease staging in multiple sclerosis. *Ann. Neurol.* **2013**, *73*, 729–740.

40. Lorenzi, J.C.; Brum, D.G.; Zanette, D.L.; de Paula Alves Souza, A.; Barbuzano, F.G.; Dos Santos, A.C.; Barreira, A.A.; da Silva, W.A. *miR-15a* and *16-1* are downregulated in CD4+ T cells of multiple sclerosis relapsing patients. *Int. J. Neurosci.* **2012**, *122*, 466–471.

41. Sievers, C.; Meira, M.; Hoffmann, F.; Fontoura, P.; Kappos, L.; Lindberg, R.L. Altered microRNA expression in B lymphocytes in multiple sclerosis: Towards a better understanding of treatment effects. *Clin. Immunol.* **2012**, *144*, 70–79.

42. Smith, K.M.; Guerau-de-Arellano, M.; Costinean, S.; Williams, J.L.; Bottoni, A.; Mavrikis Cox, G.; Satoskar, A.R.; Croce, C.M.; Racke, M.K.; Lovett-Racke, A.E.; *et al.* miR-29ab1 deficiency identifies a negative feedback loop controlling Th1 bias that is dysregulated in multiple sclerosis. *J. Immunol.* **2012**, *189*, 1567–1576.

43. Ma, F.; Xu, S.; Liu, X.; Zhang, Q.; Xu, X.; Liu, M.; Hua, M.; Li, N.; Yao, H.; Cao, X. The microRNA *miR-29* controls innate and adaptive immune responses to intracellular bacterial infection by targeting interferon-gamma. *Nat. Immunol.* **2011**, *12*, 861–869.

44. Steiner, D.F.; Thomas, M.F.; Hu, J.K.; Yang, Z.; Babiarz, J.E.; Allen, C.D.; Matloubian, M.; Blelloch, R.; Ansel, K.M. *MicroRNA-29* regulates T-box transcription factors and interferon-gamma production in helper T cells. *Immunity* **2011**, *35*, 169–181.

45. Mott, J.L.; Kurita, S.; Cazanave, S.C.; Bronk, S.F.; Werneburg, N.W.; Fernandez-Zapico, M.E. Transcriptional suppression of *mir-29b-1/mir-29a* promoter by c-Myc, hedgehog, and NF-kappaB. *J. Cell Biochem.* **2010**, *110*, 1155–1164.

46. Verrier, J.D.; Lau, P.; Hudson, L.; Murashov, A.K.; Renne, R.; Notterpek, L. Peripheral myelin protein 22 is regulated post-transcriptionally by *miRNA-29a*. *Glia* **2009**, *57*, 1265–1279.

47. Martinelli-Boneschi, F.; Fenoglio, C.; Brambilla, P.; Sorosina, M.; Giacalone, G.; Esposito, F.; Serpente, M.; Cantoni, C.; Ridolfi, E.; Rodegher, M.; *et al.* MicroRNA and mRNA expression profile screening in multiple sclerosis patients to unravel novel pathogenic steps and identify potential biomarkers. *Neurosci. Lett.* **2012**, *508*, 4–8.
48. Junker, A.; Krumbholz, M.; Eisele, S.; Mohan, H.; Augstein, F.; Bittner, R.; Lassmann, H.; Wekerle, H.; Hohlfeld, R.; Meinl, E. MicroRNA profiling of multiple sclerosis lesions identifies modulators of the regulatory protein CD47. *Brain* **2009**, *132*, 3342–3352.
49. Haghikia, A.; Haghikia, A.; Hellwig, K.; Baraniskin, A.; Holzmann, A.; Décard, B.F.; Thum, T.; Gold, R. Regulated microRNAs in the CSF of patients with multiple sclerosis: A case-control study. *Neurology* **2012**, *79*, 2166–2170.
50. Lindberg, R.L.; Hoffmann, F.; Mehling, M.; Kuhle, J.; Kappos, L. Altered expression of *miR-17-5p* in CD4+ lymphocytes of relapsing-remitting multiple sclerosis patients. *Eur. J. Immunol.* **2010**, *40*, 888–898.
51. Otaegui, D.; Baranzini, S.E.; Armañanzas, R.; Calvo, B.; Muñoz-Culla, M.; Khankhanian, P.; Inza, I.; Lozano, J.A.; Castillo-Triviño, T.; Asensio, A.; *et al.* Differential micro RNA expression in PBMC from multiple sclerosis patients. *PLoS One* **2009**, *4*, e6309.
52. Ovcharenko, D.; Kelnar, K.; Johnson, C.; Leng, N.; Brown, D. Genome-scale microRNA and small interfering RNA screens identify small RNA modulators of TRAIL-induced apoptosis pathway. *Cancer Res.* **2007**, *67*, 10782–10788.
53. Keller, A.; Leidinger, P.; Lange, J.; Borries, A.; Schroers, H.; Scheffler, M.; Lenhof, H.P.; Ruprecht, K.; Meese, E. Multiple sclerosis: MicroRNA expression profiles accurately differentiate patients with relapsing-remitting disease from healthy controls. *PLoS One* **2009**, *4*, e7440.
54. Ridolfi, E.; Fenoglio, C.; Cantoni, C.; Calvi, A.; de Riz, M.; Pietroboni, A.; Villa, C.; Serpente, M.; Bonsi, R.; Vercellino, M.; *et al.* Expression and genetic analysis of microRNAs involved in multiple sclerosis. *Int. J. Mol. Sci.* **2013**, *14*, 4375–4384.
55. Li, T.; Morgan, M.J.; Choksi, S.; Zhang, Y.; Kim, Y.S.; Liu, Z.G. MicroRNAs modulate the noncanonical transcription factor NF-kappaB pathway by regulating expression of the kinase IKKalpha during macrophage differentiation. *Nat. Immunol.* **2010**, *11*, 799–805.
56. Dweep, H.; Sticht, C.; Pandey, P.; Gretz, N. miRWalk—Database: Prediction of possible miRNA binding sites by "walking" the genes of three genomes. *J. Biomed. Inform.* **2011**, *44*, 839–847.
57. Hsu, S.D.; Lin, F.M.; Wu, W.Y.; Liang, C.; Huang, W.C.; Chan, W.L.; Tsai, W.T.; Chen, G.Z.; Lee, C.J.; Chiu, C.M.; *et al.* miRTarBase: A database curates experimentally validated microRNA-target interactions. *Nucleic Acids Res.* **2011**, *39*, D163–D169.
58. Abbas, Y.M.; Pichlmair, A.; Górna, M.W.; Superti-Furga, G.; Nagar, B. Structural basis for viral 5'-PPP-RNA recognition by human IFIT proteins. *Nature* **2013**, *494*, 60–64.
59. Katibah, G.E.; Lee, H.J.; Huizar, J.P.; Vogan, J.M.; Alber, T.; Collins, K. tRNA binding, structure, and localization of the human interferon-induced protein IFIT5. *Mol. Cell* **2013**, *49*, 743–750.
60. Lai, X.; Schmitz, U.; Gupta, S.K.; Bhattacharya, A.; Kunz, M.; Wolkenhauer, O.; Vera, J. Computational analysis of target hub gene repression regulated by multiple and cooperative miRNAs. *Nucleic Acids Res.* **2012**, *40*, 8818–8834.
61. Leaman, D.W.; Chawla-Sarkar, M.; Vyas, K.; Reheman, M.; Tamai, K.; Toji, S.; Borden, E.C. Identification of X-linked inhibitor of apoptosis-associated factor-1 as an interferon-stimulated gene that augments TRAIL Apo2L-induced apoptosis. *J. Biol. Chem.* **2002**, *277*, 28504–28511.

62. Elefant, N.; Berger, A.; Shein, H.; Hofree, M.; Margalit, H.; Altuvia, Y. RepTar: A database of predicted cellular targets of host and viral miRNAs. *Nucleic Acids Res.* **2011**, *39*, D188–D194.
63. Hundeshagen, A.; Hecker, M.; Paap, B.K.; Angerstein, C.; Kandulski, O.; Fatum, C.; Hartmann, C.; Koczan, D.; Thiesen, H.J.; Zettl, U.K. Elevated type I interferon-like activity in a subset of multiple sclerosis patients: Molecular basis and clinical relevance. *J. Neuroinflammation* **2012**, *9*, 140.
64. Hausser, J.; Berninger, P.; Rodak, C.; Jantscher, Y.; Wirth, S.; Zavolan, M. MirZ: An integrated microRNA expression atlas and target prediction resource. *Nucleic Acids Res.* **2009**, *37*, W266–W272.
65. Landgraf, P.; Rusu, M.; Sheridan, R.; Sewer, A.; Iovino, N.; Aravin, A.; Pfeffer, S.; Rice, A.; Kamphorst, A.O.; Landthaler, M.; *et al.* A mammalian microRNA expression atlas based on small RNA library sequencing. *Cell* **2007**, *129*, 1401–1414.
66. Alexiou, P.; Vergoulis, T.; Gleditzsch, M.; Prekas, G.; Dalamagas, T.; Megraw, M.; Grosse, I.; Sellis, T.; Hatzigeorgiou, A.G. miRGen 2.0: A database of microRNA genomic information and regulation. *Nucleic Acids Res.* **2010**, *38*, D137–D141.
67. Kai, Z.S.; Pasquinelli, A.E. MicroRNA assassins: Factors that regulate the disappearance of miRNAs. *Nat. Struct. Mol. Biol.* **2010**, *17*, 5–10.
68. Hentze, M.W.; Preiss, T. Circular RNAs: Splicing's enigma variations. *EMBO J.* **2013**, *32*, 923–925.
69. Valadi, H.; Ekström, K.; Bossios, A.; Sjöstrand, M.; Lee, J.J.; Lötvall, J.O. Exosome-mediated transfer of mRNAs and microRNAs is a novel mechanism of genetic exchange between cells. *Nat. Cell Biol.* **2007**, *9*, 654–659.
70. Ferrari, F.; Bortoluzzi, S.; Coppe, A.; Sirota, A.; Safran, M.; Shmoish, M.; Ferrari, S.; Lancet, D.; Danieli, G.A.; Bicciato, S. Novel definition files for human GeneChips based on GeneAnnot. *BMC Bioinf.* **2007**, *8*, 446.
71. Burns, M.; Valdivia, H. Modelling the limit of detection in real-time quantitative PCR. *Eur. Food Res. Technol.* **2008**, *226*, 1513–1524.
72. Smoot, M.E.; Ono, K.; Ruscheinski, J.; Wang, P.L.; Ideker, T. Cytoscape 2.8: New features for data integration and network visualization. *Bioinformatics* **2011**, *27*, 431–432.

7

Long Non-Coding RNAs and Complex Human Diseases

Jing Li [1,†], Zhenyu Xuan [2,*] and Changning Liu [1,*]

[1] Bioinformatics Research Group, Key Laboratory of Intelligent Information Processing, Advanced Computer Research Center, Institute of Computing Technology, Chinese Academy of Sciences, Beijing 100190, China; E-Mail: lij.thu@gmail.com

[2] Department of Molecular and Cell Biology, Center for Systems Biology, University of Texas at Dallas, 800 W Campbell Road, Richardson, TX 75080, USA

[†] Present address: School of Medicine, Tsinghua University, Beijing 100084, China.

* Authors to whom correspondence should be addressed: E-Mails: zhenyu.xuan@utdallas.edu (Z.X.); lcn@ict.ac.cn (C.L.);

Abstract: Long non-coding RNAs (lncRNAs) are a heterogeneous class of RNAs that are generally defined as non-protein-coding transcripts longer than 200 nucleotides. Recently, an increasing number of studies have shown that lncRNAs can be involved in various critical biological processes, such as chromatin remodeling, gene transcription, and protein transport and trafficking. Moreover, lncRNAs are dysregulated in a number of complex human diseases, including coronary artery diseases, autoimmune diseases, neurological disorders, and various cancers, which indicates their important roles in these diseases. Here, we reviewed the current understanding of lncRNAs, including their definition and subclassification, regulatory functions, and potential roles in different types of complex human diseases.

Keywords: non-coding RNA; long non-coding RNA; complex human disease

---

1. Introduction

While about 20,000 protein-coding genes, representing less than 2% of the human genome, have been reported [1,2], a large part of the genome can be transcribed into non-coding RNAs (ncRNAs), which have little or no protein-coding capability [3,4]. Besides many widely studied classes of short ncRNAs, such as microRNAs (miRNA) and Piwi-interacting RNAs (piRNA) [5,6], one class of heterogeneous ncRNAs with lengths longer than 200 nucleotides, recently designated as long non-coding RNAs (lncRNAs), is increasingly attracting the attention of ncRNA researchers [7,8].

Over the past decade, furthered by the rapid progress in high-throughput sequencing technology, thousands of lncRNAs have been identified in mammalian transcriptomes [9,10]. A number of studies have revealed that lncRNAs can participate in various critical biological processes, such as chromatin remodeling, gene transcription, and protein transport and trafficking [11,12], implicating their impact on a wide range of complex human diseases [13,14]. However, despite some well-characterized lncRNAs, such as Xist and HOTAIR [15,16], little is known about the general features of most lncRNAs, such as gene structure, transcriptional regulation, and functional domains, and even less is known about their possible molecular mechanisms in different human diseases. Understanding the function of lncRNAs remains a significant challenge. Here, we review the current literature reporting on lncRNAs, including their definition, subclassification, regulatory functions, and roles in different types of complex human diseases.

2. Long Non-Coding RNAs: A Heterogeneous Class of RNAs

Long non-coding RNAs are most commonly defined as an RNA transcript more than 200 nucleotides (nt) long that cannot be translated into a protein [17]. However, this length threshold is not strict; it may vary 100 to 200 nt, or even longer. In different biochemical fractionation protocols, this threshold is primarily used to exclude most of the categories of small RNAs, such as small nucleolar RNA (snoRNA), microRNA (miRNA), Piwi-interacting RNA (piRNA), transfer RNA (tRNA), and small nuclear RNA (snRNA). Therefore, according to these simple criteria of transcript size and protein-coding capability, the designated lncRNAs contain a group of structurally and functionally heterogeneous RNAs, having a length that varies from approximately 200 nt to over 100 kb. They can undergo either splicing or not, with cellular locations in either nucleus or cytoplasm. They can be transcribed by RNA polymerase II or III, and they can play different functional and structural roles in different biological processes [18]. Based on these features, lncRNAs can be further categorized into different subgroups, as listed in Table 1. In past years, along with in-depth studies of lncRNAs, several lncRNA databases have been constructed (Table 2). These databases can facilitate further functional research on lncRNAs.

Table 1. Types of long non-coding RNAs.

Long non-coding RNA	Symbol	References
Long intergenic non-coding RNA	LincRNA	[19,20]
Long intronic non-coding RNA		[14,21]
Natural antisense transcript	NAT	[22–24]
Promoter-associated long RNA	PALR	[25]
Promoter upstream transcript	PROMPT	[26]
Repetitive element-associated non-coding RNA		[27–29]
Transcribed pseudogene		[30,31]
Transcribed ultraconserved region	T-UCR	[32]
Enhancer-like non-coding RNA	eRNA	[33]

Table 2. Public lncRNA databases.

Name	Website	Reference
ChIPBase	http://deepbase.sysu.edu.cn/chipbase/	[34]
fRNAdb	http://www.ncrna.org/frnadb/	[35]
LNCipedia	http://www.lncipedia.org/	[36]
lncRNAdb	http://www.lncrnadb.org/	[37]
NONCODE	http://www.noncode.org/	[9]
NRED	http://nred.matticklab.com/cgi-bin/ncrnadb.pl	[38]

Genome-wide transcriptional maps have shown that lncRNAs are pervasively transcribed throughout mammalian genomes [39,40]. Clusters of overlapping sense and antisense transcripts can be found inside of known genes, as well as in intergenic regions. Natural antisense transcripts (NATs), which have been largely discovered in human, mouse, and many other species, are endogenous RNA molecules that exhibit partial or complete complementarities to other transcripts [22–24]. NATs may regulate the expression level of their sense counterparts. Several plausible regulation models, like blocking translation by sense-antisense pairing or antisense RNA-directed chromatin remodeling, have been proposed [13,41]. However, little mechanistic information has supported these suppositions, and more intensive studies are needed. Unlike NATs, long intergenic non-coding RNAs (lincRNAs) are large multiexonic RNAs, which are transcribed from intergenic regions, and may act in trans within large ribonucleoprotein complexes [10,19,42]. For example, Huarte *et al.* have reported that lincRNA-p21, which can physically interact with hnRNP-K, serves as a repressor in p53-dependent transcriptional responses by modulating hnRNP-K localization to chromatin [43]. A recent comprehensive screen has identified dozens of lincRNAs, which bind to multiple chromatin regulatory proteins, to affect related gene expression procedures and function critically in the pathway controlling pluripotent embryonic stem cell (ES) state [20]. Long non-coding RNAs have also been transcribed from widespread repetitive elements. For instance, human Alu and mouse B2 RNAs are originally derived from short interspersed repeat elements (SINEs). They are transcribed by RNA polymerase III in response to environmental stresses, such as heat shock, and act as transcriptional repressors by directly targeting RNA polymerase II [27–29]. Another interesting example is lncRNA PTENP1, a biologically active pseudogene of the tumor suppressor gene Phosphatase and Tensin Homolog (*PTEN*). PTENP1 performs a tumor suppressive function by acting as an "endogenous miRNA sponge", which can positively regulate PTEN protein level via competing for PTEN-targeting miRNA binding [30,31].

Unlike protein-coding genes, which are usually conserved across species, most lncRNAs are poorly conserved and have been taken for transcriptional noise [44]. However, lack of conservation does not mean lack of function [45,46]. For example, two lncRNAs, Antisense Igf2r RNA (Air) and X Specific Transcript (Xist), are poorly conserved, but are well functional [47,48]. Their subjection to a series of recent and rapid adaptive selections may provide one explanation for the poor conservation of lncRNAs. For example, Highly Accelerated Region 1F (HAR1F), an lncRNA which is exclusively expressed in Cajal-Retzius neurons in the human neocortex, has undergone rapid evolutionary change in the human lineage since our last common ancestor of chimpanzee [49]. Moreover, for lncRNAs, which exert functions by secondary structures or short sequence motifs, we may only find small conserved regions interspersed in long poorly conserved transcripts [50]. Besides those poorly conserved lncRNAs, non-coding transcripts can also be transcribed from ultraconserved genomic regions (UCRs). UCRs were first discovered in the sequence comparison of mouse, rat, and human genomes [51]. They are genomic elements longer than 200 bp with 100% identity between orthologous regions in these three genomes [52,53]. Genomic variations in UCRs, such as single nucleotide polymorphisms (SNPs), have been reported to be associated with increased cancer risk [54]. Genome-wide expression profiling has revealed that a large fraction of UCRs are transcribed (transcribed-UCRs, T-UCRs) with significant alteration at both DNA and RNA levels in adult chronic lymphocytic leukemias, as well as colorectal and hepatocellular carcinomas [32,55]. Recent study in neuroblastoma has discovered the relevance between expression levels of specific T-UCRs and important clinical-genetic parameters, suggesting that T-UCRs may be used as signatures associated with cancer diagnosis, prognosis, and treatment [55].

As a heterogeneous class of RNAs, lncRNAs have been implicated in the regulation of a number of complicated biological processes. A study in human cell lines suggests that about 30% of lncRNAs are specifically expressed in the nucleus [25]. Many of them are involved in chromatin remodeling complexes and mediate genomic silencing [10]. One of the most well-known examples is the participation of lncRNAs in X-chromosome inactivation (XCI), a process by which one of the two copies of the X chromosome present in female mammals is inactivated [15]. During XCI, Inactive X Specific Transcript (Xist), a 17-kb lncRNA transcribed from the XCI center, will accumulate on the inactive X chromosome and recruit Polycomb complexes for subsequent epigenetic modifications. Its antisense counterpart, Tsix, which is another lncRNA specifically expressed from the other X chromosome, can also interact with Polycomb complexes and maintain the activity of X chromosome [48]. Apart from in *trans* regulation, some lncRNAs can directly regulate gene expression in *cis*. By using knockdown approaches and reporter assays, Orom *et al.* have discovered an enhancer-like effect for a set of lncRNAs in human cell lines [56]. Depletion of these lncRNAs leads to decreased expression of their neighboring protein-coding genes. More interestingly, a recent genome-wide study of transcriptional enhancers in mouse has shown that some lncRNAs, termed as enhancer-RNAs (eRNAs), are transcribed from functional enhancers [33]. Although the function of eRNAs remains largely unclear, their close correlation with active enhancers suggests an important role of eRNAs in transcriptional activation. Long non-coding RNAs are also actively involved in diverse cytoplasmic processes. One of the well-studied examples is noncoding repressor of *NFAT* (NRON), an lncRNA repressor of nuclear factor of activated T cells (*NFAT*). Using an arrayed library of short hairpin RNAs and cell-based assays, Willingham *et al.* have identified that NRON interacts with multiple proteins, including members of the importin-beta superfamily, and possibly functions as a specific regulator of *NFAT* nuclear trafficking [57]. Moreover, lncRNAs are probably involved in stress-related signaling pathways. Utilizing whole-genome tiling arrays, Silva *et al.* identified a new class of long stress responsive non-coding transcripts (LSINCTs), which have increased expression in response to DNA damage induced by the tobacco carcinogen 4-(methylnitrosamino)-1-(3-pyridyl)-1-butanone (NNK) [58]. Interestingly, LSINCTs also have increased expression in a number of cancer-derived cell lines, indicating its stress response under a carcinogenic environment.

3. Long Non-Coding RNAs and Complex Human Diseases

Complex diseases are multifactorial or polygenic disorders of the body. They are likely caused by multiple genetic variants with low penetrance in combination with various environmental and lifestyle factors, and they do not simply obey the standard Mendelian patterns of inheritance [59,60]. Coronary artery diseases, autoimmune diseases, neurological disorders, various cancers, and many other diseases all belong within this classification. The recent discovery that lncRNAs can participate in a wide range of biological processes has attracted substantial scientific interest in their potential impact on these complex diseases. It has been reported that lncRNAs are dysregulated in a variety of complex human diseases and are closely associated with disease development and progression (Table 3). Here, we describe some of the well-characterized lncRNAs associated with different types of complex human diseases.

Table 3. Examples of lncRNAs dysregulated in complex human diseases.

LncRNA	Disease	References
aHIF	Multiple cancers	[61,62]
AK023948	Papillary thyroid carcinoma	[63]
ANRIL	Coronary artery disease; Multiple cancers	[64–68]
ASFMR1	Fragile X syndrome; Fragile X tremor ataxia syndrome	[69]
ATXN8OS	Spinocerebellar ataxia type 8	[70]
BACE1-AS	Alzheimer's disease	[71]
BC200	Alzheimer's disease; Multiple cancers	[72–74]
BIC	B-cell lymphoma	[75]
CUDR	Squamous carcinoma	[76]
DD3	Prostate cancer	[77,78]
FMR4	Fragile X syndrome; Fragile X tremor ataxia syndrome	[79]
GAS5	Breast cancer	[80]
H19	Multiple cancers	[81–85]
HOTAIR	Multiple cancers	[16,86]
HULC	Multiple cancers	[87,88]
Kcnq1ot1	Colon cancer	[89]
Kras1p	Prostate cancer	[30]
Linc-p21	Lung cancer	[43]
LOC285194	Osteosarcoma	[90]
MALAT-1	Multiple cancers	[91–93]
MEG3	Multiple cancers	[94–97]
MIAT	Myocardial infarction	[98]
ncRAN	Neuroblastoma	[99,100]
NDM29	Neuroblastoma	[101]
PCGEM1	Prostate cancer	[102]
PRINS	Psoriasis	[103]
PRNCR1	Prostate cancer	[104]
PTENP1	Prostate cancer	[30]
RMRP	Leukemia and lymphoma	[105]
SAS-ZFAT	Autoimmune thyroid disease	[106]

Table 3. *Cont.*

LncRNA	Disease	References
SPRY4-IT1	Melanoma	[107]
SRA-1	Breast cancer	[108,109]
TERC	Multiple cancers	[110]
Ube3a-as	Angelman syndrome	[111,112]
uc.73A	Colon cancer	[32]
UCA1	Bladder cancer	[113–115]
Zfas1	Breast cancer	[116]

3.1. LncRNAs in Coronary Artery Diseases

Facilitated by single nucleotide polymorphism (SNP) array, genome-wide association study (GWAS) is becoming one of the most powerful approaches to identify genetic variants susceptible to common diseases [117]. From these studies, a large amount of disease-associated SNPs are found to be mapped to non-coding genomic regions. While some of these SNPs could be associated with enhancers, it would not be surprising that many others are associated with lncRNAs. By using 52,608 haplotype-based SNP markers, the lncRNA called Myocardial Infarction Associated Transcript (MIAT) was first identified in a large-scale case-control association study of the samples from 3435 MI patients and 3774 controls [98]. MIAT dwells at a susceptible locus for myocardial infarction (MI) on chromosome 22q12.1. This study discovered six SNPs showing significant association with MI in this locus. The MIAT transcript is approximately 10 kb in length and has five exons. No translational product is encoded in MIAT based on *in vitro* translation assay, which indicates that it is an lncRNA. From *in vitro* functional analyses, MIAT transcription is increased by the minor variant of one SNP in exon 5. In contrast to the non-risk allele, the risk allele has more intense binding of nuclear protein(s). The study concluded that MIAT may play some roles in the pathogenesis of MI with altered expression by SNP [98].

Recent GWASs have identified a region on chromosome 9p that is associated with coronary artery disease (CAD) [118,119]. A long non-coding antisense RNA gene, named as Antisense non-coding RNA in the *INK4* locus (ANRIL), is a prime candidate for the chromosome 9p CAD locus [64]. Several recent GWASs showed that ANRIL has increased susceptibility to intracranial aneurysm, breast cancer, glioma, and basal cell carcinomas [65–68]. ANRIL is located in the *INK4b/ARF/INK4a* locus, and it is coregulated with *INK4a*, *INK4b* and *ARF*. Expression studies have confirmed that ANRIL is expressed in multiple atherosclerosis-related cell lines, including vascular endothelial cell, monocyte-derived macrophages and coronary smooth muscle cells [64,120]. Moreover, a mouse model study has demonstrated the pivotal role of ANRIL in the regulation of *INK4a/b* expression through a *cis*-acting mechanism and its implication in proliferation and senescence [121,122]. Interestingly, several studies discovered that Polycomb complexes are able to bind the *INK4/ARF* locus and alter expression of *INK4a* and *INK4b* [123,124]. Similar to lncRNA XIST, the ANRIL gene presents unusual masses of repetitive elements, as well as many binding sites for repressive transcription factors [120]. All these observations suggest that ANRIL, much like XIST, may regulate the expression of the INK4a/b transcript by recruiting Polycomb complex to the *INK4/ARF* locus and imposing a repressive chromatin state.

3.2. LncRNAs in Autoimmune Diseases

Long non-coding RNAs may also function in the regulation of downstream protein-coding genes, thus forming a complicated mutual regulation network with both coding and non-coding genes [125,126]. Recent studies have shown that autoimmune diseases, which result from an inappropriate immune response of the body against substances and tissues normally present in the body, have a complex genetic context that involves multiple protein-coding and non-coding genes. For example, in association study of 515 affected individuals and 526 controls, Shirasawa *et al.* discovered that the T allele of SNP Ex9b-SNP10 is correlated to increased risk for autoimmune thyroid disease (AITD) [106]. The Ex9b-SNP10 resides in intron 9 of the protein-coding gene *ZFAT* and the promoter region of an lncRNA, SAS-ZFAT, which is an antisense transcript of the ZFAT gene. With the existence of SNP Ex9b-SNP10, SAS-ZFAT expression is evidently upregulated, which, in turn, downregulates the expression level of its sense counterpart—truncated *ZFAT*. Since SAS-ZFAT is exclusively expressed in CD19+ B cells in peripheral blood lymphocytes, these results implicated that SAS-ZFAT might play a critical role in B cell function and determine susceptibility to AITD.

Another example is lncRNA PRINS, a Psoriasis Susceptibility-related non-coding RNA gene that harbors two Alu elements [103]. PRINS is transcribed by RNA polymerase II and is expressed at different levels in various human tissues. Real-time RT-PCR analysis showed that PRINS has higher expression in the uninvolved epidermis of psoriatic patients than in both psoriatic lesional and healthy epidermis, suggesting that PRINS plays a role in psoriasis susceptibility. *In silico* structural and homology studies have suggested that PRINS acts as a non-coding RNA. The RNA expression level of PRINS is decreased in the uninvolved psoriatic, but not healthy, epidermis with treatment of T-lymphokines that are known to precipitate psoriatic symptoms. Moreover, downregulating the RNA level of PRINS by RNAi can impair cell viability after serum starvation, but not under normal serum conditions. It was also discovered that PRINS could function as a "riboregulator" to regulate the expression of other genes involved in the proliferation and survival of cells.

3.3. LncRNAs in Neurological Disorders

Neurological disorders are diseases of the body's nervous systems, which include the central nervous system, the peripheral nervous system, and the autonomic nervous system. Previous transcriptome studies have shown a number of lncRNAs in the mammalian brain, and most of them exhibit particular expression profiles within specific neuroanatomical regions, cell types, or subcellular compartments, implicating that lncRNAs probably have a significant impact on neurological regulation [127].

Long non-coding RNAs may participate in the pathogenesis of fragile X syndrome (FXS) and fragile X tremor ataxia syndrome (FXTAS), both of which are caused by the aberrant expansion of CGG trinucleotide, repeat in the 5' UTR of protein-coding fragile-X mental retardation 1 gene (*FMR1*) [128,129]. Studies have shown that two lncRNAs, FMR4 and ASFMR1, are expressed from the *FMR1* locus. FMR4 is a primate-specific lncRNA that likely shares a bidirectional promoter with *FMR1* [79], while ASFMR1 is a spliced and poly-adenylated antisense transcript that overlaps the 5' UTR CGG repeat region of *FMR1* [69]. *In vitro* studies of FMR4 have shown that it may function to prevent neurons or their progenitors from apoptosis during the progress of development in human. FMR4 and

ASFMR1, as well as *FMR1*, may participate synergistically in neurological regulation in a RNA-protein interacting manner, since they are all silenced in FXS or upregulated in FXTAS patients. Dysregulation of these delicate interactions may result in various neurological disorders [69,79].

Long non-coding RNAs have also been reported to be dysregulated in different types of neurodegenerative diseases, such as spinocerebellar ataxia type 8 (SCA8) and Alzheimer's disease (AD). SCA8 is an autosomal dominant disorder caused by repeat expansion [130]. Two bidirectionally transcribed genes are located within the SCA8 expansion region: the protein-coding gene ataxin 8 (*ATXN8*), with a CAG expansion that encodes a polyglutamine expansion tract protein, and ataxin 8 opposite strand (ATXN8OS) which is an lncRNA with a CUG repeat. In studies of transgenic mice expressing SCA8 expansion, it was found that the ATXN8OS mutant is overexpressed and co-localized with Muscleblind-like splicing regulator 1 (*MBNL1*) in neurons, which can lead to dysregulation of MBNL1-mediated alternative splicing, loss of GABAergic inhibition within the granular cell layer, and set the stage for the occurrence of disease [70]. AD is a form of dementia, which is believed to be caused by the formation of amyloid plaques in neurons [131]. Studies have shown that BACE1-AS, an antisense lncRNA counterpart of protein-coding gene *BACE1*, is highly expressed in tissues from AD patients [71]. *BACE1* is an enzyme that is responsible for amyloid precursor protein (APP) cleavage into amyloid β peptides, which form amyloid plaques in the neurons of AD patients [132]. Upregulation of BACE1-AS promotes the stabilization of *BACE1* mRNA and boosts the expression of *BACE1* protein, which leads to the production of pathogenic amyloid β peptides and thus may speed up the pathogenesis of AD [71]. Another lncRNA involved in AD is BC200, which is expressed almost exclusively in neuronal cells. In AD patients, the level of BC200 becomes upregulated. The increased expression of BC200 was found to be correlated with the severity of AD [74]. In addition, studies of BC1, the mouse functional homolog of BC200, have shown that BC1 knockout mice exhibit behavioral changes, thus demonstrating an important role for BC1 in brain function [133]. All these results suggest that dysregulation of BC200 may contribute to AD susceptibility.

3.4. LncRNAs in Cancers

Cancer is a broad group of various diseases in which abnormal cells divide uncontrollably and tend to invade other tissues. Up to now, although hundreds of oncogenes and tumor suppressor genes have been identified, the exact cause of most cancers remains unknown or poorly understood. In recent years, researchers have increasingly come to recognize lncRNAs as major mediators in cancer pathogenesis [134]. Thus far, no concrete evidence has surfaced to indicate any lncRNAs as causal factors in cancer. However, many lncRNAs have been found to be differentially expressed in a variety of cancers and may act as either oncogenes, such as MALAT-1, HOTAIR, and ANRIL, or tumor suppressor genes, such as MEG3, lincRNA-p21, and PTENP1, in cancer development. Here, we discuss some examples of such lncRNAs.

Like protein-coding oncogenes, some lncRNAs can promote cell proliferation and induce tumorigenesis. Metastasis-associated lung adenocarcinoma transcript 1 (MALAT-1), which correlates with high metastasis and poor prognosis in non-small-cell lung cancer, is an abundant 8.7-kb lncRNA encoded in the human chromosome 11q13 [93]. MALAT-1 is broadly expressed in normal human tissues and is found to be upregulated in many solid tumors, such as lung, breast, prostate, liver, and colon tumors [91,92,135]. MALAT-1 is believed to play a vital role in cell proliferation, migration, and invasion. By interacting with serine-arginine-rich splicing factor (*SR*), which is responsible for alternative splicing (AS) in a concentration- and phosphorylation-dependent manner, studies have shown that MALAT-1 can modulate the phosphorylation of *SR* proteins and thus regulate AS of selective pre-mRNAs [136]. MALAT1 is also involved in the regulation of cell mobility. RNAi-mediated silencing of MALAT1 impaired the *in vitro* migration of lung adenocarcinoma cells and reduced cell proliferation and invasive potential in a cervical cancer cell line [92]. Another onco-lncRNA example is *HOX* Antisense Intergenic RNA (HOTAIR). HOTAIR, a 2.2-kb spliced and poly-adenylated lncRNA, is transcribed from the antisense strand of the Homeobox C (*HOXC*) gene cluster on chromosome 12 [16]. Studies have shown that *HOTAIR* is overexpressed in breast tumors, hepatocellular carcinoma, and colorectal cancer [137–139]. A high level of HOTAIR expression is directly correlated with poor patient prognosis and metastasis. Recent studies revealed that HOTAIR is likely to work as a molecular scaffold to bind two distinct histone modification complexes, the Polycomb repressor complex 2 (PRC2) and the histone demethylase LSD1, facilitating their genome-wide retargeting to specific regions for coupled histone H3K27 methylation and H3K4 demethylation [140]. *In vitro* studies have shown that overexpression of HOTAIR in cell lines leads to the recruitment of PRC2 and LSD1 to over 800 additional loci, including those of tumor suppressor genes [16]. These observations indicate that dysregulation of HOTAIR may reprogram the epigenetic information to promote tumor cell invasion and subsequent metastasis.

Long non-coding RNAs can also act as tumor suppressor genes. One example is maternally expressed gene 3 (MEG3), a maternally imprinted RNA gene of approximately 1700 nucleotides [94]. Studies have revealed that MEG3 is expressed in many normal tissues, but not in the majority of human meningiomas or human meningioma cell lines [141]. Moreover, ectopic expression of MEG3 was found to suppress the growth of several human cancer cell lines, further supporting the effect of MEG3 on tumor suppression [95]. MEG3 was found to be a positive regulator of *p53*, a tumor suppressor protein [142]. In cells that are transfected with MEG3, *p53* protein level increases significantly, which results in dramatically stimulating the transcription of p53-dependent genes from a p53-responsive promoter. Studies have shown that MEG3 is also capable of inhibiting cell proliferation in the absence of *p53* [143]. These data suggest that MEG3 can function as a tumor suppressor through both p53-dependent and p53-independent pathways. MEG3 has a total of twelve isoforms from alternative splicing, all of which contain three distinct secondary folding motifs (M1, M2, and M3). Deletion analysis indicates that motifs M2 and M3 are important for *p53* activation. Furthermore, a hybrid MEG3 RNA, which contains a piece of unrelated sequence, but preserves the original secondary structure, retained the functions of both *p53* activation and growth suppression [144]. As a regulatory lncRNA, all of these experiments demonstrated that the proper conformation of MEG3 is critical to its biological functions.

4. Conclusions

Long non-coding RNAs are rapidly becoming a focal point for intensified research in the biological and medical sciences. Increasing evidence has indicated that lncRNAs play important roles in various critical biological processes and that they add a new layer of complexity to already complex human diseases. We believe that the further functional and mechanistic studies of these versatile macromolecules will expand our understanding of general principles in biological systems and provide new approaches to the diagnosis and treatment of complex human diseases.

Acknowledgments

This work was supported by National Natural Science Foundation of China (Grant No. 31000586) and The University of Texas at Dallas Startup Grant.

References

1. Consortium, E.P.; Dunham, I.; Kundaje, A.; Aldred, S.F.; Collins, P.J.; Davis, C.A.; Doyle, F.; Epstein, C.B.; Frietze, S.; Harrow, J.; *et al.* An integrated encyclopedia of DNA elements in the human genome. *Nature* **2012**, *489*, 57–74.
2. Lander, E.S.; Linton, L.M.; Birren, B.; Nusbaum, C.; Zody, M.C.; Baldwin, J.; Devon, K.; Dewar, K.; Doyle, M.; FitzHugh, W.; *et al.* Initial sequencing and analysis of the human genome. *Nature* **2001**, *409*, 860–921.
3. Liu, C.; Bai, B.; Skogerbo, G.; Cai, L.; Deng, W.; Zhang, Y.; Bu, D.; Zhao, Y.; Chen, R. Noncode: An integrated knowledge database of non-coding RNAs. *Nucleic Acids Res.* **2005**, *33*, D112–D115.
4. Carninci, P.; Kasukawa, T.; Katayama, S.; Gough, J.; Frith, M.C.; Maeda, N.; Oyama, R.; Ravasi, T.; Lenhard, B.; Wells, C.; *et al.* The transcriptional landscape of the mammalian genome. *Science* **2005**, *309*, 1559–1563.
5. Lau, N.C.; Seto, A.G.; Kim, J.; Kuramochi-Miyagawa, S.; Nakano, T.; Bartel, D.P.; Kingston, R.E. Characterization of the pirna complex from rat testes. *Science* **2006**, *313*, 363–367.
6. David, R. Small RNAs: MiRNA machinery disposal. *Nat. Rev.* **2013**, *14*, 4–5.
7. Wilusz, J.E.; Sunwoo, H.; Spector, D.L. Long noncoding RNAs: Functional surprises from the RNA world. *Genes Dev.* **2009**, *23*, 1494–1504.
8. Liao, Q.; Liu, C.; Yuan, X.; Kang, S.; Miao, R.; Xiao, H.; Zhao, G.; Luo, H.; Bu, D.; Zhao, H.; *et al.* Large-scale prediction of long non-coding RNA functions in a coding-non-coding gene co-expression network. *Nucleic Acids Res.* **2011**, *39*, 3864–3878.
9. Bu, D.; Yu, K.; Sun, S.; Xie, C.; Skogerbo, G.; Miao, R.; Xiao, H.; Liao, Q.; Luo, H.; Zhao, G.; *et al.* Noncode v3.0: Integrative annotation of long noncoding RNAs. *Nucleic Acids Res.* **2012**, *40*, D210–D215.
10. Khalil, A.M.; Guttman, M.; Huarte, M.; Garber, M.; Raj, A.; Rivea Morales, D.; Thomas, K.; Presser, A.; Bernstein, B.E.; van Oudenaarden, A.; *et al.* Many human large intergenic noncoding RNAs associate with chromatin-modifying complexes and affect gene expression. *Proc. Natl. Acad. Sci. USA* **2009**, *106*, 11667–11672.
11. Mercer, T.R.; Mattick, J.S. Structure and function of long noncoding RNAs in epigenetic regulation. *Nat. Struct. Mol. Biol.* **2013**, *20*, 300–307.
12. Wang, K.C.; Chang, H.Y. Molecular mechanisms of long noncoding RNAs. *Mol. Cell* **2011**, *43*, 904–914.

13. Tufarelli, C.; Stanley, J.A.; Garrick, D.; Sharpe, J.A.; Ayyub, H.; Wood, W.G.; Higgs, D.R. Transcription of antisense RNA leading to gene silencing and methylation as a novel cause of human genetic disease. *Nat. Genet.* **2003**, *34*, 157–165.
14. Tahira, A.C.; Kubrusly, M.S.; Faria, M.F.; Dazzani, B.; Fonseca, R.S.; Maracaja-Coutinho, V.; Verjovski-Almeida, S.; Machado, M.C.; Reis, E.M. Long noncoding intronic RNAs are differentially expressed in primary and metastatic pancreatic cancer. *Mol. Cancer* **2011**, *10*, 141.
15. Chow, J.; Heard, E. X inactivation and the complexities of silencing a sex chromosome. *Curr. Opin. Cell Biol.* **2009**, *21*, 359–366.
16. Gupta, R.A.; Shah, N.; Wang, K.C.; Kim, J.; Horlings, H.M.; Wong, D.J.; Tsai, M.C.; Hung, T.; Argani, P.; Rinn, J.L.; *et al.* Long non-coding RNA hotair reprograms chromatin state to promote cancer metastasis. *Nature* **2010**, *464*, 1071–1076.
17. Kung, J.T.; Colognori, D.; Lee, J.T. Long noncoding RNAs: Past, present, and future. *Genetics* **2013**, *193*, 651–669.
18. Ponting, C.P.; Oliver, P.L.; Reik, W. Evolution and functions of long noncoding RNAs. *Cell* **2009**, *136*, 629–641.
19. Guttman, M.; Amit, I.; Garber, M.; French, C.; Lin, M.F.; Feldser, D.; Huarte, M.; Zuk, O.; Carey, B.W.; Cassady, J.P.; *et al.* Chromatin signature reveals over a thousand highly conserved large non-coding RNAs in mammals. *Nature* **2009**, *458*, 223–227.
20. Guttman, M.; Donaghey, J.; Carey, B.W.; Garber, M.; Grenier, J.K.; Munson, G.; Young, G.; Lucas, A.B.; Ach, R.; Bruhn, L.; *et al.* LincRNAs act in the circuitry controlling pluripotency and differentiation. *Nature* **2011**, *477*, 295–300.
21. Louro, R.; Smirnova, A.S.; Verjovski-Almeida, S. Long intronic noncoding RNA transcription: Expression noise or expression choice? *Genomics* **2009**, *93*, 291–298.
22. Yin, Y.; Zhao, Y.; Wang, J.; Liu, C.; Chen, S.; Chen, R.; Zhao, H. Anticode: A natural sense-antisense transcripts database. *BMC Bioinf.* **2007**, *8*, 319.
23. Katayama, S.; Tomaru, Y.; Kasukawa, T.; Waki, K.; Nakanishi, M.; Nakamura, M.; Nishida, H.; Yap, C.C.; Suzuki, M.; Kawai, J.; *et al.* Antisense transcription in the mammalian transcriptome. *Science* **2005**, *309*, 1564–1566.
24. Lehner, B.; Williams, G.; Campbell, R.D.; Sanderson, C.M. Antisense transcripts in the human genome. *Trends Genet.* **2002**, *18*, 63–65.
25. Kapranov, P.; Cheng, J.; Dike, S.; Nix, D.A.; Duttagupta, R.; Willingham, A.T.; Stadler, P.F.; Hertel, J.; Hackermuller, J.; Hofacker, I.L.; *et al.* RNA maps reveal new RNA classes and a possible function for pervasive transcription. *Science* **2007**, *316*, 1484–1488.
26. Preker, P.; Nielsen, J.; Kammler, S.; Lykke-Andersen, S.; Christensen, M.S.; Mapendano, C.K.; Schierup, M.H.; Jensen, T.H. RNA exosome depletion reveals transcription upstream of active human promoters. *Science* **2008**, *322*, 1851–1854.
27. Liu, W.M.; Chu, W.M.; Choudary, P.V.; Schmid, C.W. Cell stress and translational inhibitors transiently increase the abundance of mammalian sine transcripts. *Nucleic Acids Res.* **1995**, *23*, 1758–1765.
28. Espinoza, C.A.; Goodrich, J.A.; Kugel, J.F. Characterization of the structure, function, and mechanism of b2 RNA, an ncRNA repressor of RNA polymerase II transcription. *RNA* **2007**, *13*, 583–596.

29. Mariner, P.D.; Walters, R.D.; Espinoza, C.A.; Drullinger, L.F.; Wagner, S.D.; Kugel, J.F.; Goodrich, J.A. Human Alu RNA is a modular transacting repressor of mRNA transcription during heat shock. *Mol. Cell* **2008**, *29*, 499–509.
30. Poliseno, L.; Salmena, L.; Zhang, J.; Carver, B.; Haveman, W.J.; Pandolfi, P.P. A coding-independent function of gene and pseudogene mRNAs regulates tumour biology. *Nature* **2010**, *465*, 1033–1038.
31. Salmena, L.; Poliseno, L.; Tay, Y.; Kats, L.; Pandolfi, P.P. A ceRNA hypothesis: The rosetta stone of a hidden RNA language? *Cell* **2011**, *146*, 353–358.
32. Calin, G.A.; Liu, C.G.; Ferracin, M.; Hyslop, T.; Spizzo, R.; Sevignani, C.; Fabbri, M.; Cimmino, A.; Lee, E.J.; Wojcik, S.E.; *et al.* Ultraconserved regions encoding ncRNAs are altered in human leukemias and carcinomas. *Cancer Cell* **2007**, *12*, 215–229.
33. Kim, T.K.; Hemberg, M.; Gray, J.M.; Costa, A.M.; Bear, D.M.; Wu, J.; Harmin, D.A.; Laptewicz, M.; Barbara-Haley, K.; Kuersten, S.; *et al.* Widespread transcription at neuronal activity-regulated enhancers. *Nature* **2010**, *465*, 182–187.
34. Yang, J.H.; Li, J.H.; Jiang, S.; Zhou, H.; Qu, L.H. Chipbase: A database for decoding the transcriptional regulation of long non-coding RNA and microRNA genes from chip-seq data. *Nucleic Acids Res.* **2013**, *41*, D177–D187.
35. Mituyama, T.; Yamada, K.; Hattori, E.; Okida, H.; Ono, Y.; Terai, G.; Yoshizawa, A.; Komori, T.; Asai, K. The functional RNA database 3.0: Databases to support mining and annotation of functional RNAs. *Nucleic Acids Res.* **2009**, *37*, D89–D92.
36. Volders, P.J.; Helsens, K.; Wang, X.; Menten, B.; Martens, L.; Gevaert, K.; Vandesompele, J.; Mestdagh, P. Lncipedia: A database for annotated human lncRNA transcript sequences and structures. *Nucleic Acids Res.* **2013**, *41*, D246–D251.
37. Amaral, P.P.; Clark, M.B.; Gascoigne, D.K.; Dinger, M.E.; Mattick, J.S. LncRNAdb: A reference database for long noncoding RNAs. *Nucleic Acids Res.* **2011**, *39*, D146–D151.
38. Dinger, M.E.; Pang, K.C.; Mercer, T.R.; Crowe, M.L.; Grimmond, S.M.; Mattick, J.S. Nred: A database of long noncoding RNA expression. *Nucleic Acids Res.* **2009**, *37*, D122–D126.
39. Okazaki, Y.; Furuno, M.; Kasukawa, T.; Adachi, J.; Bono, H.; Kondo, S.; Nikaido, I.; Osato, N.; Saito, R.; Suzuki, H.; *et al.* Analysis of the mouse transcriptome based on functional annotation of 60,770 full-length cDNAs. *Nature* **2002**, *420*, 563–573.
40. Bertone, P.; Stolc, V.; Royce, T.E.; Rozowsky, J.S.; Urban, A.E.; Zhu, X.; Rinn, J.L.; Tongprasit, W.; Samanta, M.; Weissman, S.; *et al.* Global identification of human transcribed sequences with genome tiling arrays. *Science* **2004**, *306*, 2242–2246.
41. Knee, R.; Murphy, P.R. Regulation of gene expression by natural antisense RNA transcripts. *Neurochem. Inter.* **1997**, *31*, 379–392.
42. Cabili, M.N.; Trapnell, C.; Goff, L.; Koziol, M.; Tazon-Vega, B.; Regev, A.; Rinn, J.L. Integrative annotation of human large intergenic noncoding RNAs reveals global properties and specific subclasses. *Genes Dev.* **2011**, *25*, 1915–1927.
43. Huarte, M.; Guttman, M.; Feldser, D.; Garber, M.; Koziol, M.J.; Kenzelmann-Broz, D.; Khalil, A.M.; Zuk, O.; Amit, I.; Rabani, M., *et al.* A large intergenic noncoding RNA induced by p53 mediates global gene repression in the p53 response. *Cell* **2010**, *142*, 409–419.

44. Wang, J.; Zhang, J.; Zheng, H.; Li, J.; Liu, D.; Li, H.; Samudrala, R.; Yu, J.; Wong, G.K. Mouse transcriptome: Neutral evolution of "non-coding" complementary DNAs. *Nature* **2004**, *431*, doi:10.1038/nature03016.
45. Kutter, C.; Watt, S.; Stefflova, K.; Wilson, M.D.; Goncalves, A.; Ponting, C.P.; Odom, D.T.; Marques, A.C. Rapid turnover of long noncoding RNAs and the evolution of gene expression. *PLoS Genet.* **2012**, *8*, e1002841.
46. Pang, K.C.; Frith, M.C.; Mattick, J.S. Rapid evolution of noncoding RNAs: Lack of conservation does not mean lack of function. *Trends Genet.* **2006**, *22*, 1–5.
47. Nagano, T.; Mitchell, J.A.; Sanz, L.A.; Pauler, F.M.; Ferguson-Smith, A.C.; Feil, R.; Fraser, P. The air noncoding RNA epigenetically silences transcription by targeting g9a to chromatin. *Science* **2008**, *322*, 1717–1720.
48. Leeb, M.; Steffen, P.A.; Wutz, A. X chromosome inactivation sparked by non-coding RNAs. *RNA Biol.* **2009**, *6*, 94–99.
49. Pollard, K.S.; Salama, S.R.; Lambert, N.; Lambot, M.A.; Coppens, S.; Pedersen, J.S.; Katzman, S.; King, B.; Onodera, C.; Siepel, A.; *et al.* An RNA gene expressed during cortical development evolved rapidly in humans. *Nature* **2006**, *443*, 167–172.
50. Leontis, N.B.; Westhof, E. Analysis of RNA motifs. *Curr. Opin. Struct. Biol.* **2003**, *13*, 300–308.
51. Bejerano, G.; Pheasant, M.; Makunin, I.; Stephen, S.; Kent, W.J.; Mattick, J.S.; Haussler, D. Ultraconserved elements in the human genome. *Science* **2004**, *304*, 1321–1325.
52. Chen, C.T.; Wang, J.C.; Cohen, B.A. The strength of selection on ultraconserved elements in the human genome. *Am. J. Hum. Genet.* **2007**, *80*, 692–704.
53. Katzman, S.; Kern, A.D.; Bejerano, G.; Fewell, G.; Fulton, L.; Wilson, R.K.; Salama, S.R.; Haussler, D. Human genome ultraconserved elements are ultraselected. *Science* **2007**, *317*, 915.
54. Catucci, I.; Verderio, P.; Pizzamiglio, S.; Manoukian, S.; Peissel, B.; Barile, M.; Tizzoni, L.; Bernard, L.; Ravagnani, F.; Galastri, L.; *et al.* Snps in ultraconserved elements and familial breast cancer risk. *Carcinogenesis* **2009**, *30*, 544–545.
55. Scaruffi, P.; Stigliani, S.; Moretti, S.; Coco, S.; de Vecchi, C.; Valdora, F.; Garaventa, A.; Bonassi, S.; Tonini, G.P. Transcribed-ultra conserved region expression is associated with outcome in high-risk neuroblastoma. *BMC Cancer* **2009**, *9*, 441.
56. Orom, U.A.; Derrien, T.; Beringer, M.; Gumireddy, K.; Gardini, A.; Bussotti, G.; Lai, F.; Zytnicki, M.; Notredame, C.; Huang, Q.; *et al.* Long noncoding RNAs with enhancer-like function in human cells. *Cell* **2010**, *143*, 46–58.
57. Willingham, A.T.; Orth, A.P.; Batalov, S.; Peters, E.C.; Wen, B.G.; Aza-Blanc, P.; Hogenesch, J.B.; Schultz, P.G. A strategy for probing the function of noncoding RNAs finds a repressor of nfat. *Science* **2005**, *309*, 1570–1573.
58. Silva, J.M.; Perez, D.S.; Pritchett, J.R.; Halling, M.L.; Tang, H.; Smith, D.I. Identification of long stress-induced non-coding transcripts that have altered expression in cancer. *Genomics* **2010**, *95*, 355–362.
59. Hunter, D.J. Gene-environment interactions in human diseases. *Nat. Rev.* **2005**, *6*, 287–298.
60. Liu, L.; Li, Y.; Tollefsbol, T.O. Gene-environment interactions and epigenetic basis of human diseases. *Curr. Issues Mol. Biol.* **2008**, *10*, 25–36.
61. Thrash-Bingham, C.A.; Tartof, K.D. Ahif: A natural antisense transcript overexpressed in human renal cancer and during hypoxia. *J. Natl. Cancer Inst.* **1999**, *91*, 143–151.

62. Cayre, A.; Rossignol, F.; Clottes, E.; Penault-Llorca, F. Ahif but not hif-1alpha transcript is a poor prognostic marker in human breast cancer. *Breast Cancer Res.* **2003**, *5*, R223–R230.
63. He, H.; Nagy, R.; Liyanarachchi, S.; Jiao, H.; Li, W.; Suster, S.; Kere, J.; de la Chapelle, A. A susceptibility locus for papillary thyroid carcinoma on chromosome 8q24. *Cancer Res.* **2009**, *69*, 625–631.
64. Broadbent, H.M.; Peden, J.F.; Lorkowski, S.; Goel, A.; Ongen, H.; Green, F.; Clarke, R.; Collins, R.; Franzosi, M.G.; Tognoni, G.; *et al.* Susceptibility to coronary artery disease and diabetes is encoded by distinct, tightly linked snps in the anril locus on chromosome 9p. *Hum. Mol. Genet.* **2008**, *17*, 806–814.
65. Yasuno, K.; Bilguvar, K.; Bijlenga, P.; Low, S.K.; Krischek, B.; Auburger, G.; Simon, M.; Krex, D.; Arlier, Z.; Nayak, N.; *et al.* Genome-wide association study of intracranial aneurysm identifies three new risk loci. *Nat. Genet.* **2010**, *42*, 420–425.
66. Shete, S.; Hosking, F.J.; Robertson, L.B.; Dobbins, S.E.; Sanson, M.; Malmer, B.; Simon, M.; Marie, Y.; Boisselier, B.; Delattre, J.Y.; *et al.* Genome-wide association study identifies five susceptibility loci for glioma. *Nat. Genet.* **2009**, *41*, 899–904.
67. Stacey, S.N.; Sulem, P.; Masson, G.; Gudjonsson, S.A.; Thorleifsson, G.; Jakobsdottir, M.; Sigurdsson, A.; Gudbjartsson, D.F.; Sigurgeirsson, B.; Benediktsdottir, K.R.; *et al.* New common variants affecting susceptibility to basal cell carcinoma. *Nat. Genet.* **2009**, *41*, 909–914.
68. Turnbull, C.; Ahmed, S.; Morrison, J.; Pernet, D.; Renwick, A.; Maranian, M.; Seal, S.; Ghoussaini, M.; Hines, S.; Healey, C.S.; *et al.* Genome-wide association study identifies five new breast cancer susceptibility loci. *Nat. Genet.* **2010**, *42*, 504–507.
69. Ladd, P.D.; Smith, L.E.; Rabaia, N.A.; Moore, J.M.; Georges, S.A.; Hansen, R.S.; Hagerman, R.J.; Tassone, F.; Tapscott, S.J.; Filippova, G.N. An antisense transcript spanning the cgg repeat region of fmr1 is upregulated in premutation carriers but silenced in full mutation individuals. *Hum. Mol. Genet.* **2007**, *16*, 3174–3187.
70. Moseley, M.L.; Zu, T.; Ikeda, Y.; Gao, W.; Mosemiller, A.K.; Daughters, R.S.; Chen, G.; Weatherspoon, M.R.; Clark, H.B.; Ebner, T.J.; *et al.* Bidirectional expression of cug and cag expansion transcripts and intranuclear polyglutamine inclusions in spinocerebellar ataxia type 8. *Nat. Genet.* **2006**, *38*, 758–769.
71. Faghihi, M.A.; Modarresi, F.; Khalil, A.M.; Wood, D.E.; Sahagan, B.G.; Morgan, T.E.; Finch, C.E.; St Laurent, G., III; Kenny, P.J.; Wahlestedt, C. Expression of a noncoding RNA is elevated in Alzheimer's disease and drives rapid feed-forward regulation of beta-secretase. *Nat. Med.* **2008**, *14*, 723–730.
72. Chen, W.; Bocker, W.; Brosius, J.; Tiedge, H. Expression of neural bc200 RNA in human tumours. *J. Pathol.* **1997**, *183*, 345–351.
73. Iacoangeli, A.; Lin, Y.; Morley, E.J.; Muslimov, I.A.; Bianchi, R.; Reilly, J.; Weedon, J.; Diallo, R.; Bocker, W.; Tiedge, H. Bc200 RNA in invasive and preinvasive breast cancer. *Carcinogenesis* **2004**, *25*, 2125–2133.
74. Mus, E.; Hof, P.R.; Tiedge, H. Dendritic bc200 RNA in aging and in Alzheimer's disease. *Proc. Natl. Acad. Sci. USA* **2007**, *104*, 10679–10684.

75. Eis, P.S.; Tam, W.; Sun, L.; Chadburn, A.; Li, Z.; Gomez, M.F.; Lund, E.; Dahlberg, J.E. Accumulation of mir-155 and bic RNA in human b cell lymphomas. *Proc. Natl. Acad. Sci. USA* **2005**, *102*, 3627–3632.
76. Tsang, W.P.; Wong, T.W.; Cheung, A.H.; Co, C.N.; Kwok, T.T. Induction of drug resistance and transformation in human cancer cells by the noncoding RNA cudr. *RNA* **2007**, *13*, 890–898.
77. Bussemakers, M.J.; van Bokhoven, A.; Verhaegh, G.W.; Smit, F.P.; Karthaus, H.F.; Schalken, J.A.; Debruyne, F.M.; Ru, N.; Isaacs, W.B. Dd3: A new prostate-specific gene, highly overexpressed in prostate cancer. *Cancer Res.* **1999**, *59*, 5975–5979.
78. De Kok, J.B.; Verhaegh, G.W.; Roelofs, R.W.; Hessels, D.; Kiemeney, L.A.; Aalders, T.W.; Swinkels, D.W.; Schalken, J.A. Dd3(pca3), a very sensitive and specific marker to detect prostate tumors. *Cancer Res.* **2002**, *62*, 2695–2698.
79. Khalil, A.M.; Faghihi, M.A.; Modarresi, F.; Brothers, S.P.; Wahlestedt, C. A novel RNA transcript with antiapoptotic function is silenced in fragile x syndrome. *PLoS One* **2008**, *3*, e1486.
80. Mourtada-Maarabouni, M.; Pickard, M.R.; Hedge, V.L.; Farzaneh, F.; Williams, G.T. Gas5, a non-protein-coding RNA, controls apoptosis and is downregulated in breast cancer. *Oncogene* **2009**, *28*, 195–208.
81. Brannan, C.I.; Dees, E.C.; Ingram, R.S.; Tilghman, S.M. The product of the h19 gene may function as an RNA. *Mol. Cell. Biol.* **1990**, *10*, 28–36.
82. Gabory, A.; Jammes, H.; Dandolo, L. The h19 locus: Role of an imprinted non-coding RNA in growth and development. *BioEssays* **2010**, *32*, 473–480.
83. Matouk, I.J.; DeGroot, N.; Mezan, S.; Ayesh, S.; Abu-lail, R.; Hochberg, A.; Galun, E. The h19 non-coding RNA is essential for human tumor growth. *PLoS One* **2007**, *2*, e845.
84. Yang, F.; Bi, J.; Xue, X.; Zheng, L.; Zhi, K.; Hua, J.; Fang, G. Up-regulated long non-coding RNA h19 contributes to proliferation of gastric cancer cells. *FEBS J.* **2012**, *279*, 3159–3165.
85. Luo, M.; Li, Z.; Wang, W.; Zeng, Y.; Liu, Z.; Qiu, J. Long non-coding RNA h19 increases bladder cancer metastasis by associating with ezh2 and inhibiting e-cadherin expression. *Cancer Lett.* **2013**, *333*, 213–221.
86. Rinn, J.L.; Kertesz, M.; Wang, J.K.; Squazzo, S.L.; Xu, X.; Brugmann, S.A.; Goodnough, L.H.; Helms, J.A.; Farnham, P.J.; Segal, E.; *et al.* Functional demarcation of active and silent chromatin domains in human hox loci by noncoding RNAs. *Cell* **2007**, *129*, 1311–1323.
87. Panzitt, K.; Tschernatsch, M.M.; Guelly, C.; Moustafa, T.; Stradner, M.; Strohmaier, H.M.; Buck, C.R.; Denk, H.; Schroeder, R.; Trauner, M.; *et al.* Characterization of hulc, a novel gene with striking up-regulation in hepatocellular carcinoma, as noncoding RNA. *Gastroenterology* **2007**, *132*, 330–342.
88. Matouk, I.J.; Abbasi, I.; Hochberg, A.; Galun, E.; Dweik, H.; Akkawi, M. Highly upregulated in liver cancer noncoding RNA is overexpressed in hepatic colorectal metastasis. *Eur. J. Gastroenterol. Hepatol.* **2009**, *21*, 688–692.
89. Tanaka, K.; Shiota, G.; Meguro, M.; Mitsuya, K.; Oshimura, M.; Kawasaki, H. Loss of imprinting of long qt intronic transcript 1 in colorectal cancer. *Oncology* **2001**, *60*, 268–273.
90. Pasic, I.; Shlien, A.; Durbin, A.D.; Stavropoulos, D.J.; Baskin, B.; Ray, P.N.; Novokmet, A.; Malkin, D. Recurrent focal copy-number changes and loss of heterozygosity implicate two noncoding RNAs and one tumor suppressor gene at chromosome 3q13.31 in osteosarcoma. *Cancer Res.* **2010**, *70*, 160–171.

91. Ying, L.; Chen, Q.; Wang, Y.; Zhou, Z.; Huang, Y.; Qiu, F. Upregulated malat-1 contributes to bladder cancer cell migration by inducing epithelial-to-mesenchymal transition. *Mol. Biosyst.* **2012**, *8*, 2289–2294.
92. Tano, K.; Mizuno, R.; Okada, T.; Rakwal, R.; Shibato, J.; Masuo, Y.; Ijiri, K.; Akimitsu, N. Malat-1 enhances cell motility of lung adenocarcinoma cells by influencing the expression of motility-related genes. *FEBS Lett.* **2010**, *584*, 4575–4580.
93. Ji, P.; Diederichs, S.; Wang, W.; Boing, S.; Metzger, R.; Schneider, P.M.; Tidow, N.; Brandt, B.; Buerger, H.; Bulk, E.; *et al.* Malat-1, a novel noncoding RNA, and thymosin beta4 predict metastasis and survival in early-stage non-small cell lung cancer. *Oncogene* **2003**, *22*, 8031–8041.
94. Miyoshi, N.; Wagatsuma, H.; Wakana, S.; Shiroishi, T.; Nomura, M.; Aisaka, K.; Kohda, T.; Surani, M.A.; Kaneko-Ishino, T.; Ishino, F. Identification of an imprinted gene, meg3/gtl2 and its human homologue meg3, first mapped on mouse distal chromosome 12 and human chromosome 14q. *Genes Cells* **2000**, *5*, 211–220.
95. Zhang, X.; Zhou, Y.; Mehta, K.R.; Danila, D.C.; Scolavino, S.; Johnson, S.R.; Klibanski, A. A pituitary-derived meg3 isoform functions as a growth suppressor in tumor cells. *J. Clin. Endocrinol. Metab.* **2003**, *88*, 5119–5126.
96. Braconi, C.; Kogure, T.; Valeri, N.; Huang, N.; Nuovo, G.; Costinean, S.; Negrini, M.; Miotto, E.; Croce, C.M.; Patel, T. MicroRNA-29 can regulate expression of the long non-coding RNA gene meg3 in hepatocellular cancer. *Oncogene* **2011**, *30*, 4750–4756.
97. Ying, L.; Huang, Y.; Chen, H.; Wang, Y.; Xia, L.; Chen, Y.; Liu, Y.; Qiu, F. Downregulated meg3 activates autophagy and increases cell proliferation in bladder cancer. *Mol. Biosyst.* **2013**, *9*, 407–411.
98. Ishii, N.; Ozaki, K.; Sato, H.; Mizuno, H.; Saito, S.; Takahashi, A.; Miyamoto, Y.; Ikegawa, S.; Kamatani, N.; Hori, M.; *et al.* Identification of a novel non-coding RNA, miat, that confers risk of myocardial infarction. *J. Hum. Genet.* **2006**, *51*, 1087–1099.
99. Yu, M.; Ohira, M.; Li, Y.; Niizuma, H.; Oo, M.L.; Zhu, Y.; Ozaki, T.; Isogai, E.; Nakamura, Y.; Koda, T.; *et al.* High expression of ncran, a novel non-coding RNA mapped to chromosome 17q25.1, is associated with poor prognosis in neuroblastoma. *Int. J. Oncol.* **2009**, *34*, 931–938.
100. Zhu, Y.; Yu, M.; Li, Z.; Kong, C.; Bi, J.; Li, J.; Gao, Z.; Li, Z. Ncran, a newly identified long noncoding RNA, enhances human bladder tumor growth, invasion, and survival. *Urology* **2011**, *77*, e511–e515.
101. Castelnuovo, M.; Massone, S.; Tasso, R.; Fiorino, G.; Gatti, M.; Robello, M.; Gatta, E.; Berger, A.; Strub, K.; Florio, T.; *et al.* An alu-like RNA promotes cell differentiation and reduces malignancy of human neuroblastoma cells. *FASEB J.* **2010**, *24*, 4033–4046.
102. Petrovics, G.; Zhang, W.; Makarem, M.; Street, J.P.; Connelly, R.; Sun, L.; Sesterhenn, I.A.; Srikantan, V.; Moul, J.W.; Srivastava, S., Elevated expression of pcgem1, a prostate-specific gene with cell growth-promoting function, is associated with high-risk prostate cancer patients. Oncogene 2004, 23, 605-611.
103. Sonkoly, E.; Bata-Csorgo, Z.; Pivarcsi, A.; Polyanka, H.; Kenderessy-Szabo, A.; Molnar, G.; Szentpali, K.; Bari, L.; Megyeri, K.; Mandi, Y.; *et al.* Identification and characterization of a novel, psoriasis susceptibility-related noncoding RNA gene, prins. *J. Biol. Chem.* **2005**, *280*, 24159–24167.

104. Chung, S.; Nakagawa, H.; Uemura, M.; Piao, L.; Ashikawa, K.; Hosono, N.; Takata, R.; Akamatsu, S.; Kawaguchi, T.; Morizono, T.; *et al.* Association of a novel long non-coding RNA in 8q24 with prostate cancer susceptibility. *Cancer Sci.* **2011**, *102*, 245–252.
105. Maida, Y.; Yasukawa, M.; Furuuchi, M.; Lassmann, T.; Possemato, R.; Okamoto, N.; Kasim, V.; Hayashizaki, Y.; Hahn, W.C.; Masutomi, K. An RNA-dependent RNA polymerase formed by tert and the rmrp RNA. *Nature* **2009**, *461*, 230–235.
106. Shirasawa, S.; Harada, H.; Furugaki, K.; Akamizu, T.; Ishikawa, N.; Ito, K.; Ito, K.; Tamai, H.; Kuma, K.; Kubota, S.; *et al.* Snps in the promoter of a b cell-specific antisense transcript, sas-zfat, determine susceptibility to autoimmune thyroid disease. *Hum. Mol. Genet.* **2004**, *13*, 2221–2231.
107. Khaitan, D.; Dinger, M.E.; Mazar, J.; Crawford, J.; Smith, M.A.; Mattick, J.S.; Perera, R.J. The melanoma-upregulated long noncoding RNA spry4-it1 modulates apoptosis and invasion. *Cancer Res.* **2011**, *71*, 3852–3862.
108. Leygue, E. Steroid receptor RNA activator (sra1): Unusual bifaceted gene products with suspected relevance to breast cancer. *Nucl. Recept. Signaling* **2007**, *5*, e006.
109. Chooniedass-Kothari, S.; Hamedani, M.K.; Troup, S.; Hube, F.; Leygue, E. The steroid receptor RNA activator protein is expressed in breast tumor tissues. *Int. J. Cancer* **2006**, *118*, 1054–1059.
110. Cao, Y.; Bryan, T.M.; Reddel, R.R. Increased copy number of the tert and terc telomerase subunit genes in cancer cells. *Cancer Sci.* **2008**, *99*, 1092–1099.
111. Mishra, A.; Godavarthi, S.K.; Jana, N.R. Ube3a/e6-ap regulates cell proliferation by promoting proteasomal degradation of p27. *Neurobiol. Dis.* **2009**, *36*, 26–34.
112. Numata, K.; Kohama, C.; Abe, K.; Kiyosawa, H. Highly parallel snp genotyping reveals high-resolution landscape of mono-allelic ube3a expression associated with locus-wide antisense transcription. *Nucleic Acids Res.* **2011**, *39*, 2649–2657.
113. Wang, F.; Li, X.; Xie, X.; Zhao, L.; Chen, W. Uca1, a non-protein-coding RNA up-regulated in bladder carcinoma and embryo, influencing cell growth and promoting invasion. *FEBS Lett.* **2008**, *582*, 1919–1927.
114. Wang, X.S.; Zhang, Z.; Wang, H.C.; Cai, J.L.; Xu, Q.W.; Li, M.Q.; Chen, Y.C.; Qian, X.P.; Lu, T.J.; Yu, L.Z.; *et al.* Rapid identification of uca1 as a very sensitive and specific unique marker for human bladder carcinoma. *Clin. Cancer Res.* **2006**, *12*, 4851–4858.
115. Yang, C.; Li, X.; Wang, Y.; Zhao, L.; Chen, W. Long non-coding RNA uca1 regulated cell cycle distribution via creb through pi3-k dependent pathway in bladder carcinoma cells. *Gene* **2012**, *496*, 8–16.
116. Askarian-Amiri, M.E.; Crawford, J.; French, J.D.; Smart, C.E.; Smith, M.A.; Clark, M.B.; Ru, K.; Mercer, T.R.; Thompson, E.R.; Lakhani, S.R.; *et al.* Snord-host RNA zfas1 is a regulator of mammary development and a potential marker for breast cancer. *RNA* **2011**, *17*, 878–891.
117. Visscher, P.M.; Brown, M.A.; McCarthy, M.I.; Yang, J. Five years of gwas discovery. *Am. J. Hum. Genet.* **2012**, *90*, 7–24.
118. McPherson, R.; Pertsemlidis, A.; Kavaslar, N.; Stewart, A.; Roberts, R.; Cox, D.R.; Hinds, D.A.; Pennacchio, L.A.; Tybjaerg-Hansen, A.; Folsom, A.R.; *et al.* A common allele on chromosome 9 associated with coronary heart disease. *Science* **2007**, *316*, 1488–1491.

119. Samani, N.J.; Erdmann, J.; Hall, A.S.; Hengstenberg, C.; Mangino, M.; Mayer, B.; Dixon, R.J.; Meitinger, T.; Braund, P.; Wichmann, H.E.; *et al*. Genomewide association analysis of coronary artery disease. *N. Engl. J. Med.* **2007**, *357*, 443–453.
120. Jarinova, O.; Stewart, A.F.; Roberts, R.; Wells, G.; Lau, P.; Naing, T.; Buerki, C.; McLean, B.W.; Cook, R.C.; Parker, J.S.; *et al*. Functional analysis of the chromosome 9p21.3 coronary artery disease risk locus. *Arterioscler. Thromb. Vasc. Biol.* **2009**, *29*, 1671–1677.
121. Pasmant, E.; Sabbagh, A.; Vidaud, M.; Bieche, I. Anril, a long, noncoding RNA, is an unexpected major hotspot in gwas. *FASEB J.* **2011**, *25*, 444–448.
122. Visel, A.; Zhu, Y.; May, D.; Afzal, V.; Gong, E.; Attanasio, C.; Blow, M.J.; Cohen, J.C.; Rubin, E.M.; Pennacchio, L.A. Targeted deletion of the 9p21 non-coding coronary artery disease risk interval in mice. *Nature* **2010**, *464*, 409–412.
123. Yap, K.L.; Li, S.; Munoz-Cabello, A.M.; Raguz, S.; Zeng, L.; Mujtaba, S.; Gil, J.; Walsh, M.J.; Zhou, M.M. Molecular interplay of the noncoding RNA anril and methylated histone h3 lysine 27 by polycomb cbx7 in transcriptional silencing of ink4a. *Mol. Cell* **2010**, *38*, 662–674.
124. El Messaoudi-Aubert, S.; Nicholls, J.; Maertens, G.N.; Brookes, S.; Bernstein, E.; Peters, G. Role for the mov10 RNA helicase in polycomb-mediated repression of the ink4a tumor suppressor. *Nat. Struct. Mol. Biol.* **2010**, *17*, 862–868.
125. He, S.; Su, H.; Liu, C.; Skogerbo, G.; He, H.; He, D.; Zhu, X.; Liu, T.; Zhao, Y.; Chen, R. MicroRNA-encoding long non-coding RNAs. *BMC Genom.* **2008**, *9*, 236.
126. Zhao, Y.; He, S.; Liu, C.; Ru, S.; Zhao, H.; Yang, Z.; Yang, P.; Yuan, X.; Sun, S.; Bu, D.; *et al.* MicroRNA regulation of messenger-like noncoding RNAs: A network of mutual microRNA control. *Trends Genet.* **2008**, *24*, 323–327.
127. Mercer, T.R.; Dinger, M.E.; Sunkin, S.M.; Mehler, M.F.; Mattick, J.S. Specific expression of long noncoding RNAs in the mouse brain. *Proc. Natl. Acad. Sci. USA* **2008**, *105*, 716–721.
128. Hagerman, P.J.; Hagerman, R.J. Fragile x-associated tremor/ataxia syndrome—An older face of the fragile x gene. *Nat. Clin. Pract. Neurol.* **2007**, *3*, 107–112.
129. Jacquemont, S.; Hagerman, R.J.; Hagerman, P.J.; Leehey, M.A. Fragile-x syndrome and fragile x-associated tremor/ataxia syndrome: Two faces of fmr1. *Lancet Neurol.* **2007**, *6*, 45–55.
130. Nemes, J.P.; Benzow, K.A.; Moseley, M.L.; Ranum, L.P.; Koob, M.D. The sca8 transcript is an antisense RNA to a brain-specific transcript encoding a novel actin-binding protein (klhl1). *Hum. Mol. Genet.* **2000**, *9*, 1543–1551.
131. Berchtold, N.C.; Cotman, C.W. Evolution in the conceptualization of dementia and Alzheimer's disease: Greco-roman period to the 1960s. *Neurobiol. Aging* **1998**, *19*, 173–189.
132. Fleck, D.; Garratt, A.N.; Haass, C.; Willem, M. Bace1 dependent neuregulin processing: Review. *Curr. Alzheimer Res.* **2012**, *9*, 178–183.
133. Lewejohann, L.; Skryabin, B.V.; Sachser, N.; Prehn, C.; Heiduschka, P.; Thanos, S.; Jordan, U.; Dell'Omo, G.; Vyssotski, A.L.; Pleskacheva, M.G.; *et al.* Role of a neuronal small non-messenger RNA: Behavioural alterations in bc1 RNA-deleted mice. *Behav. Brain Res.* **2004**, *154*, 273–289.
134. Qiu, M.T.; Hu, J.W.; Yin, R.; Xu, L. Long noncoding RNA: An emerging paradigm of cancer research. *Tumour Biol.* **2013**, *34*, 613–620.
135. Xu, C.; Yang, M.; Tian, J.; Wang, X.; Li, Z. Malat-1: A long non-coding RNA and its important 3' end functional motif in colorectal cancer metastasis. *Int. J. Oncol.* **2011**, *39*, 169–175.

136. Tripathi, V.; Ellis, J.D.; Shen, Z.; Song, D.Y.; Pan, Q.; Watt, A.T.; Freier, S.M.; Bennett, C.F.; Sharma, A.; Bubulya, P.A.; *et al.* The nuclear-retained noncoding RNA malat1 regulates alternative splicing by modulating sr splicing factor phosphorylation. *Mol. Cell* **2010**, *39*, 925–938.
137. Lu, L.; Zhu, G.; Zhang, C.; Deng, Q.; Katsaros, D.; Mayne, S.T.; Risch, H.A.; Mu, L.; Canuto, E.M.; Gregori, G.; *et al.* Association of large noncoding RNA hotair expression and its downstream intergenic cpg island methylation with survival in breast cancer. *Breast Cancer Res. Treat.* **2012**, *136*, 875–883.
138. Ishibashi, M.; Kogo, R.; Shibata, K.; Sawada, G.; Takahashi, Y.; Kurashige, J.; Akiyoshi, S.; Sasaki, S.; Iwaya, T.; Sudo, T.; *et al.* Clinical significance of the expression of long non-coding RNA hotair in primary hepatocellular carcinoma. *Oncol. Rep.* **2013**, *29*, 946–950.
139. Kogo, R.; Shimamura, T.; Mimori, K.; Kawahara, K.; Imoto, S.; Sudo, T.; Tanaka, F.; Shibata, K.; Suzuki, A.; Komune, S.; *et al.* Long noncoding RNA hotair regulates polycomb-dependent chromatin modification and is associated with poor prognosis in colorectal cancers. *Cancer Res.* **2011**, *71*, 6320–6326.
140. Tsai, M.C.; Manor, O.; Wan, Y.; Mosammaparast, N.; Wang, J.K.; Lan, F.; Shi, Y.; Segal, E.; Chang, H.Y. Long noncoding RNA as modular scaffold of histone modification complexes. *Science* **2010**, *329*, 689–693.
141. Zhang, X.; Gejman, R.; Mahta, A.; Zhong, Y.; Rice, K.A.; Zhou, Y.; Cheunsuchon, P.; Louis, D.N.; Klibanski, A. Maternally expressed gene 3, an imprinted noncoding RNA gene, is associated with meningioma pathogenesis and progression. *Cancer Res.* **2010**, *70*, 2350–2358.
142. Cho, Y.; Gorina, S.; Jeffrey, P.D.; Pavletich, N.P. Crystal structure of a p53 tumor suppressor-DNA complex: Understanding tumorigenic mutations. *Science* **1994**, *265*, 346–355.
143. Zhou, Y.; Zhong, Y.; Wang, Y.; Zhang, X.; Batista, D.L.; Gejman, R.; Ansell, P.J.; Zhao, J.; Weng, C.; Klibanski, A. Activation of p53 by meg3 non-coding RNA. *J. Biol. Chem.* **2007**, *282*, 24731–24742.
144. Zhang, X.; Rice, K.; Wang, Y.; Chen, W.; Zhong, Y.; Nakayama, Y.; Zhou, Y.; Klibanski, A. Maternally expressed gene 3 (meg3) noncoding ribonucleic acid: Isoform structure, expression, and functions. *Endocrinology* **2010**, *151*, 939–947.

8

Non-Coding RNA in Oncogenesis: A New Era of Identifying Key Players

Guorui Deng [1,2] **and Guangchao Sui** [1,*]

[1] Department of Cancer Biology and Comprehensive Cancer Center, Wake Forest University School of Medicine, Winston-Salem, NC 27157, USA

[2] Hypertension and Vascular Research Center, Wake Forest University School of Medicine, Winston-Salem, NC 27157, USA; E-Mail: gdeng@wakehealth.edu

* Author to whom correspondence should be addressed; E-Mail: gsui@wakehealth.edu;

Abstract: New discoveries and accelerating progresses in the field of noncoding RNAs (ncRNAs) continuously challenges our deep-rooted doctrines in biology and sometimes our imagination. A growing body of evidence indicates that ncRNAs are important players in oncogenesis. While a stunning list of ncRNAs has been discovered, only a small portion of them has been examined for their biological activities and very few have been characterized for the molecular mechanisms of their action. To date, ncRNAs have been shown to regulate a wide range of biological processes, including chromatin remodeling, gene transcription, mRNA translation and protein function. Dysregulation of ncRNAs contributes to the pathogenesis of a variety of cancers and aberrant ncRNA expression has a high potential to be prognostic in some cancers. Thus, a new cancer research era has begun to identify novel key players of ncRNAs in oncogenesis. In this review, we will first discuss the function and regulation of miRNAs, especially focusing on the interplay between miRNAs and several key cancer genes, including p53, PTEN and c-Myc. We will then summarize the research of long ncRNAs (lncRNAs) in cancers. In this part, we will discuss the lncRNAs in four categories based on their activities, including regulating gene expression, acting as miRNA decoys, mediating mRNA translation, and modulating protein activities. At the end, we will also discuss recently unraveled activities of circular RNAs (circRNAs).

Keywords: noncoding RNA; microRNA; gene expression; cancer; tumor suppressor; oncogene

---

1. Introduction

The central dogma of biology dictates that genetic information flows in a unidirectional fashion of DNA-mRNA-proteins. Thus, the majority of the early research in molecular and cellular biology was focused on the protein-coding genes and their transcripts, messenger RNAs (mRNAs). However, in the past two decades, noncoding RNAs (ncRNAs) have garnered increased appreciation for their important roles in regulating various biological processes. In the whole human genome, more than 90% of the DNA sequence can be transcribed, but only about 2% of it encodes proteins [1–5]. Recent studies reveal that most of these excess transcripts are not transcriptional noise, but rather serve as functional ncRNAs regulating chromatin modifications, gene transcription, mRNA translation and protein function [5–7].

The discovery of the first human oncogene RAS [8–10] and the first tumor suppressor Retinoblastoma (Rb) in the 1980s invigorated the field of cancer research [11,12]. The fact that a simple point mutation in the RAS gene could transform a normal growth-regulating gene into a cancer-causing factor inspired cancer researchers to seek out additional cancer-regulating genes. In the decades that followed, the search for novel oncogenes and tumor suppressors has dominated the field of cancer research, leading to the discovery of many well-characterized tumor suppressors, such as p53, and oncogenes, such as c-Myc. With the discovery of ncRNAs and their involvement in oncogenesis, a new surge of research in determining the oncogenic and tumor suppressive roles of these noncoding transcripts has taken place. To date, multiple classes of ncRNAs have been demonstrated to regulate different biological or physiological processes. It is possible that some ncRNAs may not have a discernible function; however, the enormous number of the identified noncoding transcripts and their differential expression profiles in cancers indicate that we are still at the beginning of this exciting era to explore the functions and regulatory mechanisms of ncRNAs in cancers.

NcRNAs can be divided into three categories based on length or number of nucleotides (nts) [13]. (1) Short ncRNAs: these RNAs are 17–30 nts in length and include microRNAs (miRNAs), piwi-interacting RNAs (piRNAs), and transcription initiation RNAs (tiRNAs). MiRNAs will be one of the major focuses of this review; (2) Middle-size ncRNAs: these RNAs have variable sizes but are typically between 20 and 200 nts in length. The small nucleolar RNAs (snoRNAs) belong to this category; (3) Long ncRNAs (lncRNAs): these RNAs are over 200 nts. This category includes several well-characterized ncRNA, such as MALAT1 and HOTAIR. Circular RNAs (circRNAs) have recently been discovered and can be grouped here based on their sizes [14]. Many members in this category were identified in the last few years, which will be extensively discussed in this review.

Due to rapid expanding of the ncRNA field, it is difficult to cover all aspects of ncRNA-related studies in a single review article. There are many excellent reviews discussing the activities of ncRNAs from different perspectives. In this review, we will discuss the biological function of miRNAs and lncRNAs in cancers, with particular attention paid to their interplay with p53, c-Myc and PTEN. We will also summarize recent studies of circRNAs that have returned to the arena with a recently unraveled function since their discovery many years ago.

2. MicroRNAs

2.1. Introduction of MicroRNAs

MiRNAs are small and evolutionarily conserved noncoding RNAs that are typically 18 to 25 nts in length. The biogenesis of miRNAs begins with the synthesis of the primary transcripts (pri-miRNAs) by RNA polymerase II in the nucleus (Figure 1). Like the transcripts of protein-coding genes, pri-miRNAs contain a 5' cap structure, a poly(A) tail, and sometimes intron sequences [15,16]. A stem loop structure can be formed by regions of partial complementary sequences in each pri-miRNA. The nuclear ribonuclease DROSHA and its partner DGCR8 recognize this stem-loop structure, and subsequently crop the pri-miRNA to generate the pre-miRNA intermediate. After the pre-miRNA is exported to the cytoplasm by Exportin-5/Ran-GTP, it is further processed by DICER1, another ribonuclease, to generate a double-stranded miRNA molecule. Both strands can act as mature miRNAs; however, if only one strand (guide strand) becomes a functional miRNA, the other strand (passenger strand) will be quickly degraded [17,18].

A mature miRNA can associate with Argonaute proteins to form a RNA-induced silencing complex (RISC), in which the miRNA guides the complex to the 3'-untranslated region (3'-UTR) of the target mRNA to block translational protein synthesis and/or cause its degradation (Figure 1). Each miRNA contains a seed sequence (7 nts; from the nucleotides 2 to 8) at its 5'-end with conserved complementarity to perfectly pair the ‘seed-match’ sequence at the 3'-UTR of its target mRNA [19,20]. Since only a 7-nucleotide match between a target 3'-UTR and a miRNA seed sequence will theoretically make the mRNA a potential target of the miRNA, miRNAs do not have a high specificity for their targets like small interference RNAs (siRNAs) have. MiRNAs can be divided into different families. A miRNA family consists miRNAs that share the same seed sequence and thus may target the same set of genes [19]. MiRNA coding regions, or their genes, can be located in either protein-coding or noncoding regions of transcription units in the human genome. Each of these regions may encode one or a cluster of mRNAs. A miRNA cluster is a set of two or more miRNAs that are transcribed from physically adjacent miRNA genes. Thus, the miRNAs in a cluster are transcribed by the same promoter, in the same direction, and as an unseparated transcription unit. One cluster typically contains two or three miRNAs, but large clusters do exist, such as the miRNA-17-92 cluster consisting of seven miRNAs. Interestingly, some members in one miRNA family are encoded by one miRNA gene cluster, such as miR-15a/miR-16-1 and miR-34b/miR-34c. Different miRNA clusters may belong to the same miRNA family, such as miR-15a/miR-16-1 and miR-497/miR-195.

Figure 1. MicroRNA biogenesis and their role in inhibiting gene expression. A miRNA is transcribed by RNA polymerase II (Pol II) mediated by other transcription factors, such as p53 and c-Myc. The generated pri-miRNA is processed by the DROSHA complex to become a pre-miRNA, which is transported from nucleus to cytoplasm with a complex consisting of Exportin-5 (Exp-5) and Ran-GTP. In cytoplasm, the pre-miRNA is further processed by a Dicer-1 complex to become a duplex that consists of a guide strand (to become a mature miRNA) and a passenger strand. The guide strand associates with the RNA-induced silencing complex (RISC) and guides it to the target site on the 3'-UTR of an mRNA to inhibit the mRNA translation and cause mRNA degradation. For some miRNA-duplexes post Dicer-1 processing, both strands can become mature miRNAs; the one at the 5'-end of pre-miRNA is suffixed by "-5p", while the 3'-end one is suffixed by "-3", such as miR-17-5p and miR-17-3p.

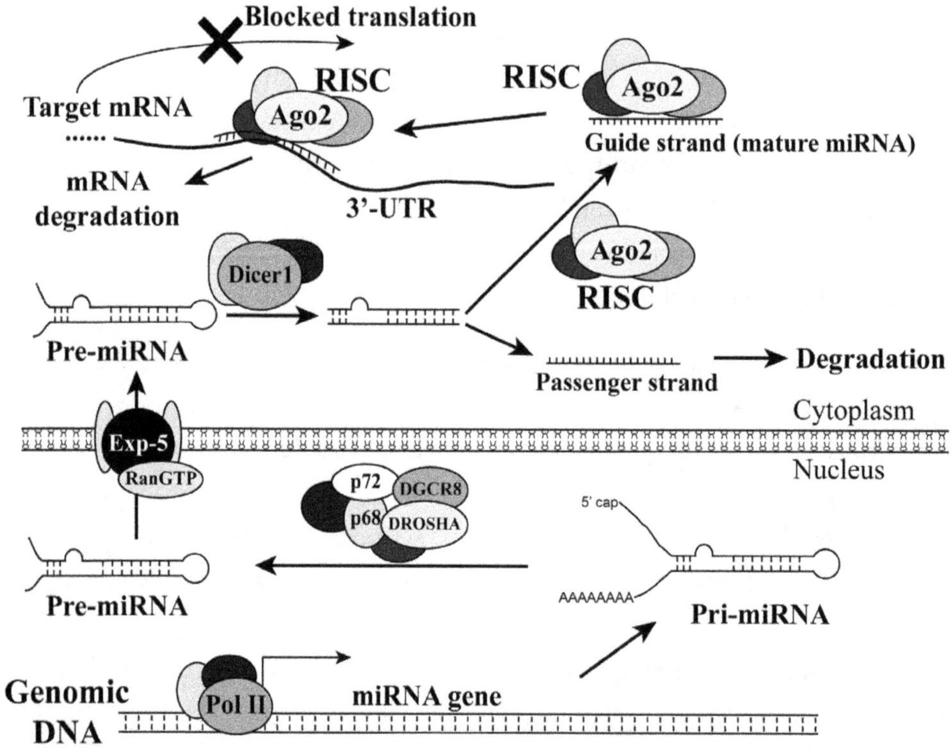

To date, over 2,500 human microRNAs have been identified based on the microRNA database (miRBase) at the Sanger Institute (http://www.mirbase.org/). Bioinformatical analyses suggest that miRNAs may regulate over 5,300 human genes, which represents 30% of human genes [19] and each miRNA regulates about 200 genes [21]. Thus, altered expression of microRNAs in cancer can cause significant perturbation of gene expression with a profound effect on malignant transformation and cancer progression [20].

There are much debate and investigation regarding the mechanisms of miRNA-repressed protein expression. Initially, miRNA-mediated gene silencing was portrayed as a different mechanism from that of siRNA-mediated mRNA degradation, based on the observation that the lin-4 miRNA decreased lin-14 protein expression without affecting its mRNA levels in *C. elegans* [22,23]. This was supported by multiple studies showing that miRNA-mediated gene silencing could be achieved with no or minor change of the target mRNA levels in *Drosophila* [24], *Arabidopsis* [25] and mammalian cells [26,27].

However, several later studies suggested that microRNAs can both block translation and trigger target mRNA degradation [28,29]. Notably, a recent report by Djuranovic *et al.* provided new insights into the kinetics of miRNA-mediated gene silencing in *Drosophila* cells. The authors demonstrated that miR-9b and miR-279 first repress the translation of the target mRNAs and then cause mRNA deadenylation and degradation [30]. Whether this mechanism is applicable to all miRNAs or other species remains to be determined.

Through the ability to repress the expression of multiple genes, miRNAs play an important role in regulating many cellular activities such as proliferation, differentiation and apoptosis. Calin *et al.* provided the evidence showing the deletion of the miR-15/miR-16 cluster at 13q14 and its downregulation in patients with B cell chronic lymphocytic leukemia [31]. This was the first study suggesting that noncoding genes correlate with and may even contribute to oncogenesis. The same group further investigated the loci of 186 miRNAs in the human genome to evaluate their potential involvement in cancer pathogenesis [32]. They discovered that over 50% of these miRNA genes are present in the genomic regions with reported alterations in cancers. These are highly instable loci, including fragile sites, minimal heterozygous deletion regions, frequently amplified sections and common breakpoints. However, the instability of these miRNA-coding regions does not generate frequent miRNA somatic mutations within their seed sequences in cancers but rather changes their expression. A more recent study indicated that the genes encoding oncogenic miRNAs are mainly located in the amplified regions in human cancers, whereas the majority of genes for tumor suppressive miRNAs are in the deleted regions [33]. Interestingly, many oncogenes can produce alternative mRNA isoforms with shorter 3'-UTR sequences through a mechanism involved in alternative cleavage and polyadenylation [34]. A short mRNA isoform of an oncogene can avoid miRNA-mediated inhibition, which consequently increases its stability and typically produces ten-fold more protein. Overall, miRNA expression is globally reduced in tumors compared to their matched normal tissues [35]. Thus, numerous studies have demonstrated the potential of using the expression profiles of single or multiple miRNAs as biomarkers to classify tumor origins, stages and clinical outcomes [36–39]. Currently, microRNA expression detection is not used clinically in cancer diagnosis and prognosis, but it bears great promise in multiple applications of cancer therapies. Especially, due to the high stability of miRNAs and advance of RNA purification techniques, miRNAs can be extracted from not only tumors samples exposed to variable treatments, including formalin-fixing and paraffin-embedding [40], but also serum and urine [41–44].

2.2. MicroRNAs in Oncogenesis

Many miRNAs have been shown to regulate cell survival and proliferation, angiogenesis, and epithelial-mesenchymal transition (EMT). The miRNAs associated with oncogenesis are also known as "oncomirs". Depending on their major targets, oncomirs can be classified into the oncogenic and tumor suppressive miRNA groups. As the classification implies, tumor suppressive miRNAs repress protein-coding oncogenes whereas oncogenic miRNAs repress protein-coding tumor suppressors. Some miRNAs display both oncogenic and tumor suppressive activities, depending on the tissue and tumor contexts. Wang *et al.* demonstrated that oncogenic and tumor suppressor miRNAs show clearly different patterns in many aspects, including evolutionary rates, expression patterns, chromosome distribution, molecule sizes and of course targets [33].

An example of miRNAs that have opposing effects is the miRNA-17-92 cluster that has been implicated in regulating cellular survival. The polycistronic miRNA-17-92 cluster (miR-17-92) is located on chromosome 13 open reading frame 25 (c13orf25) in the human genome and contains seven miRNAs (*miR-92-1, miR-19a, miR-20a, miR-19b, miR-18a, miR-17-5p and miR-17-3p*) [45]. Ota *et al.* reported that the chromosomal region 13q31-q32, where the miR-17-92 cluster resides, is amplified in malignant lymphoma [46]. He *et al.* observed that five miRNAs (*miR-92-1, miR-19a, miR-20a, miR-19b and miR-17-5p*) from this cluster are up-regulated in human B-cell lymphoma and cell lines [47]. Ectopic expression of this miRNA cluster promotes c-Myc-induced lymphomagenesis in mice suggesting the cooperation between mir-17-19b and c-Myc in oncogenesis. In another study, O'Donnell *et al.* identified an additional miRNA, miR-18a, at the miRNA-17-92 locus and demonstrated that c-Myc transcriptionally activates the expression of these 6 miRNAs (*miR-92-1, miR-19a, miR-20a, miR-19b, miR-17-5p* and miR-18a) [48]. Interestingly, miR17-5p and miR-20a negatively regulate E2F1 expression, which is also activated by c-Myc. E2F1 is a crucial driver of cell cycle progression from the G1 to S phase, but its expression can induce either pro-metastatic activity or cell apoptosis depending on its molecular contexts [49,50]. Thus, the ectopic miR-17-92 expression in a c-Myc background confers an overall pro-survival trait to the cells through eliminating E2F1-mediated apoptotic potential. This is corroborated by the aforementioned studies of He *et al.*, in which c-Myc and miR-19b overexpression promoted murine lymphomas without causing any indicative sign of apoptosis [47]. The presence of this interactive regulatory network suggests a tight control of c-Myc-mediated proliferation signal during oncogenesis.

Data from several other groups also suggest tumor suppressive activities of the miRNAs in this cluster. Hossain *et al.* reported that miR-17-5p acts as a tumor suppressor through repressing the expression of AIB1 (or NCOA3) and consequently reducing the proliferation of breast cancer cells [51]. AIB1 was named for its "amplified in breast cancer" and has been suggested to play an oncogenic role in mammary oncogenesis [52,53]. Interestingly, AIB1 functions as a coactivator of E2F1 to promote breast cancer cell proliferation [52]. Thus, through repressing both AIB1 and E2F1, miR-17-5p may "kill two birds with one stone" in reducing breast cancer cell proliferation.

Yu *et al.* reported a negative feedback regulation between miR-17-5p/miR-20a and cyclin D1 [54], an oncoprotein in breast cancer. Cyclin D1 inversely correlated with miR-17-5p/miR-20a levels in breast cancer samples and cell lines. In a negative regulatory loop, cyclin D1 induces miR-17-5p/miR-20a expression, while these two miRNAs target the cyclin D1 3'-UTR to limit its proliferative activities. The same group also reported that miR-17-5p/miR-20a repress cytokeratin 8 through inhibiting cyclin D1 [55]. Additionally, these two miRNAs target the 3'-UTR of the interleukin-8 mRNA. Thus, miR-17-5p/miR-20a regulates both cellular secretion and tumor microenvironment to block migration and invasion of neighboring cells in breast cancer [55]. The seventh member of this cluster, miR-17-3p, is the "passenger strand" of miR-17-5p that is also processed into a mature miRNA. Recent studies indicate that it represses the expression of vimentin and Mdm2, suggesting a tumor suppressive role in oncogenesis [56,57].

Recent studies revealed a tumor suppressive role of miR-101 in a variety of cancers, including prostate, bladder, breast and gastric cancers [58–62]. MiR-101 is frequently downregulated in these cancers compared to their normal adjacent tissues. On the other hand, ectopic expression of miR-101 in various cancer cell lines resulted in decreased cellular proliferation, motility and invasiveness, indicating its antitumoral activities. The tumor suppressive function of miR-101 has been corroborated by its inhibitory effect on EZH2, an oncogene of many solid tumors [63]. Varambally *et al.* reported the miR-101 loss resulted in increased EZH2 expression [62]. Importantly, genomic loss of the miR-101 locus correlates with EZH2 overexpression in solid tumors. Cao *et al.* further demonstrated that miR-101 negatively regulates EZH2 expression in prostate cancer cells, while miR-101 expression is modulated by androgen receptor and HIF-1α/HIF-1β [58]. The negative regulation of EZH2 by miR-101 has been confirmed by the studies from several other groups [64–66]. Consistent with these observations, a recent report suggested a positive regulation of E-cadherin by miR-101 [67]. E-cadherin is frequently downregulated in aggressive cancers and its loss increases cell dissemination and cancer cell invasiveness. In the study by Qazi *et al.*, ectopic miR-101 restored E-cadherin expression in pancreatic ductal adenocarcinoma cells by reducing EZH2-mediated histone H3-K27 methylation. Overall, these studies suggest that miR-101 primarily targets oncogene EZH2 and its deletion contributes to oncogenesis through aberrant epigenetics caused by EZH2 overexpression.

The let-7 family was one of the first mammalian miRNAs to be discovered and consists of 13 members located on different genomic loci that are often lost in human cancers [68]. Reduced expression of the let-7 members is associated with more dedifferentiated and aggressive cancers [69]. These observations suggested a possible tumor suppressive role of the let-7 family members. Recent studies indicated that let-7 can target oncogenes RAS and c-Myc that have multiple potential binding sites of let-7 miRNAs in their 3'-UTRs. Johnson *et al.* detected reduced let-7 expression in tumor tissues compared to normal adjacent tissues in lung cancer samples [70]. Interestingly, let-7 miRNA levels inversely correlated with RAS protein expression but not its mRNA levels, suggesting a mechanism of let-7-mediated translational inhibition without mRNA degradation. Overall, let-7 miRNAs may target several key oncogenes, such as RAS and c-Myc, to suppress the proliferative signals from these two oncogenes.

Many other miRNAs have been demonstrated to play a role in cancer development and progression. It is noteworthy that reports from different groups may present paradoxical results. These discrepancies could result from different experimental settings and also reflect the complexity of miRNA-regulated network in cancers. As discussed above, each miRNA potentially target hundreds genes and its ectopic overexpression may perturb multiple cellular processes leading to artificial phenotypic changes. Thus, studies only using miRNA overexpression or reporter assay without any miRNA depletion or tumor sample correlation studies may not truly represent physiological relevance. Nevertheless, currently available literature unequivocally indicates the prognostic potential and biological activities of miRNAs in different cancers.

2.3. P53 Is a Key Regulator of MicroRNAs Biogenesis

The tumor suppressor p53 is one of the best-studied proteins in the field of cancer research. Owing its well-characterized regulation, the role of p53 in the miRNA network has been extensively explored in the past decade. As a transcription factor, p53 forms tetramers to bind to its consensus sites on target genes and mediates their transcription. Increasing evidence shows that p53 exerts its antiproliferative activities at least partially through the transcriptional regulation of miRNA expression (Figure 2), in addition to its canonical tumor suppressive role [20,71]. The best characterized p53 target is the miR-34 family [72]. The miR-34 family includes miR-34a, -34b and -34c. MiR-34a is transcribed by chromosome 1, and miR-34b and -34c are transcribed by two proximal loci on chromosome 11 and controlled by the same promoter. These miRNAs exert tumor suppressive activities through repressing proliferative genes, such as c-Myc and BCL2, and their overall function is inducing apoptosis, cell cycle arrest or senescence [73–75]. Consistently, they are frequently silenced by promoter methylation in tumors [76]. Thus, the p53-promoted expression of the miR-34 family extended its activated tumor suppressive network.

Figure 2. The interplay between miRNAs and p53. While p53 regulates the gene expression of many miRNAs, its expression is also inhibited by miRNAs. When cells are exposed to genotoxic stresses, p53 activates four miRNAs that repress Mdm2 expression, which leads to p53 accumulation and cell cycle arrest or apoptosis. P53 protein also associates with the DROSHA complex to directly regulate miRNA maturation.

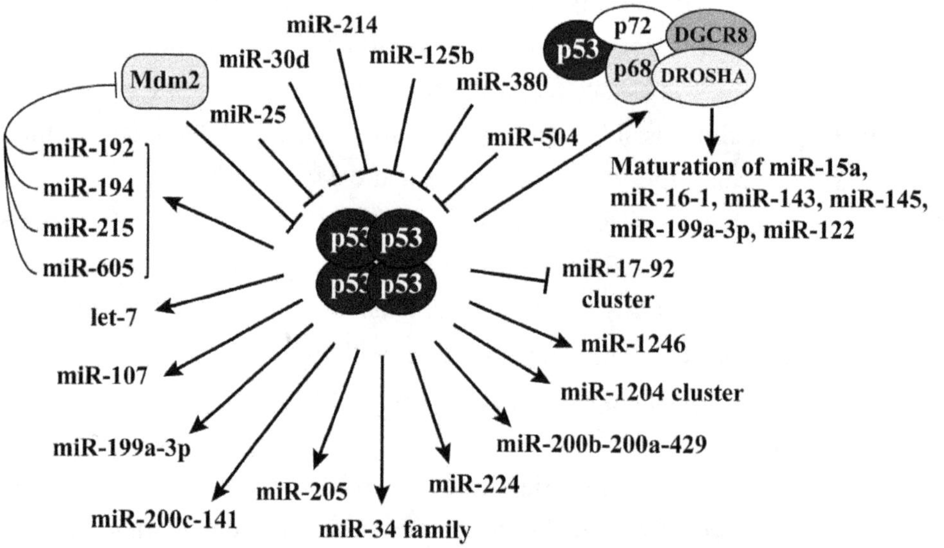

P53 also stimulates the expression of many other miRNAs that have antiproliferative activities. For example, p53-mediated miRNA expression plays a role in hypoxia. During hypoxia of tumor cells, the upregulated hypoxia inducing factor (HIF)-1α forms a heterodimer with HIF-1β to become a transcription factor HIF-1 that activates pro-angiogenic genes such as vascular endothelial growth factor A (VEGFA), to promote angiogenesis for tumor growth and metastasis [77,78]. Yamakuchi *et al.* reported that miR-107 reduces hypoxia signaling by inhibiting HIF-1α expression in human colon cancer cells [79]. As a transactivator, p53 promotes the expression of miR-107 to reduce HIF-1α levels, which quenches the hypoxic signal to block tumor angiogenesis. Among other p53-activated miRNAs, miR-145 represses c-Myc expression [80], and miR-200c/miR-141 and miR-200b/miR-200a/miR-429 inhibit EMT through downregulating ZEB1 and ZEB2 [81,82]. As discussed above, miR-17-92 cluster exhibits proliferative activities based on most literature. P53 represses the expression of the miR-17-92 cluster, opposite to its transactivating effects on most other miRNAs (Figure 2), and this regulation likely contributes to the p53-induced apoptosis [83].

In addition to transcriptional regulation, p53 is directly involved in the maturation process of miRNAs. DROSHA is a major component of the complex that processes pri-miRNA into pre-miRNA. P53 associates with DROSHA in a RNA dependent manner and facilitates DROSHA-mediated pri-miRNA processing [84]. Interestingly, although this process is independent of its transcriptional activity, transcriptionally inactive p53 mutants do not show this capability. Actually, these p53 mutants interfere with the assembly of the DROSHA complex and consequently attenuate miRNA processing. Currently, p53 has been reported to regulate the maturation of at least 6 miRNAs (miR-15a, miR-16-1, miR-143, miR-145, miR-199a, and miR-122) [84–87]. Whether p53 is involved in the processing of other miRNAs remains to be determined.

Conversely, p53 expression can also be regulated by miRNAs (Figure 2). The functions of these miRNAs are not uniformly oncogenic; thus, whether their role in repressing p53 expression is physiologically significant is unclear. There is a negative feedback loop between p53 and the ubiquitin E3 ligase Mdm2. While Mdm2 stimulates p53 ubiquitination and degradation, p53 activates Mdm2 gene expression. This feedback regulation is important to maintain p53 homeostasis in normal cells, which is disrupted when exposed to genotoxic stresses [88]. Recent studies reveal that some miRNAs can break this negative feedback loop leading to p53 accumulation. Pichiorri *et al.* reported that p53 activates the expression of miR-192, miR-194 and miR-215 and these miRNAs target the Mdm2 3'-UTR to repress its expression; thus, downregulation of p53-inducible miR-192, miR-194 and miR-215 causes aberrant Mdm2 increase and p53 downregulation in multiple myeloma [89]. Consistently, miR-192, miR-194 and miR-215 are downregulated in multiple myeloma and renal cancers [89–91]. Xiao *et al.* demonstrated that p53 activates miR-605 that also represses Mdm2 expression [92]. Thus, p53 activates these four miRNAs to downregulate its negative regulator Mdm2 leading to p53 accumulation in response to stress.

2.4. C-Myc Regulates the Synthesis of miRNAs

The oncogene c-Myc is a transcription factor and mediates target gene expression through recruiting various chromatin modifiers [93]. Recent studies reveal its role in regulating miRNA synthesis. Interestingly, the effects of c-Myc-mediated transcription of many miRNAs are opposite to those regulated by p53.

C-Myc represses the expression of multiple miRNAs (Figure 3). As discussed above, p53 transactivates the miR-34 family that represses c-Myc expression [72,75]. On the other hand, c-Myc represses the expression of the miRNAs in the miR-34a family [94]. C-Myc also recruits HDAC3 to downregulate the expression of miR-15a/miR-16-1 [95] that block the expression of multiple oncogenes, such as BCL2 and cyclin D1 [27,96]. Similarly, c-Myc also recruits HDAC3 and EZH2 to silence the expression of miR-29 [97] that has been defined as a tumor suppressor [98]. Among other c-Myc repressed miRNAs, let-7 inhibits androgen receptor and KRAS [68,99], miR-23a/miR-23b block glutaminase [100], and miR-26 targets EZH2 and cyclin D2 [58]. Thus, c-Myc-mediated repression of these miRNAs can release the expression of multiple oncogenes or proliferative genes to promote oncogenesis.

Figure 3. The interplay between miRNAs and c-Myc. While c-Myc regulates the expression of multiple miRNAs, its expression is inhibited by different miRNAs. C-Myc also activates the gene expression of DROSHA to directly promote miRNA procession.

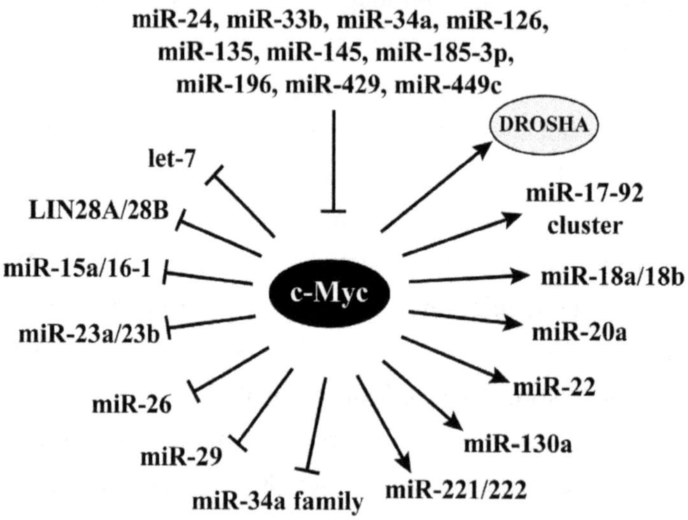

c-Myc also activates the expression of multiple miRNAs. Similar to its role in antagonizing p53-mediated miR-34 miRNA expression, c-Myc promotes the transcription of miR-17-92 cluster [101] that is downregulated by p53. Kim *et al.* carried out a combined analysis of mRNA and miRNA expression profiles that revealed multiple c-Myc-induced miRNAs and their downstream targets [102]. Among their discovered miRNAs, miR-221 and miR-222 have been shown to target p27, p57 and PTEN, and exhibit proliferative activities [103–106]. For the other c-Myc-activated miRNAs, the authors predicted that miR-20a targets p21, RB1, PTEN and interferon regulatory factor 1 (IRF-1), and miR-130a inhibits tuberous sclerosis 1 (TSC1) and CYLD [102]. While p53 protein direct associates with the miRNA processing machinery, c-Myc activates DROSHA expression through binding to the E-box in its promoter and consequently facilitates miRNA processing [107].

While c-Myc acts as a regulator of miRNAs, its expression has been reported to be modulated by a number of miRNAs (Figure 3). Among them, miR-34a, miR-126 and miR-145 have been documented as tumor suppressors [33].

3. Long Noncoding RNA

Long noncoding RNAs (lncRNAs) were previously defined as RNA molecules longer than 200 nucleotides that are not translated into proteins [6,108]. Recently, Spizzo *et al.* amended this definition linking to the biological functions and described that lncRNAs are a class of RNA molecules that do not fit into any known class of small and structural RNAs, and possess regulatory roles in their primary or spliced form [109]. The GENCODE Consortium is a part of the ENCODE (ENCyclopedia Of DNA Elements) project and aims to identify all gene features in the human genome. The GENCODE 7 release in 2012 indicates that the human genome contains at least 9,640 long noncoding RNA loci that can potentially encode 15,512 transcripts [110]. LncRNA production is regulated by mechanisms similar to these of protein-coding genes, such as histone modifications and RNA splicing, and their expression shows tissue-specific patterns [111]. Most lncRNAs are localized in nucleus and associated with chromatin, and some of them are preferentially processed into small RNAs [111]. Banfai *et al.* demonstrated that lncRNAs are rarely translated in two tested human cell lines, suggesting that ribosomes can differentiate the coding and noncoding transcripts for translation [112].

Recent studies continue elucidating novel and unpredicted biological activities of the lncRNAs, which were previously ascribed as protein functions. Based on the regulatory mechanisms, the lncRNAs can be divided into four categories, including the lncRNAs that (1) regulate gene expression, (2) act as miRNA decoys to free target mRNAs, (3) regulate mRNA translation, and (4) regulate protein activities.

3.1. LncRNAs Regulating Gene Expression

The X-inactive-specific transcript (Xist) was one of the first lncRNAs discovered in mammals [113]. Xist is encoded by the inactive X chromosome (Xi) and the genomic locus can transcribe a 17- to 20-kb RNA. This lncRNA binds the Xi *in cis* to induce chromosome X silencing by recruiting the polycomb repressive complex 2 (PRC2) that induces histone H3-K27 methylation, a hallmark of gene inactivation [114,115]. In this process, the transcription repressor Yin Yang 1 (YY1) confers allele-specific binding of Xist to the Xi with the involvement of two other noncoding RNAs, Jpx [116] and Ftx [117]. There is another lncRNA, Tsix, that is transcribed at the same locus of Xist but in the reverse direction and thus antisense to Xist [118]. Tsix regulates imprinted and random X inactivation in development [119]. An additional Xist-related RNA transcript is Xite that also plays a role in the X chromosome inactivation [120].

The association of epigenetic silencing complexes with Xist to induce transcriptional silencing has been extended to several recently characterized lncRNAs, which also associate with the PRC2 and other chromatin repressive complexes [121,122]. One of these lncRNAs is HOX Antisense Intergenic RNA (HOTAIR) that is a 2.2-kb transcript located at the HOXC gene cluster on chromosome 12 [122]. HOTAIR also modulates gene expression through epigenetic regulation. Unlike Xist that acts *in cis*, HOTAIR functions *in trans* to recruit the PRC2 to the *HOXD* locus on chromosome 2 to induce transcriptional silencing [122]. In addition to associating with PRC2, HOTAIR also interacts with the LSD1/CoREST/REST histone modification complex, leading to both histone H3-K27 methylation and H3-K4 demethylation [123]. Since HOTAIR plays an important role in the epigenetic regulation of its target genes, it is not surprising that its deregulation has been observed in different types of cancers. Recent studies suggest that HOTAIR overexpression is positively associated with increased tumor cell malignancy. Gupta *et al.* reported that HOTAIR is overexpressed in both primary and metastatic breast cancer tissues, and its levels in the primary tumors could be used as a significant predictor of subsequent tumor metastasis and survival of the patients [124]. Ectopically expressed HOTAIR could confer the breast epithelial cancer cells with invasive and metastatic potential while its depletion in breast cancer cells abrogated these activities [124]. The role of HOTAIR in promoting oncogenesis has also been reported in other cancers. Yang *et al.* compared HOTAIR levels between tumorigenic and adjacent non-tumorigenic tissues of hepatocellular carcinoma (HCC) samples and found that this lncRNA was expressed at higher levels in malignant tissues [125]. In addition, the survival analysis of a cohort consisting of 60 HCC patients revealed that high HOTAIR expression could serve as an independent prognostic marker for disease recurrence and reduced patient survival [125]. Another

HCC-related study suggested that HOTAIR expression is a potential biomarker for lymph node metastasis from the primary tumors [126]. Furthermore, HOTAIR upregulation was also observed in a cohort of patients diagnosed with the stage IV colorectal cancer (CRC) [127]. In this study, Kogo *et al.* indicated that high HOTAIR expression showed significantly positive correlation with the liver metastasis and poor patient outcome, further supporting that HOTAIR expression is a potential prognostic marker of multiple cancers. In pancreatic cancer, HOTAIR levels were also increased in tumorigenic tissues compared to the non-tumorigenic tissues, and associated with a more aggressive phenotype [128]. The oncogenic role of HOTAIR in pancreatic cancer cell invasion was validated by its siRNA-mediated knockdown and overexpression studies [128], consistent with the observations in the other aforementioned cancers. Interestingly, gene array studies showed only a small overlap of HOTAIR-regulated genes between pancreatic cancer and breast cancer [128], suggesting that HOTAIR may regulate different sets of target genes in a cell type-specific manner.

EZH2 is a core component of the PRC2. EZH2 knockdown followed by chromatin immunoprecipitation demonstrated that HOTAIR-mediated gene repression could be either PRC2-depedent or -independent in pancreatic cancer cells, although the PRC2 is necessary for HOTAIR target gene repression in breast cancer cell [124,128]. This discrepancy between the two cancer types suggests the presence of yet unidentified epigenetic mechanisms regulating HOTAIR-mediated transcriptional silencing.

Wang *et al.* identified the lincRNA HOTTIP that is transcribed from the 5' tip of the HOXA locus and modulates the activity of the WDR5-MLL complex, in which the WD40-repeat protein WDR5 binds the MLL complex to activate its histone H3-K4 methyltransferase activity [129]. Chromosomal looping can make HOTTIP stay in the vicinity of its target genes and let it bind WDR5 to promote the WDR5/MLL complexes-mediated histone H3-K4 methylation, which leads to target gene activation. Thus, HOTTIP serves as a key intermediate to transmit information from higher order chromosomal looping into chromatin modifications [129]. Another HOXA-related lncRNA is HOX antisense intergenic RNA myeloid 1 (HOTAIRM1) that is transcribed at a direction antisense to the HOXA gene [130]. The knockdown of HOTAIRM1 reduced the expression of HOXA1 and HOXA4 during the myeloid differentiation in promyelocytic leukemia cells. Whether this lncRNA acts as a miRNA decoy to promote the expression of these HOX genes remains to be determined.

Metastasis-associated lung adenocarcinoma transcript 1 (MALAT1), also known as nuclear-enriched abundant transcript 2 (NEAT2), is one of the first identified cancer-associated lncRNAs [131]. MALAT1 is a highly conserved noncoding transcript of over 8000 nts encoded by a locus on chromosome 11. It was initially recognized as a prognostic marker of increased metastatic risk for the patients of non-small cell lung carcinoma (NSCLC). Subsequent studies revealed that MALAT1 is localized in nuclear structures enriched with splicing and transcription factors, known as nuclear speckles, suggesting that this lncRNA may modulate alternative splicing of target genes [132,133]. Using RNAi-mediated depletion for the components of the nuclear speckles, Miyagawa *et al.* demonstrated that the pre-mRNA splicing activator RNPS1, the splicing coactivator SRm160 and the spliceosomal intron binding protein IBP160 promote MALAT1 localization to the nuclear speckles [134]. Furthermore, MALAT1 depletion and delocalization from the nuclear speckles resulted in downregulation of two interferon-induced genes (OASL and IFI44) and a potential celiac disease susceptibility gene, SPINK4 [134].

Despite these studies suggesting a role of MALAT1 in RNA splicing, a recent study by Gutschner *et al.* showed that this lncRNA regulates gene expression but not alternative splicing in lung cancer cells [135]. MALAT1 knockout was achieved by genomic integration of RNA destabilizing elements using zinc finger nucleases, leading to 1000-fold MALAT1 reduction. Using these MALAT1-knockout cells, the authors demonstrated that MALAT1 represses anti-metastatic genes and activates pro-metastatic genes in lung cancer cells, but does not affect genes regulating cell growth. In a xenograft mouse model, the MALAT1-depleted lung cancer cells showed reduced tumor formation compared to the cells with the intact MALAT1. This observation is in contrast to the findings reported by Yang *et al.* showing that MALAT1 cooperates with Polycomb 2 protein (Pc2) in regulating the activation of the growth-control gene program in 293T cells [136]. The different cell types employed in their experiments might contribute to the discrepancy between the two studies. Consistent with this prediction, MALAT1 is widely expressed in most normal human tissues, such as pancreas and lung, but absent in several other tissues including skin, stomach, bone marrow and uterus [131]. This suggests that this lncRNA may possess tissue specific functions. In addition to its role in regulating lung cancer metastasis, MALAT1 is upregulated in uterine endometrial stromal sarcoma [137], cervical cancer [4] and hepatocellular carcinoma [138], while its expression in the corresponding healthy tissues is undetectable or intermediate [131]. Although the oncogenic role of MALAT1 in different cancers has been demonstrated by correlational and functional studies, the molecular mechanisms underlying its activities in regulating gene expression and RNA splicing remain undetermined. To date, several studies suggested the essential role of the 3' end sequence and structure to its metastasis-promoting function, nuclear localization and stability [139–142].

ANRIL is a large antisense ncRNA of the INK4b/ARF/INK4a locus [143]. Yap *et al.* demonstrated that ANRIL binds chromobox 7 (CBX7), a component of the polycomb repressive complex 1. This interaction contributes to the role of CBX in promoting EZH2-mediated H3-K27 methylation at the INK4b/ARF/INK4a locus and consequently represses the tumor suppresser INK4a gene. Consistently, both CBX7 and ANRIL are increasingly expressed in prostate cancer [144].

GAS5 (growth arrest-specific transcript 5) is a lncRNA regulating growth arrest of T-cells and lymphocytes [145]. Ectopic GAS5 increases apoptosis and reduces cell cycle progression. Consistently, its downregulation inhibits apoptosis and promotes cell cycle. Kino *et al.* investigated the mechanism underlying the growth suppressive activities of GAS5 and discovered its role in blocking gene expression mediated by glucocorticoid receptor (GR) [146]. GAS5 binds to the DNA-binding domain of GR and thus prevents its association with the glucocorticoid response element of the GR target genes with anti-apoptotic activities, such as inhibitor of apoptosis 2 (cIAP2). Thus, abundantly expressed GAS5 during starvation can sensitize the cells to apoptosis. The nonsense-mediated mRNA decay (NMD) is a system that controls the quality of gene transcripts and reduces errors in gene expression by eliminating RNAs with premature stop codons [147]. Meanwhile, this mechanism also regulates the abundance of cellular transcripts, including ncRNAs. Recently, Zhang *et al.* demonstrated a reciprocally negative regulation between GAS5 and miR-21 [148]. While miR-21 represses GAS5 by targeting a sequence encoded by its exon 4, GAS5 inhibits miR-21 expression. Thus, GAS5 antagonizes the oncogenic activity of miR-21 [149] through reducing its cellular levels. Consistently, miR-21 and GAS5 showed negative correlation in breast cancer specimens [148]. The tumor suppressive role of GAS5 is supported by the identification of genetic susceptibility of its genomic locus, 1q25, to several cancers, including melanoma [150], prostate cancer [151], breast cancer [152,153], colorectal cancer [154] and B-cell lymphoma [155]. Additionally, Tani *et al.* indicated that GAS5 can be stabilized with the depletion of UPF1, an essential component of NMD or during starvation [156] and the GAS5 introns encode multiple snoRNAs [157–160].

Several other lncRNAs have also been demonstrated to regulate chromatin remodeling and gene transcription. PTENP1 is the PTEN pseudogene and encodes two antisense RNA (asRNA) transcripts, asRNA α and β [161]. The α asRNA isoform can recruit DNMT3A, EZH2 and G9A to the PTEN promoter and repress its transcription. This will be further discussed below with other regulatory mechanisms of PTENP1. Evf2 is a polyadenylated lncRNA identified in embryonic brain cells [162]. This lncRNA regulates the transcription of homeodomain transcription factors DLX5 and DLX6 through recruiting DLX and MECP2 to the DNA regulatory elements in the intergenic region of these two genes. As a p53 transactivated lncRNA, lincRNA-p21 is a key mediator of p53-dependent gene repression through a mechanism of recruiting heterogeneous nuclear ribonucleoprotein K (HNRNP) to these p53 target genes [163]. Thus, inhibition of lincRNA-p21 affects the expression of a number of p53 repressed genes. Sheik *et al.* identified an Oct4-activated lncRNA, AK028326, and discovered that this lncRNA regulates pluripotency in mouse embryonic stem cells [164]. Interestingly, AK028326 activates Oct4 expression in a regulatory feedback loop.

3.2. LncRNAs Acting as miRNA Decoys to Free Target mRNAs

Since the regulation of gene expression by miRNAs was revealed, researchers have been using miRNA sponges, RNA molecules containing the target sequence or reverse complementary sequence of a miRNA to be sponged, as a tool to inhibit the function of miRNAs and release their target gene expression [165]. Recent studies suggest that this approach naturally exists in cancers to modulate tumor suppressor and oncogene levels. In some literature, these decoy RNAs are also named as competitive endogenous RNAs (ceRNAs).

PTEN (phosphatase and tensin homolog) is a well characterized tumor suppressor with phosphatase activity. It is encoded at the 10q23.3 locus on chromosome 10 and frequently inactivated through diverse mechanisms in human cancers, highlighting its crucial role in oncogenesis. The 3'-UTR of the PTEN mRNA has 3,329 nts, markedly longer than the average 3'-UTR length (740 nts) of eukaryotic mRNAs [166], implicating its vulnerability as a target of miRNAs. Thus, while PTEN inactivation can be achieved by gene deletion and epigenetic silencing in cancers, its expression is also regulated by multiple miRNAs. For example, the PTEN 3'-UTR contains potential binding sites for over 10 miRNAs overexpressed in glioblastoma multiforme, which is more than 2 times higher than any other tumor suppressor [167]. PTEN expression can be repressed by miR-21, miR-221 and miR-222 [106,168]. Recent studies revealed novel mechanisms regulating PTEN expression through its pseudogene PTENP1 (also called PTH2 or ψPTEN). Pseudogenes are dysfunctional relatives of their cognate genes but have lost the protein-coding ability due to premature stop codons, deletions/insertions or frameshift mutations, and thus cannot be translated into functional proteins [169]. The PTEN pseudogene PTENP1 is highly transcribed in certain tissues and cells, suggesting that this lncRNA may have biological activities [170]. The functional relationship between the PTEN and its pseudogene was first discovered by Poliseno *et al.* [171]. In their study, the authors demonstrated that PTENP1 modulates endogenous PTEN transcript levels by acting as a molecular sponge for

PTEN-targeting miRNAs; thus, the PTENP1 transcript serves as an effective decoy and exerts tumor suppressive functions (Figure 4). This novel regulatory role of the PTENP1 can be extended to KRAS1P, the pseudogene of the oncogene KRAS. In a study by Poliseno *et al.*, the overexpression of KRAS1P led to increased KRAS mRNA levels in prostate cancer DU145 cells through a mechanism of sequestering KRAS-targeting microRNAs, and consequently promoted cell proliferation [171].

Figure 4. PTEN expression is regulated by multiple ncRNAs. (**A**) The PTEN pseudogene, PTENP1, can transcribe into the PTENP1 lncRNA that acts as a decoy to sponge the miRNAs targeting at the 3'-UTR of the PTEN mRNA; (**B**,**C**) The locus of the PTEN pseudogene can also transcribe from the reverse direction to make two antisense RNA (asRNA) isoforms, α and β; (**B**) The α asRNA isoform binds the PTEN promoter and recruits DNMT3A, EZH2 and G9A, which causes epigenetic silencing of the PTEN gene; (**C**) The β asRNA isoform associates with the PTENP1 lncRNA to increase its stability and miRNA decoy activity; (**D**) The ZEB2 mRNA has a long 3'-UTR with many binding sites of miRNAs that also potentially target the PTEN 3'-UTR. Thus, the ZEB2 mRNA acts as a decoy to sponge many miRNAs and consequently promotes PTEN expression.

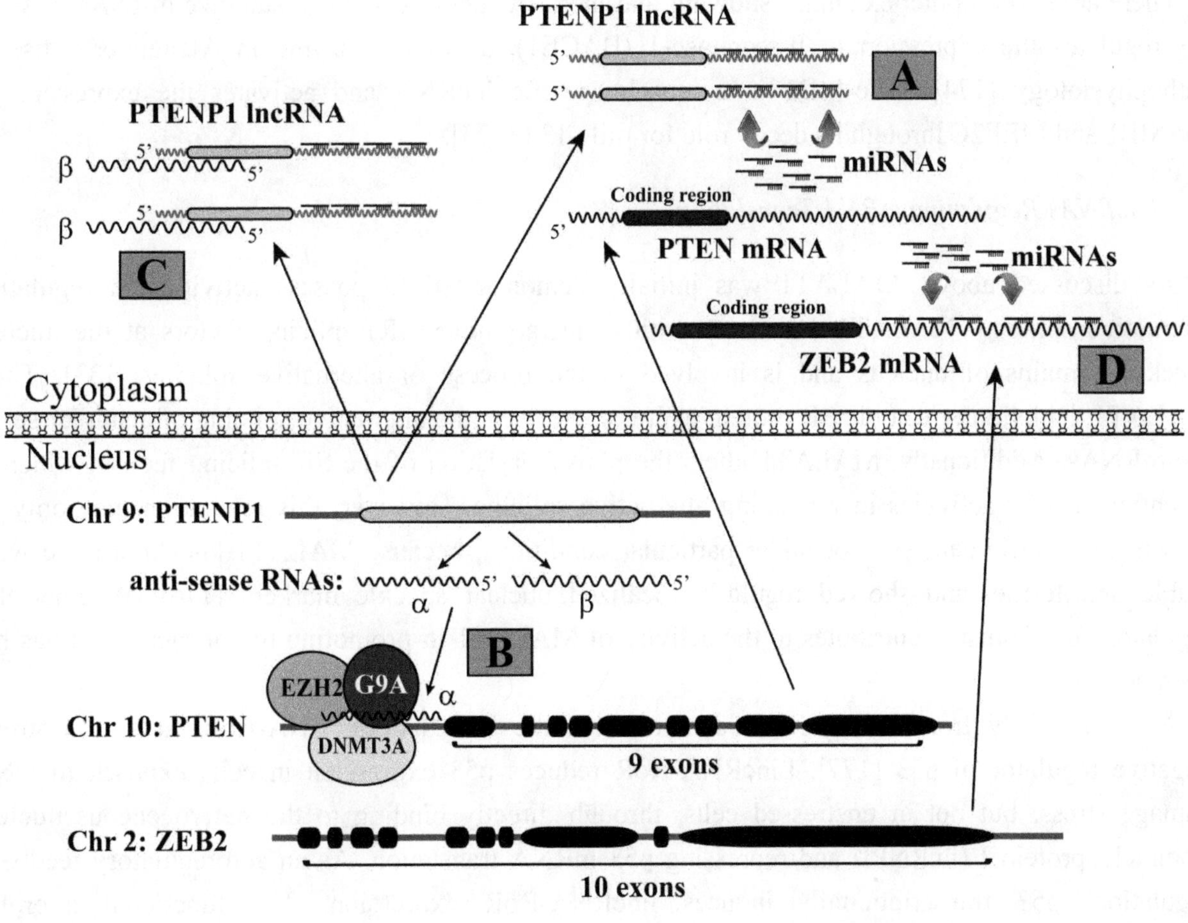

As briefly discussed above, a recent study by Johnsson *et al.* provided evidence of another regulatory mechanism of PTEN by its pseudogene PTENP1 (Figure 4). The authors discovered that the PTENP1 locus can be transcribed from a reverse direction to create two isoforms of an antisense RNA (asRNA), α and β [161]. The α isoform of the asRNA binds the PTEN promoter and recruits

DNMT3A and two histone methyltransferases EZH2 and G9A, which mediate the methylation of histone H3-K27 and H3-K9, respectively, two well-characterized markers of gene repression [172]; thus, the PTENP1 α asRNA negatively regulates PTEN gene expression through promoting the epigenetic silencing of its promoter. The β asRNA isoform interacts with the PTENP1 lncRNA through RNA-RNA pairing, which can maintain stable PTENP1 lncRNA levels in cytoplasm, increase its stability and facilitate its role as a microRNA sponge; thus, the PTENP1 β asRNA activates PTEN expression through facilitating the decoy activity of the PTENP1 lncRNA. Overall, the PTENP1 α and β asRNAs exhibit oncogenic and tumor suppressive roles, respectively, based on their effects on PTEN expression (Figure 4). Whether the two asRNA isoforms are differentially expressed in human cancers remains to be determined.

Interestingly, PTEN expression can also be regulated at the translational level by the sponge effect of the transcript from another gene. The full length ZEB2 mRNA has a very long 3'-UTR (over 5,000 nts) and contains multiple potential binding sites of miRNAs that can also target the PTEN 3'-UTR, such as miR-181a and miR-200/miR-141 [173]. Thus, the ZEB2 mRNA serves as a decoy or ceRNA of the PTEN mRNA (Figure 4) and reduced ZEB2 expression activates the PI3K/AKT pathway through downregulating PTEN.

There are several other examples showing that lncRNAs act as decoys to stabilize mRNAs. BACE-AS regulates the expression of β-secretase-1 (BACE1), a crucial enzyme in Alzheimer's disease pathophysiology [174]. Linc-MD1 is a muscle-specific lncRNA and activates the expression of MAML1 and MEF2C through its decoy role for miR-133 [175].

3.3. LncRNAs Regulating mRNA Translation

As discussed above, MALAT1 was initially demonstrated to possess activities in regulating alternative splicing. MALAT1 associates with serine/arginine (SR) splicing factors at the nuclear speckle domains of nucleus and is involved in the process of alternative splicing [133]. Thus, MALAT1 depletion or ectopic SR protein expression affects the alternative splicing of a similar set of pre-mRNAs. Additionally, MALAT1 alters the phosphorylation of the SR splicing factors, which is essential to their activities in regulating alternative splicing. However, this regulation may only be present in specific cell types or under particular conditions, because MALAT1-knockout mice were viable and fertile, and showed regularly localized nuclear speckle markers [176]. Whether this regulatory mechanism contributes to the activity of MALAT1 in promoting tumor metastasis has not been determined.

A recent study from Zhang *et al.* demonstrated the lncRNA, lincRNA-RoR, acts as a strong negative regulator of p53 [177]. LincRNA-RoR reduces p53 expression in cells exposed to DNA damage stress, but not in unstressed cells, through directly binding to the heterogeneous nuclear ribonucleoprotein I (hnRNPI) and repressing p53 mRNA translation. As an autoregulatory feedback regulation, p53 transcriptionally induces lincRNA-RoR expression. The functional interplay between lincRNA-RoR and p53 may serve as additional surveillance to mediate cell response to genotoxic stresses.

As discussed above, another p53-transactivated lncRNA, lincRNA-p21, is involved in p53-dependent gene transrepression [163]. A recent study from Yoon *et al.* demonstrated that lincRNA-p21 also modulates translation [178]. This lncRNA can associate with JunB and β-catenin mRNAs to reduce their translation rates. A RNA-binding protein, HuR, can recruit let-7/AGO2 to lincRNA-p21 to reduce its stability; thus, elevated HuR releases lincRNA-p21-mediated repression of JunB and β-catenin expression. Since lincRNA-p21 is transactivated by p53, the negative regulation of JunB and β-catenin by lincRNA-p21 is consistent to the tumor suppressive role of p53.

3.4. LncRNAs Regulating Protein Activities

Telomeres are the DNA-protein complexes at the end of eukaryotic chromosomes and essential to chromosome stability. As a frequently activated reverse transcriptase, telomerase adds DNA sequence repeats ("TTAGGG" in vertebrates) to the telomere regions of chromosome to maintain the telomere length. Azzalin *et al.* discovered that telomeres can be transcribed into telomeric repeat-containing RNA (TERRA) [179]. These molecules have different lengths and were predicted to play a role in the maintenance of telomere integrity through an unclear mechanism. A later study from Redon *et al.* provided a possible mechanism underlying this activity of TERRA [180]. The authors demonstrated that TERRA binds telomerase and thus acts as a potent competitive inhibitor for the telomeric DNA. Consistently, TERRA expression is significantly downregulated in multiple tumor cell lines, which plays a role in telomere maintenance and cell immortalization of cancer cells.

Wang *et al.* demonstrated an indirect regulation of the histone acetyltransferase activities of CBP and p300 by ncRNAs [181]. The 5'-regulatory region of the cyclin D1 gene encodes at least four ncRNAs. In response to ionizing radiation, these ncRNAs bind TLS (translocated in liposarcoma) protein at the chromatin of the cyclin D1 promoter region and cause an allosteric effect on this protein. This TLS conformational change promotes its activity of inhibiting CBP/p300-mediated histone acetylation and consequently silences cyclin D1 gene. These data indicate that ncRNAs encoded by a promoter can act as selective ligands to modulate the activities of transcription cofactors in response to genotoxic stresses.

LncRNAs can also regulate protein activity through altering their subcellular localization. Nuclear factor of activated T-cells (NFAT) represents a family of transcription factors regulating immune response. Some of NFAT proteins, such as NFAT1 and NFAT5, contribute to tumor metastasis and cell motility [182,183]. The noncoding repressor of NFAT (NRON) is a ncRNA associated with multiple proteins [184]. NRON binds phosphorylated NFAT1 to sequester NFAT in the cytoplasm. The depletion of NRON leads to NFAT dephosphorylation and nuclear import. Thus, ncRNAs can be a part of a scaffold to trap a latent transcription factor [185] and regulate the expression of its target genes.

4. Circular RNAs

Recent advances in high-throughput sequencing coupled with powerful computational analyses of expression data have allowed researchers to identify and characterize new RNA species, one of which is a new class of RNA molecules, circular RNAs (circRNAs), present in humans and animals. CircRNAs were first discovered in plants and considered as viroids due to their predicted role as subviral agents [186]. These covalently linked and single-stranded ncRNAs have molecular weights of over 100,000 Da and are highly thermal stable. Initially, most circRNA molecules were dismissed as transcription noise or by-products generated by cellular splicing machineries. However, this notion has been challenged by emerging studies suggesting that some circRNAs are evolutionary conserved in human and mice [14,187]. The exact mechanism of circRNA biogenesis remains to be elucidated. Salzman *et al.* indicated that circRNAs are produced by a non-canonical mode of RNA splicing [188]. Several other studies suggested that circRNA may be at least partially contributed by exon-skipping events, which can create an exon-containing lariat and then possibly undergo internal splicing to generate an exon circle [187,189,190].

In 2011, Hansen *et al.* reported the miRNA decoy function of a noncoding circRNA [191]. In this study, the authors demonstrated that miR-671 directly targets and cleaves a circular antisense transcript of the cerebellar degeneration-related protein 1 (CDR1) in an AGO2-dependent manner. In 2013, thousands of well-expressed and stable circRNAs with tissue- and developmental stage-specific expression were identified [14]. The antisense to the CDR1 transcript (CDR1as) is densely bound by Argonaute proteins and contains 63 conserved binding sites of miR-7, suggesting that this circRNA functions as a miRNA sponge of miR-7 to release its target gene expression [14,190]. Indeed, using a zebrafish model, Memczak *et al.* demonstrated that introduction of human CDR1as resulted in a phenotype similar to that of miR-7 knockdown, while injection of the miR-7 precursor partially reversed this phenotype, further implicating CDR1as as an antagonist of miR-7 [14]. The results of this study provided evidence that circRNAs may function as regulators of gene expression at the post-transcriptional level. In addition to these studies, Jeck *et al.* recently discovered over 25,000 circRNAs in human fibroblasts [187]. Due to the lack of exposed 5' and 3' ends, circRNAs are predicted to be resistant to degradation by cellular enzymes such as ribonucleases and thus more stable than linear RNAs. The increased stability of circRNAs makes these novel RNA molecules act as efficient miRNA sponges in modulating gene expression.

5. Conclusions

NcRNAs have gained intensive and growing attention for their potential as both regulators and biomarkers of cancers in the past two decades. We are now witnessing the beginning of a new era of identifying key players and determining their underlying mechanisms during oncogenesis. In this review, we are only able to summarize some studies of miRNAs, lncRNAs and circRNAs. We did not intentionally ignore other excellent reports, but just could not include them due to the enormous amounts of ncRNA-related studies and the limitation of this article. Many review papers have summarized the roles of ncRNAs in different diseases including cancers and some of them also discussed several other types of ncRNAs, such as snoRNA and small nuclear RNAs (snRNA).

Currently, many thousands of ncRNAs have been identified and their differential expression profiles in a variety of cancers or between normal and tumorigenic specimens have been demonstrated; however, only a small number of ncRNAs, especially lncRNAs, have been well characterized. Thus, what we have discovered is just a tip of the iceberg in this area. Understanding the functions and regulatory mechanisms of ncRNAs in cancer pathogenesis remains a fertile research field to be explored at least in next decade. Future studies are needed to dissect the upstream mechanisms regulating ncRNA expression and processing, and the downstream proteins or pathways mediated by ncRNAs at different stages of cancer development. These efforts may lead to the discovery of entirely novel regulatory mechanisms or advance our understanding for the currently recognized signaling pathways. The achievement of these ncRNA studies in cancers will definitely provide insights into discovering new biomarkers for cancer diagnosis and prognosis, unraveling novel therapeutic targets, and developing unconventional therapeutic modalities to reduce cancer-related mortality.

Acknowledgments

We thank Steven J Kridel for critical reading of the manuscript. GD is supported by the Randi B. Weiss Cancer Fund. GS is supported by American Cancer Society (116403-RSG-09-082-01-MGO) and NIH R01 (5R01CA106314).

References

1. Bertone, P.; Stolc, V.; Royce, T.E.; Rozowsky, J.S.; Urban, A.E.; Zhu, X.; Rinn, J.L.; Tongprasit, W.; Samanta, M.; Weissman, S.; *et al.* Global identification of human transcribed sequences with genome tiling arrays. *Science* **2004**, *306*, 2242–2246.
2. Birney, E.; Stamatoyannopoulos, J.A.; Dutta, A.; Guigo, R.; Gingeras, T.R.; Margulies, E.H.; Weng, Z.; Snyder, M.; Dermitzakis, E.T.; Thurman, R.E.; *et al.* Identification and analysis of functional elements in 1% of the human genome by the ENCODE pilot project. *Nature* **2007**, *447*, 799–816.
3. Cheng, J.; Kapranov, P.; Drenkow, J.; Dike, S.; Brubaker, S.; Patel, S.; Long, J.; Stern, D.; Tammana, H.; Helt, G.; *et al.* Transcriptional maps of 10 human chromosomes at 5-nucleotide resolution. *Science* **2005**, *308*, 1149–1154.
4. Kapranov, P.; Cheng, J.; Dike, S.; Nix, D.A.; Duttagupta, R.; Willingham, A.T.; Stadler, P.F.; Hertel, J.; Hackermuller, J.; Hofacker, I.L.; *et al.* RNA maps reveal new RNA classes and a possible function for pervasive transcription. *Science* **2007**, *316*, 1484–1488.
5. Wilusz, J.E.; Sunwoo, H.; Spector, D.L. Long noncoding RNAs: Functional surprises from the RNA world. *Genes Dev.* **2009**, *23*, 1494–1504.
6. Mercer, T.R.; Dinger, M.E.; Mattick, J.S. Long non-coding RNAs: Insights into functions. *Nat. Rev. Genet.* **2009**, *10*, 155–159.
7. Wang, K.C.; Chang, H.Y. Molecular mechanisms of long noncoding RNAs. *Mol. Cell* **2011**, *43*, 904–914.
8. Cooper, G.M. Cellular transforming genes. *Science* **1982**, *217*, 801–806.
9. Santos, E.; Tronick, S.R.; Aaronson, S.A.; Pulciani, S.; Barbacid, M. T24 human bladder carcinoma oncogene is an activated form of the normal human homologue of BALB- and Harvey-MSV transforming genes. *Nature* **1982**, *298*, 343–347.
10. Parada, L.F.; Tabin, C.J.; Shih, C.; Weinberg, R.A. Human EJ bladder carcinoma oncogene is homologue of Harvey sarcoma virus ras gene. *Nature* **1982**, *297*, 474–478.
11. Godbout, R.; Dryja, T.P.; Squire, J.; Gallie, B.L.; Phillips, R.A. Somatic inactivation of genes on chromosome 13 is a common event in retinoblastoma. *Nature* **1983**, *304*, 451–453.
12. Murphree, A.L.; Benedict, W.F. Retinoblastoma: Clues to human oncogenesis. *Science* **1984**, *223*, 1028–1033.
13. Esteller, M. Non-coding RNAs in human disease. *Nat. Rev. Genet.* **2011**, *12*, 861–874.

14. Memczak, S.; Jens, M.; Elefsinioti, A.; Torti, F.; Krueger, J.; Rybak, A.; Maier, L.; Mackowiak, S.D.; Gregersen, L.H.; Munschauer, M.; *et al.* Circular RNAs are a large class of animal RNAs with regulatory potency. *Nature* **2013**, *495*, 333–338.
15. Kim, V.N. MicroRNA biogenesis: Coordinated cropping and dicing. *Nat. Rev. Mol. Cell Biol.* **2005**, *6*, 376–385.
16. Czech, B.; Hannon, G.J. Small RNA sorting: Matchmaking for Argonautes. *Nat. Rev. Genet.* **2011**, *12*, 19–31.
17. Bartel, D.P. MicroRNAs: Target recognition and regulatory functions. *Cell* **2009**, *136*, 215–233.
18. Guo, L.; Lu, Z. The fate of miRNA* strand through evolutionary analysis: Implication for degradation as merely carrier strand or potential regulatory molecule? *PLoS One* **2010**, *5*, e11387.
19. Lewis, B.P.; Burge, C.B.; Bartel, D.P. Conserved seed pairing, often flanked by adenosines, indicates that thousands of human genes are microRNA targets. *Cell* **2005**, *120*, 15–20.
20. Lujambio, A.; Lowe, S.W. The microcosmos of cancer. *Nature* **2012**, *482*, 347–355.
21. Krek, A.; Grun, D.; Poy, M.N.; Wolf, R.; Rosenberg, L.; Epstein, E.J.; MacMenamin, P.; da Piedade, I.; Gunsalus, K.C.; Stoffel, M.; *et al.* Combinatorial microRNA target predictions. *Nat. Genet.* **2005**, *37*, 495–500.
22. Lee, R.C.; Feinbaum, R.L.; Ambros, V. The C. elegans heterochronic gene lin-4 encodes small RNAs with antisense complementarity to lin-14. *Cell* **1993**, *75*, 843–854.
23. Wightman, B.; Ha, I.; Ruvkun, G. Posttranscriptional regulation of the heterochronic gene lin-14 by lin-4 mediates temporal pattern formation in *C. elegans*. *Cell* **1993**, *75*, 855–862.
24. Brennecke, J.; Hipfner, D.R.; Stark, A.; Russell, R.B.; Cohen, S.M. Bantam encodes a developmentally regulated microRNA that controls cell proliferation and regulates the proapoptotic gene hid in Drosophila. *Cell* **2003**, *113*, 25–36.
25. Chen, X. A microRNA as a translational repressor of APETALA2 in Arabidopsis flower development. *Science* **2004**, *303*, 2022–2025.
26. Poy, M.N.; Eliasson, L.; Krutzfeldt, J.; Kuwajima, S.; Ma, X.; Macdonald, P.E.; Pfeffer, S.; Tuschl, T.; Rajewsky, N.; Rorsman, P.; *et al.* A pancreatic islet-specific microRNA regulates insulin secretion. *Nature* **2004**, *432*, 226–230.
27. Cimmino, A.; Calin, G.A.; Fabbri, M.; Iorio, M.V.; Ferracin, M.; Shimizu, M.; Wojcik, S.E.; Aqeilan, R.I.; Zupo, S.; Dono, M.; *et al.* miR-15 and miR-16 induce apoptosis by targeting BCL2. *Proc. Natl. Acad. Sci. USA* **2005**, *102*, 13944–13949.
28. Valencia-Sanchez, M.A.; Liu, J.; Hannon, G.J.; Parker, R. Control of translation and mRNA degradation by miRNAs and siRNAs. *Genes Dev.* **2006**, *20*, 515–524.
29. Huntzinger, E.; Izaurralde, E. Gene silencing by microRNAs: Contributions of translational repression and mRNA decay. *Nat. Rev. Genet.* **2011**, *12*, 99–110.
30. Djuranovic, S.; Nahvi, A.; Green, R. miRNA-mediated gene silencing by translational repression followed by mRNA deadenylation and decay. *Science* **2012**, *336*, 237–240.
31. Calin, G.A.; Dumitru, C.D.; Shimizu, M.; Bichi, R.; Zupo, S.; Noch, E.; Aldler, H.; Rattan, S.; Keating, M.; Rai, K.; *et al.* Frequent deletions and down-regulation of micro- RNA genes miR15 and miR16 at 13q14 in chronic lymphocytic leukemia. *Proc. Natl. Acad. Sci. USA* **2002**, *99*, 15524–15529.

32. Calin, G.A.; Sevignani, C.; Dumitru, C.D.; Hyslop, T.; Noch, E.; Yendamuri, S.; Shimizu, M.; Rattan, S.; Bullrich, F.; Negrini, M.; *et al.* Human microRNA genes are frequently located at fragile sites and genomic regions involved in cancers. *Proc. Natl. Acad. Sci. USA* **2004**, *101*, 2999–3004.

33. Wang, D.; Qiu, C.; Zhang, H.; Wang, J.; Cui, Q.; Yin, Y. Human microRNA oncogenes and tumor suppressors show significantly different biological patterns: From functions to targets. *PLoS One* **2010**, *5*, e13067.

34. Mayr, C.; Bartel, D.P. Widespread shortening of 3'UTRs by alternative cleavage and polyadenylation activates oncogenes in cancer cells. *Cell* **2009**, *138*, 673–684.

35. Diederichs, S.; Haber, D.A. Sequence variations of microRNAs in human cancer: Alterations in predicted secondary structure do not affect processing. *Cancer Res.* **2006**, *66*, 6097–6104.

36. Lu, J.; Getz, G.; Miska, E.A.; Alvarez-Saavedra, E.; Lamb, J.; Peck, D.; Sweet-Cordero, A.; Ebert, B.L.; Mak, R.H.; Ferrando, A.A.; *et al.* MicroRNA expression profiles classify human cancers. *Nature* **2005**, *435*, 834–838.

37. Murakami, Y.; Yasuda, T.; Saigo, K.; Urashima, T.; Toyoda, H.; Okanoue, T.; Shimotohno, K. Comprehensive analysis of microRNA expression patterns in hepatocellular carcinoma and non-tumorous tissues. *Oncogene* **2006**, *25*, 2537–2545.

38. Yanaihara, N.; Caplen, N.; Bowman, E.; Seike, M.; Kumamoto, K.; Yi, M.; Stephens, R.M.; Okamoto, A.; Yokota, J.; Tanaka, T.; *et al.* Unique microRNA molecular profiles in lung cancer diagnosis and prognosis. *Cancer Cell* **2006**, *9*, 189–198.

39. Rosenfeld, N.; Aharonov, R.; Meiri, E.; Rosenwald, S.; Spector, Y.; Zepeniuk, M.; Benjamin, H.; Shabes, N.; Tabak, S.; Levy, A.; *et al.* MicroRNAs accurately identify cancer tissue origin. *Nat. Biotechnol.* **2008**, *26*, 462–469.

40. Xi, Y.; Nakajima, G.; Gavin, E.; Morris, C.G.; Kudo, K.; Hayashi, K.; Ju, J. Systematic analysis of microRNA expression of RNA extracted from fresh frozen and formalin-fixed paraffin-embedded samples. *RNA* **2007**, *13*, 1668–1674.

41. Mitchell, P.S.; Parkin, R.K.; Kroh, E.M.; Fritz, B.R.; Wyman, S.K.; Pogosova-Agadjanyan, E.L.; Peterson, A.; Noteboom, J.; O'Briant, K.C.; Allen, A.; *et al.* Circulating microRNAs as stable blood-based markers for cancer detection. *Proc. Natl. Acad. Sci. USA* **2008**, *105*, 10513–10518.

42. Chan, M.; Liaw, C.S.; Ji, S.M.; Tan, H.H.; Wong, C.Y.; Thike, A.A.; Tan, P.H.; Ho, G.H.; Lee, A.S. Identification of circulating microRNA signatures for breast cancer detection. *Clin. Cancer Res.* **2013**, *19*, 4477–4487.

43. Chen, J.; Yao, D.; Li, Y.; Chen, H.; He, C.; Ding, N.; Lu, Y.; Ou, T.; Zhao, S.; Li, L.; *et al.* Serum microRNA expression levels can predict lymph node metastasis in patients with early-stage cervical squamous cell carcinoma. *Int. J. Mol. Med.* **2013**, *32*, 557–567.

44. Xiao, Y.F.; Yong, X.; Fan, Y.H.; Lu, M.H.; Yang, S.M.; Hu, C.J. microRNA detection in feces, sputum, pleural effusion and urine: Novel tools for cancer screening (Review). *Oncol. Rep.* **2013**, *30*, 535–544.

45. Xiang, J.; Wu, J. Feud or Friend? The role of the miR-17-92 cluster in tumorigenesis. *Curr. Genomics* **2010**, *11*, 129–135.

46. Ota, A.; Tagawa, H.; Karnan, S.; Tsuzuki, S.; Karpas, A.; Kira, S.; Yoshida, Y.; Seto, M. Identification and characterization of a novel gene, C13orf25, as a target for 13q31-q32 amplification in malignant lymphoma. *Cancer Res.* **2004**, *64*, 3087–3095.
47. He, L.; Thomson, J.M.; Hemann, M.T.; Hernando-Monge, E.; Mu, D.; Goodson, S.; Powers, S.; Cordon-Cardo, C.; Lowe, S.W.; Hannon, G.J.; *et al.* A microRNA polycistron as a potential human oncogene. *Nature* **2005**, *435*, 828–833.
48. O'Donnell, K.A.; Wentzel, E.A.; Zeller, K.I.; Dang, C.V.; Mendell, J.T. c-Myc-regulated microRNAs modulate E2F1 expression. *Nature* **2005**, *435*, 839–843.
49. Engelmann, D.; Putzer, B.M. The dark side of E2F1: In transit beyond apoptosis. *Cancer Res.* **2012**, *72*, 571–575.
50. Biswas, A.K.; Johnson, D.G. Transcriptional and nontranscriptional functions of E2F1 in response to DNA damage. *Cancer Res.* **2012**, *72*, 13–17.
51. Hossain, A.; Kuo, M.T.; Saunders, G.F. Mir-17-5p regulates breast cancer cell proliferation by inhibiting translation of AIB1 mRNA. *Mol. Cell. Biol.* **2006**, *26*, 8191–8201.
52. Louie, M.C.; Zou, J.X.; Rabinovich, A.; Chen, H.W. ACTR/AIB1 functions as an E2F1 coactivator to promote breast cancer cell proliferation and antiestrogen resistance. *Mol. Cell. Biol.* **2004**, *24*, 5157–5171.
53. Qin, L.; Liao, L.; Redmond, A.; Young, L.; Yuan, Y.; Chen, H.; O'Malley, B.W.; Xu, J. The AIB1 oncogene promotes breast cancer metastasis by activation of PEA3-mediated matrix metalloproteinase 2 (MMP2) and MMP9 expression. *Mol. Cell. Biol.* **2008**, *28*, 5937–5950.
54. Yu, Z.; Wang, C.; Wang, M.; Li, Z.; Casimiro, M.C.; Liu, M.; Wu, K.; Whittle, J.; Ju, X.; Hyslop, T.; *et al.* A cyclin Dl/microRNA 17/20 regulatory feedback loop in control of breast cancer cell proliferation. *J. Cell Biol.* **2008**, *182*, 509–517.
55. Yu, Z.; Willmarth, N.E.; Zhou, J.; Katiyar, S.; Wang, M.; Liu, Y.; McCue, P.A.; Quong, A.A.; Lisanti, M.P.; Pestell, R.G. microRNA 17/20 inhibits cellular invasion and tumor metastasis in breast cancer by heterotypic signaling. *Proc. Natl. Acad. Sci. USA* **2010**, *107*, 8231–8236.
56. Li, H.; Yang, B.B. Stress response of glioblastoma cells mediated by miR-17-5p targeting PTEN and the passenger strand miR-17-3p targeting MDM2. *Oncotarget* **2012**, *3*, 1653–1668.
57. Shan, S.W.; Fang, L.; Shatseva, T.; Rutnam, Z.J.; Yang, X.; Du, W.; Lu, W.Y.; Xuan, J.W.; Deng, Z.; Yang, B.B. Mature miR-17-5p and passenger miR-17-3p induce hepatocellular carcinoma by targeting PTEN, GalNT7 and vimentin in different signal pathways. *J. Cell Sci.* **2013**, *126*, 1517–1530.
58. Cao, P.; Deng, Z.; Wan, M.; Huang, W.; Cramer, S.D.; Xu, J.; Lei, M.; Sui, G. MicroRNA-101 negatively regulates Ezh2 and its expression is modulated by androgen receptor and HIF-1alpha/HIF-1beta. *Mol. Cancer* **2010**, *9*, doi:10.1186/1476-4598-9-108.
59. Wang, H.J.; Ruan, H.J.; He, X.J.; Ma, Y.Y.; Jiang, X.T.; Xia, Y.J.; Ye, Z.Y.; Tao, H.Q. MicroRNA-101 is down-regulated in gastric cancer and involved in cell migration and invasion. *Eur. J. Cancer* **2010**, *46*, 2295–2303.
60. Hu, Z.; Lin, Y.; Chen, H.; Mao, Y.; Wu, J.; Zhu, Y.; Xu, X.; Li, S.; Zheng, X.; Xie, L. MicroRNA-101 suppresses motility of bladder cancer cells by targeting c-Met. *Biochem. Biophys. Res. Commun.* **2013**, *435*, 82–87.

61. Wang, R.; Wang, H.B.; Hao, C.J.; Cui, Y.; Han, X.C.; Hu, Y.; Li, F.F.; Xia, H.F.; Ma, X. MiR-101 is involved in human breast carcinogenesis by targeting Stathmin1. *PLoS One* **2012**, *7*, e46173.
62. Varambally, S.; Cao, Q.; Mani, R.S.; Shankar, S.; Wang, X.; Ateeq, B.; Laxman, B.; Cao, X.; Jing, X.; Ramnarayanan, K.; *et al.* Genomic loss of microRNA-101 leads to overexpression of histone methyltransferase EZH2 in cancer. *Science* **2008**, *322*, 1695–1699.
63. Sellers, W.R.; Loda, M. The EZH2 polycomb transcriptional repressor—A marker or mover of metastatic prostate cancer? *Cancer Cell* **2002**, *2*, 349–350.
64. Friedman, J.M.; Liang, G.; Liu, C.C.; Wolff, E.M.; Tsai, Y.C.; Ye, W.; Zhou, X.; Jones, P.A. The putative tumor suppressor microRNA-101 modulates the cancer epigenome by repressing the polycomb group protein EZH2. *Cancer Res.* **2009**, *69*, 2623–2629.
65. Banerjee, R.; Mani, R.S.; Russo, N.; Scanlon, C.S.; Tsodikov, A.; Jing, X.; Cao, Q.; Palanisamy, N.; Metwally, T.; Inglehart, R.C.; *et al.* The tumor suppressor gene rap1GAP is silenced by miR-101-mediated EZH2 overexpression in invasive squamous cell carcinoma. *Oncogene* **2011**, *30*, 4339–4349.
66. Kottakis, F.; Polytarchou, C.; Foltopoulou, P.; Sanidas, I.; Kampranis, S.C.; Tsichlis, P.N. FGF-2 regulates cell proliferation, migration, and angiogenesis through an NDY1/KDM2B-miR-101-EZH2 pathway. *Mol. Cell* **2011**, *43*, 285–298.
67. Qazi, A.M.; Gruzdyn, O.; Semaan, A.; Seward, S.; Chamala, S.; Dhulipala, V.; Sethi, S.; Ali-Fehmi, R.; Philip, P.A.; Bouwman, D.L.; *et al.* Restoration of E-cadherin expression in pancreatic ductal adenocarcinoma treated with microRNA-101. *Surgery* **2012**, *152*, 704–711, discussion 711–713.
68. Nadiminty, N.; Tummala, R.; Lou, W.; Zhu, Y.; Zhang, J.; Chen, X.; eVere White, R.W.; Kung, H.J.; Evans, C.P.; Gao, A.C. MicroRNA let-7c suppresses androgen receptor expression and activity via regulation of Myc expression in prostate cancer cells. *J. Biol. Chem.* **2012**, *287*, 1527–1537.
69. Boyerinas, B.; Park, S.M.; Hau, A.; Murmann, A.E.; Peter, M.E. The role of let-7 in cell differentiation and cancer. *Endocr. Relat. Cancer* **2010**, *17*, F19–F36.
70. Johnson, S.M.; Grosshans, H.; Shingara, J.; Byrom, M.; Jarvis, R.; Cheng, A.; Labourier, E.; Reinert, K.L.; Brown, D.; Slack, F.J. RAS is regulated by the let-7 microRNA family. *Cell* **2005**, *120*, 635–647.
71. Hermeking, H. MicroRNAs in the p53 network: Micromanagement of tumour suppression. *Nat. Rev. Cancer* **2012**, *12*, 613–626.
72. Hunten, S.; Siemens, H.; Kaller, M.; Hermeking, H. The p53/microRNA network in cancer: Experimental and bioinformatics approaches. *Adv. Exp. Med. Biol.* **2013**, *774*, 77–101.
73. Yamamura, S.; Saini, S.; Majid, S.; Hirata, H.; Ueno, K.; Chang, I.; Tanaka, Y.; Gupta, A.; Dahiya, R. MicroRNA-34a suppresses malignant transformation by targeting c-Myc transcriptional complexes in human renal cell carcinoma. *Carcinogenesis* **2012**, *33*, 294–300.
74. Yamamura, S.; Saini, S.; Majid, S.; Hirata, H.; Ueno, K.; Deng, G.; Dahiya, R. MicroRNA-34a modulates c-Myc transcriptional complexes to suppress malignancy in human prostate cancer cells. *PLoS One* **2012**, *7*, e29722.
75. Cole, K.A.; Attiyeh, E.F.; Mosse, Y.P.; Laquaglia, M.J.; Diskin, S.J.; Brodeur, G.M.; Maris, J.M. A functional screen identifies miR-34a as a candidate neuroblastoma tumor suppressor gene. *Mol. Cancer Res.* **2008**, *6*, 735–742.

76. Hermeking, H. The miR-34 family in cancer and apoptosis. *Cell Death Differ.* **2010**, *17*, 193–199.
77. Wong, C.S.; Sceneay, J.; House, C.M.; Halse, H.M.; Liu, M.C.; George, J.; Hunnam, T.C.; Parker, B.S.; Haviv, I.; Ronai, Z.; *et al.* Vascular normalization by loss of Siah2 results in increased chemotherapeutic efficacy. *Cancer Res.* **2012**, *72*, 1694–1704.
78. Keith, B.; Johnson, R.S.; Simon, M.C. HIF1alpha and HIF2alpha: Sibling rivalry in hypoxic tumour growth and progression. *Nat. Rev. Cancer* **2012**, *12*, 9–22.
79. Yamakuchi, M.; Lotterman, C.D.; Bao, C.; Hruban, R.H.; Karim, B.; Mendell, J.T.; Huso, D.; Lowenstein, C.J. P53-induced microRNA-107 inhibits HIF-1 and tumor angiogenesis. *Proc. Natl. Acad. Sci. USA* **2010**, *107*, 6334–6339.
80. Sachdeva, M.; Zhu, S.; Wu, F.; Wu, H.; Walia, V.; Kumar, S.; Elble, R.; Watabe, K.; Mo, Y.Y. p53 represses c-Myc through induction of the tumor suppressor miR-145. *Proc. Natl. Acad. Sci. USA* **2009**, *106*, 3207–3212.
81. Chang, C.J.; Chao, C.H.; Xia, W.; Yang, J.Y.; Xiong, Y.; Li, C.W.; Yu, W.H.; Rehman, S.K.; Hsu, J.L.; Lee, H.H.; *et al.* p53 regulates epithelial-mesenchymal transition and stem cell properties through modulating miRNAs. *Nat. Cell Biol.* **2011**, *13*, 317–323.
82. Kim, T.; Veronese, A.; Pichiorri, F.; Lee, T.J.; Jeon, Y.J.; Volinia, S.; Pineau, P.; Marchio, A.; Palatini, J.; Suh, S.S.; *et al.* p53 regulates epithelial-mesenchymal transition through microRNAs targeting ZEB1 and ZEB2. *J. Exp. Med.* **2011**, *208*, 875–883.
83. Yan, H.L.; Xue, G.; Mei, Q.; Wang, Y.Z.; Ding, F.X.; Liu, M.F.; Lu, M.H.; Tang, Y.; Yu, H.Y.; Sun, S.H. Repression of the miR-17–92 cluster by p53 has an important function in hypoxia-induced apoptosis. *EMBO J.* **2009**, *28*, 2719–2732.
84. Suzuki, H.I.; Yamagata, K.; Sugimoto, K.; Iwamoto, T.; Kato, S.; Miyazono, K. Modulation of microRNA processing by p53. *Nature* **2009**, *460*, 529–533.
85. Kohlstedt, K.; Trouvain, C.; Boettger, T.; Shi, L.; Fisslthaler, B.; Fleming, I. AMP-activated protein kinase regulates endothelial cell angiotensin-converting enzyme expression via p53 and the post-transcriptional regulation of microRNA-143/145. *Circ. Res.* **2013**, *112*, 1150–1158.
86. Wang, J.; He, Q.; Han, C.; Gu, H.; Jin, L.; Li, Q.; Mei, Y.; Wu, M. p53-facilitated miR-199a-3p regulates somatic cell reprogramming. *Stem Cells* **2012**, *30*, 1405–1413.
87. Manfe, V.; Biskup, E.; Rosbjerg, A.; Kamstrup, M.; Skov, A.G.; Lerche, C.M.; Lauenborg, B.T.; Odum, N.; Gniadecki, R. miR-122 regulates p53/Akt signalling and the chemotherapy-induced apoptosis in cutaneous T-cell lymphoma. *PLoS One* **2012**, *7*, e29541.
88. Michael, D.; Oren, M. The p53-Mdm2 module and the ubiquitin system. *Semin. Cancer Biol.* **2003**, *13*, 49–58.
89. Pichiorri, F.; Suh, S.S.; Rocci, A.; De Luca, L.; Taccioli, C.; Santhanam, R.; Zhou, W.; Benson, D.M., Jr.; Hofmainster, C.; Alder, H.; *et al.* Downregulation of p53-inducible microRNAs 192, 194, and 215 impairs the p53/MDM2 autoregulatory loop in multiple myeloma development. *Cancer Cell* **2010**, *18*, 367–381.
90. Khella, H.W.; Bakhet, M.; Allo, G.; Jewett, M.A.; Girgis, A.H.; Latif, A.; Girgis, H.; von Both, I.; Bjarnason, G.A.; Yousef, G.M. miR-192, miR-194 and miR-215: A convergent microRNA network suppressing tumor progression in renal cell carcinoma. *Carcinogenesis* **2013**, doi:10.1093/carcin/bgt184.

91. Senanayake, U.; Das, S.; Vesely, P.; Alzoughbi, W.; Frohlich, L.F.; Chowdhury, P.; Leuschner, I.; Hoefler, G.; Guertl, B. miR-192, miR-194, miR-215, miR-200c and miR-141 are downregulated and their common target ACVR2B is strongly expressed in renal childhood neoplasms. *Carcinogenesis* **2012**, *33*, 1014–1021.
92. Xiao, J.; Lin, H.; Luo, X.; Wang, Z. miR-605 joins p53 network to form a p53:miR-605:Mdm2 positive feedback loop in response to stress. *EMBO J.* **2011**, *30*, 524–532.
93. Pelengaris, S.; Khan, M.; Evan, G. c-MYC: More than just a matter of life and death. *Nat. Rev. Cancer* **2002**, *2*, 764–776.
94. Craig, V.J.; Cogliatti, S.B.; Imig, J.; Renner, C.; Neuenschwander, S.; Rehrauer, H.; Schlapbach, R.; Dirnhofer, S.; Tzankov, A.; Muller, A. Myc-mediated repression of microRNA-34a promotes high-grade transformation of B-cell lymphoma by dysregulation of FoxP1. *Blood* **2011**, *117*, 6227–6236.
95. Zhang, X.; Chen, X.; Lin, J.; Lwin, T.; Wright, G.; Moscinski, L.C.; Dalton, W.S.; Seto, E.; Wright, K.; Sotomayor, E.; *et al.* Myc represses miR-15a/miR-16–1 expression through recruitment of HDAC3 in mantle cell and other non-Hodgkin B-cell lymphomas. *Oncogene* **2012**, *31*, 3002–3008.
96. Bonci, D.; Coppola, V.; Musumeci, M.; Addario, A.; Giuffrida, R.; Memeo, L.; D'Urso, L.; Pagliuca, A.; Biffoni, M.; Labbaye, C.; *et al.* The miR-15a-miR-16-1 cluster controls prostate cancer by targeting multiple oncogenic activities. *Nat. Med.* **2008**, *14*, 1271–1277.
97. Zhang, X.; Zhao, X.; Fiskus, W.; Lin, J.; Lwin, T.; Rao, R.; Zhang, Y.; Chan, J.C.; Fu, K.; Marquez, V.E.; *et al.* Coordinated silencing of MYC-mediated miR-29 by HDAC3 and EZH2 as a therapeutic target of histone modification in aggressive B-Cell lymphomas. *Cancer Cell* **2012**, *22*, 506–523.
98. Schmitt, M.J.; Margue, C.; Behrmann, I.; Kreis, S. MiRNA-29: A microRNA family with tumor-suppressing and immune-modulating properties. *Curr. Mol. Med.* **2013**, *13*, 572–585.
99. Chin, L.J.; Ratner, E.; Leng, S.; Zhai, R.; Nallur, S.; Babar, I.; Muller, R.U.; Straka, E.; Su, L.; Burki, E.A.; *et al.* A SNP in a let-7 microRNA complementary site in the KRAS 3' untranslated region increases non-small cell lung cancer risk. *Cancer Res.* **2008**, *68*, 8535–8540.
100. Dang, C.V.; Le, A.; Gao, P. MYC-induced cancer cell energy metabolism and therapeutic opportunities. *Clin. Cancer Res.* **2009**, *15*, 6479–6483.
101. Jung, Y.J.; Kim, J.W.; Park, S.J.; Min, B.Y.; Jang, E.S.; Kim, N.Y.; Jeong, S.H.; Shin, C.M.; Lee, S.H.; Park, Y.S.; *et al.* c-Myc-mediated overexpression of miR-17-92 suppresses replication of hepatitis B virus in human hepatoma cells. *J. Med. Virol.* **2013**, *85*, 969–978.
102. Kim, J.W.; Mori, S.; Nevins, J.R. Myc-induced microRNAs integrate Myc-mediated cell proliferation and cell fate. *Cancer Res.* **2010**, *70*, 4820–4828.
103. Medina, R.; Zaidi, S.K.; Liu, C.G.; Stein, J.L.; van Wijnen, A.J.; Croce, C.M.; Stein, G.S. MicroRNAs 221 and 222 bypass quiescence and compromise cell survival. *Cancer Res.* **2008**, *68*, 2773–2780.
104. Visone, R.; Russo, L.; Pallante, P.; de Martino, I.; Ferraro, A.; Leone, V.; Borbone, E.; Petrocca, F.; Alder, H.; Croce, C.M.; *et al.* MicroRNAs (miR)-221 and miR-222, both overexpressed in human thyroid papillary carcinomas, regulate p27Kip1 protein levels and cell cycle. *Endocr. Relat. Cancer* **2007**, *14*, 791–798.

105. Fornari, F.; Gramantieri, L.; Ferracin, M.; Veronese, A.; Sabbioni, S.; Calin, G.A.; Grazi, G.L.; Giovannini, C.; Croce, C.M.; Bolondi, L.; *et al.* MiR-221 controls CDKN1C/p57 and CDKN1B/p27 expression in human hepatocellular carcinoma. *Oncogene* **2008**, *27*, 5651–5661.
106. Garofalo, M.; di Leva, G.; Romano, G.; Nuovo, G.; Suh, S.S.; Ngankeu, A.; Taccioli, C.; Pichiorri, F.; Alder, H.; Secchiero, P.; *et al.* miR-221&222 regulate TRAIL resistance and enhance tumorigenicity through PTEN and TIMP3 downregulation. *Cancer Cell* **2009**, *16*, 498–509.
107. Wang, X.; Zhao, X.; Gao, P.; Wu, M. c-Myc modulates microRNA processing via the transcriptional regulation of Drosha. *Sci. Rep.* **2013**, *3*, doi:10.1038/srep01942.
108. Goodrich, J.A.; Kugel, J.F. Non-coding-RNA regulators of RNA polymerase II transcription. *Nat. Rev. Mol. Cell Biol.* **2006**, *7*, 612–616.
109. Spizzo, R.; Almeida, M.I.; Colombatti, A.; Calin, G.A. Long non-coding RNAs and cancer: A new frontier of translational research? *Oncogene* **2012**, *31*, 4577–4587.
110. Harrow, J.; Frankish, A.; Gonzalez, J.M.; Tapanari, E.; Diekhans, M.; Kokocinski, F.; Aken, B.L.; Barrell, D.; Zadissa, A.; Searle, S.; *et al.* GENCODE: The reference human genome annotation for The ENCODE Project. *Genome Res.* **2012**, *22*, 1760–1774.
111. Derrien, T.; Johnson, R.; Bussotti, G.; Tanzer, A.; Djebali, S.; Tilgner, H.; Guernec, G.; Martin, D.; Merkel, A.; Knowles, D.G.; *et al.* The GENCODE v7 catalog of human long noncoding RNAs: Analysis of their gene structure, evolution, and expression. *Genome Res.* **2012**, *22*, 1775–1789.
112. Banfai, B.; Jia, H.; Khatun, J.; Wood, E.; Risk, B.; Gundling, W.E., Jr.; Kundaje, A.; Gunawardena, H.P.; Yu, Y.; Xie, L.; *et al.* Long noncoding RNAs are rarely translated in two human cell lines. *Genome Res.* **2012**, *22*, 1646–1657.
113. Brown, C.J.; Hendrich, B.D.; Rupert, J.L.; Lafreniere, R.G.; Xing, Y.; Lawrence, J.; Willard, H.F. The human XIST gene: Analysis of a 17 kb inactive X-specific RNA that contains conserved repeats and is highly localized within the nucleus. *Cell* **1992**, *71*, 527–542.
114. Clemson, C.M.; McNeil, J.A.; Willard, H.F.; Lawrence, J.B. XIST RNA paints the inactive X chromosome at interphase: Evidence for a novel RNA involved in nuclear/chromosome structure. *J. Cell Biol.* **1996**, *132*, 259–275.
115. Zhao, J.; Sun, B.K.; Erwin, J.A.; Song, J.J.; Lee, J.T. Polycomb proteins targeted by a short repeat RNA to the mouse X chromosome. *Science* **2008**, *322*, 750–756.
116. Tian, D.; Sun, S.; Lee, J.T. The long noncoding RNA, Jpx, is a molecular switch for X chromosome inactivation. *Cell* **2010**, *143*, 390–403.
117. Chureau, C.; Chantalat, S.; Romito, A.; Galvani, A.; Duret, L.; Avner, P.; Rougeulle, C. Ftx is a non-coding RNA which affects Xist expression and chromatin structure within the X-inactivation center region. *Hum. Mol. Genet.* **2011**, *20*, 705–718.
118. Lee, J.T.; Davidow, L.S.; Warshawsky, D. Tsix, a gene antisense to Xist at the X-inactivation centre. *Nat. Genet.* **1999**, *21*, 400–404.
119. Lee, J.T. Disruption of imprinted X inactivation by parent-of-origin effects at Tsix. *Cell* **2000**, *103*, 17–27.
120. Ogawa, Y.; Lee, J.T. Xite, X-inactivation intergenic transcription elements that regulate the probability of choice. *Mol. Cell* **2003**, *11*, 731–743.
121. Khalil, A.M.; Guttman, M.; Huarte, M.; Garber, M.; Raj, A.; Rivea Morales, D.; Thomas, K.; Presser, A.; Bernstein, B.E.; van Oudenaarden, A.; *et al.* Many human large intergenic

noncoding RNAs associate with chromatin-modifying complexes and affect gene expression. *Proc. Natl. Acad. Sci. USA* **2009**, *106*, 11667–11672.

122. Rinn, J.L.; Kertesz, M.; Wang, J.K.; Squazzo, S.L.; Xu, X.; Brugmann, S.A.; Goodnough, L.H.; Helms, J.A.; Farnham, P.J.; Segal, E.; *et al.* Functional demarcation of active and silent chromatin domains in human HOX loci by noncoding RNAs. *Cell* **2007**, *129*, 1311–1323.
123. Tsai, M.C.; Manor, O.; Wan, Y.; Mosammaparast, N.; Wang, J.K.; Lan, F.; Shi, Y.; Segal, E.; Chang, H.Y. Long noncoding RNA as modular scaffold of histone modification complexes. *Science* **2010**, *329*, 689–693.
124. Gupta, R.A.; Shah, N.; Wang, K.C.; Kim, J.; Horlings, H.M.; Wong, D.J.; Tsai, M.C.; Hung, T.; Argani, P.; Rinn, J.L.; *et al.* Long non-coding RNA HOTAIR reprograms chromatin state to promote cancer metastasis. *Nature* **2010**, *464*, 1071–1076.
125. Yang, Z.; Zhou, L.; Wu, L.M.; Lai, M.C.; Xie, H.Y.; Zhang, F.; Zheng, S.S. Overexpression of long non-coding RNA HOTAIR predicts tumor recurrence in hepatocellular carcinoma patients following liver transplantation. *Ann. Surg. Oncol.* **2011**, *18*, 1243–1250.
126. Geng, Y.J.; Xie, S.L.; Li, Q.; Ma, J.; Wang, G.Y. Large intervening non-coding RNA HOTAIR is associated with hepatocellular carcinoma progression. *J. Int. Med. Res.* **2011**, *39*, 2119–2128.
127. Kogo, R.; Shimamura, T.; Mimori, K.; Kawahara, K.; Imoto, S.; Sudo, T.; Tanaka, F.; Shibata, K.; Suzuki, A.; Komune, S.; *et al.* Long noncoding RNA HOTAIR regulates polycomb-dependent chromatin modification and is associated with poor prognosis in colorectal cancers. *Cancer Res.* **2011**, *71*, 6320–6326.
128. Kim, K.; Jutooru, I.; Chadalapaka, G.; Johnson, G.; Frank, J.; Burghardt, R.; Kim, S.; Safe, S. HOTAIR is a negative prognostic factor and exhibits pro-oncogenic activity in pancreatic cancer. *Oncogene* **2013**, *32*, 1616–1625.
129. Wang, K.C.; Yang, Y.W.; Liu, B.; Sanyal, A.; Corces-Zimmerman, R.; Chen, Y.; Lajoie, B.R.; Protacio, A.; Flynn, R.A.; Gupta, R.A.; *et al.* A long noncoding RNA maintains active chromatin to coordinate homeotic gene expression. *Nature* **2011**, *472*, 120–124.
130. Zhang, X.; Lian, Z.; Padden, C.; Gerstein, M.B.; Rozowsky, J.; Snyder, M.; Gingeras, T.R.; Kapranov, P.; Weissman, S.M.; Newburger, P.E. A myelopoiesis-associated regulatory intergenic noncoding RNA transcript within the human HOXA cluster. *Blood* **2009**, *113*, 2526–2534.
131. Ji, P.; Diederichs, S.; Wang, W.; Boing, S.; Metzger, R.; Schneider, P.M.; Tidow, N.; Brandt, B.; Buerger, H.; Bulk, E.; *et al.* MALAT-1, a novel noncoding RNA, and thymosin beta4 predict metastasis and survival in early-stage non-small cell lung cancer. *Oncogene* **2003**, *22*, 8031–8041.
132. Hutchinson, J.N.; Ensminger, A.W.; Clemson, C.M.; Lynch, C.R.; Lawrence, J.B.; Chess, A. A screen for nuclear transcripts identifies two linked noncoding RNAs associated with SC35 splicing domains. *BMC Genomics* **2007**, *8*, 39.
133. Tripathi, V.; Ellis, J.D.; Shen, Z.; Song, D.Y.; Pan, Q.; Watt, A.T.; Freier, S.M.; Bennett, C.F.; Sharma, A.; Bubulya, P.A.; *et al.* The nuclear-retained noncoding RNA MALAT1 regulates alternative splicing by modulating SR splicing factor phosphorylation. *Mol. Cell* **2010**, *39*, 925–938.
134. Miyagawa, R.; Tano, K.; Mizuno, R.; Nakamura, Y.; Ijiri, K.; Rakwal, R.; Shibato, J.; Masuo, Y.; Mayeda, A.; Hirose, T.; *et al.* Identification of cis- and trans-acting factors involved in the localization of MALAT-1 noncoding RNA to nuclear speckles. *RNA* **2012**, *18*, 738–751.

135. Gutschner, T.; Hammerle, M.; Eissmann, M.; Hsu, J.; Kim, Y.; Hung, G.; Revenko, A.; Arun, G.; Stentrup, M.; Gross, M.; *et al.* The noncoding RNA MALAT1 is a critical regulator of the metastasis phenotype of lung cancer cells. *Cancer Res.* **2013**, *73*, 1180–1189.
136. Yang, L.; Lin, C.; Liu, W.; Zhang, J.; Ohgi, K.A.; Grinstein, J.D.; Dorrestein, P.C.; Rosenfeld, M.G. ncRNA- and Pc2 methylation-dependent gene relocation between nuclear structures mediates gene activation programs. *Cell* **2011**, *147*, 773–788.
137. Guo, F.; Li, Y.; Liu, Y.; Wang, J.; Li, G. Inhibition of metastasis-associated lung adenocarcinoma transcript 1 in CaSki human cervical cancer cells suppresses cell proliferation and invasion. *Acta Biochim. Biophys. Sin. (Shanghai)* **2010**, *42*, 224–229.
138. Yamada, K.; Kano, J.; Tsunoda, H.; Yoshikawa, H.; Okubo, C.; Ishiyama, T.; Noguchi, M. Phenotypic characterization of endometrial stromal sarcoma of the uterus. *Cancer Sci.* **2006**, *97*, 106–112.
139. Wilusz, J.E.; Freier, S.M.; Spector, D.L. 3' end processing of a long nuclear-retained noncoding RNA yields a tRNA-like cytoplasmic RNA. *Cell* **2008**, *135*, 919–932.
140. Xu, C.; Yang, M.; Tian, J.; Wang, X.; Li, Z. MALAT-1: A long non-coding RNA and its important 3' end functional motif in colorectal cancer metastasis. *Int. J. Oncol.* **2011**, *39*, 169–175.
141. Wilusz, J.E.; JnBaptiste, C.K.; Lu, L.Y.; Kuhn, C.D.; Joshua-Tor, L.; Sharp, P.A. A triple helix stabilizes the 3' ends of long noncoding RNAs that lack poly(A) tails. *Genes Dev.* **2012**, *26*, 2392–2407.
142. Brown, J.A.; Valenstein, M.L.; Yario, T.A.; Tycowski, K.T.; Steitz, J.A. Formation of triple-helical structures by the 3'-end sequences of MALAT1 and MENbeta noncoding RNAs. *Proc. Natl. Acad. Sci. USA* **2012**, *109*, 19202–19207.
143. Pasmant, E.; Laurendeau, I.; Heron, D.; Vidaud, M.; Vidaud, D.; Bieche, I. Characterization of a germ-line deletion, including the entire INK4/ARF locus, in a melanoma-neural system tumor family: Identification of ANRIL, an antisense noncoding RNA whose expression coclusters with ARF. *Cancer Res.* **2007**, *67*, 3963–3969.
144. Yap, K.L.; Li, S.; Munoz-Cabello, A.M.; Raguz, S.; Zeng, L.; Mujtaba, S.; Gil, J.; Walsh, M.J.; Zhou, M.M. Molecular interplay of the noncoding RNA ANRIL and methylated histone H3 lysine 27 by polycomb CBX7 in transcriptional silencing of INK4a. *Mol. Cell* **2010**, *38*, 662–674.
145. Mourtada-Maarabouni, M.; Hedge, V.L.; Kirkham, L.; Farzaneh, F.; Williams, G.T. Growth arrest in human T-cells is controlled by the non-coding RNA growth-arrest-specific transcript 5 (GAS5). *J. Cell Sci.* **2008**, *121*, 939–946.
146. Kino, T.; Hurt, D.E.; Ichijo, T.; Nader, N.; Chrousos, G.P. Noncoding RNA gas5 is a growth arrest- and starvation-associated repressor of the glucocorticoid receptor. *Sci. Signal* **2010**, *3*, doi:10.1126/scisignal.2000568.
147. Conti, E.; Izaurralde, E. Nonsense-mediated mRNA decay: Molecular insights and mechanistic variations across species. *Curr. Opin. Cell Biol.* **2005**, *17*, 316–325.
148. Zhang, Z.; Zhu, Z.; Watabe, K.; Zhang, X.; Bai, C.; Xu, M.; Wu, F.; Mo, Y.Y. Negative regulation of lncRNA GAS5 by miR-21. *Cell Death Differ.* **2013**, doi:10.1038/cdd.2013.110.
149. Pan, X.; Wang, Z.X.; Wang, R. MicroRNA-21: A novel therapeutic target in human cancer. *Cancer Biol. Ther.* **2010**, *10*, 1224–1232.

150. Smedley, D.; Sidhar, S.; Birdsall, S.; Bennett, D.; Herlyn, M.; Cooper, C.; Shipley, J. Characterization of chromosome 1 abnormalities in malignant melanomas. *Genes Chromosomes Cancer* **2000**, *28*, 121–125.
151. Nupponen, N.N.; Carpten, J.D. Prostate cancer susceptibility genes: Many studies, many results, no answers. *Cancer Metastasis Rev*. **2001**, *20*, 155–164.
152. Stange, D.E.; Radlwimmer, B.; Schubert, F.; Traub, F.; Pich, A.; Toedt, G.; Mendrzyk, F.; Lehmann, U.; Eils, R.; Kreipe, H.; *et al*. High-resolution genomic profiling reveals association of chromosomal aberrations on 1q and 16p with histologic and genetic subgroups of invasive breast cancer. *Clin. Cancer Res*. **2006**, *12*, 345–352.
153. Morrison, L.E.; Jewell, S.S.; Usha, L.; Blondin, B.A.; Rao, R.D.; Tabesh, B.; Kemper, M.; Batus, M.; Coon, J.S. Effects of ERBB2 amplicon size and genomic alterations of chromosomes 1, 3, and 10 on patient response to trastuzumab in metastatic breast cancer. *Genes Chromosomes Cancer* **2007**, *46*, 397–405.
154. Peters, U.; Jiao, S.; Schumacher, F.R.; Hutter, C.M.; Aragaki, A.K.; Baron, J.A.; Berndt, S.I.; Bezieau, S.; Brenner, H.; Butterbach, K.; *et al.* Identification of genetic susceptibility loci for colorectal tumors in a genome-wide meta-analysis. *Gastroenterology* **2013**, *144*, 799–807.
155. Nakamura, Y.; Takahashi, N.; Kakegawa, E.; Yoshida, K.; Ito, Y.; Kayano, H.; Niitsu, N.; Jinnai, I.; Bessho, M. The GAS5 (growth arrest-specific transcript 5) gene fuses to BCL6 as a result of t(1;3)(q25;q27) in a patient with B-cell lymphoma. *Cancer Genet Cytogenet* **2008**, *182*, 144–149.
156. Tani, H.; Torimura, M.; Akimitsu, N. The RNA degradation pathway regulates the function of GAS5 a non-coding RNA in mammalian cells. *PLoS One* **2013**, *8*, e55684.
157. Smith, C.M.; Steitz, J.A. Classification of gas5 as a multi-small-nucleolar-RNA (snoRNA) host gene and a member of the 5'-terminal oligopyrimidine gene family reveals common features of snoRNA host genes. *Mol. Cell. Biol.* **1998**, *18*, 6897–6909.
158. Tanaka, R.; Satoh, H.; Moriyama, M.; Satoh, K.; Morishita, Y.; Yoshida, S.; Watanabe, T.; Nakamura, Y.; Mori, S. Intronic U50 small-nucleolar-RNA (snoRNA) host gene of no protein-coding potential is mapped at the chromosome breakpoint t(3;6)(q27;q15) of human B-cell lymphoma. *Genes Cells* **2000**, *5*, 277–287.
159. Raho, G.; Barone, V.; Rossi, D.; Philipson, L.; Sorrentino, V. The gas 5 gene shows four alternative splicing patterns without coding for a protein. *Gene* **2000**, *256*, 13–17.
160. Hirose, T.; Steitz, J.A. Position within the host intron is critical for efficient processing of box C/D snoRNAs in mammalian cells. *Proc. Natl. Acad. Sci. USA* **2001**, *98*, 12914–12919.
161. Johnsson, P.; Ackley, A.; Vidarsdottir, L.; Lui, W.O.; Corcoran, M.; Grander, D.; Morris, K.V. A pseudogene long-noncoding-RNA network regulates PTEN transcription and translation in human cells. *Nat. Struct. Mol. Biol.* **2013**, *20*, 440–446.
162. Bond, A.M.; Vangompel, M.J.; Sametsky, E.A.; Clark, M.F.; Savage, J.C.; Disterhoft, J.F.; Kohtz, J.D. Balanced gene regulation by an embryonic brain ncRNA is critical for adult hippocampal GABA circuitry. *Nat. Neurosci*. **2009**, *12*, 1020–1027.
163. Huarte, M.; Guttman, M.; Feldser, D.; Garber, M.; Koziol, M.J.; Kenzelmann-Broz, D.; Khalil, A.M.; Zuk, O.; Amit, I.; Rabani, M.; *et al*. A large intergenic noncoding RNA induced by p53 mediates global gene repression in the p53 response. *Cell* **2010**, *142*, 409–419.

164. Sheik Mohamed, J.; Gaughwin, P.M.; Lim, B.; Robson, P.; Lipovich, L. Conserved long noncoding RNAs transcriptionally regulated by Oct4 and Nanog modulate pluripotency in mouse embryonic stem cells. *RNA* **2010**, *16*, 324–337.
165. Ebert, M.S.; Neilson, J.R.; Sharp, P.A. MicroRNA sponges: Competitive inhibitors of small RNAs in mammalian cells. *Nat. Methods* **2007**, *4*, 721–726.
166. Pesole, G.; Liuni, S.; Grillo, G.; Saccone, C. Structural and compositional features of untranslated regions of eukaryotic mRNAs. *Gene* **1997**, *205*, 95–102.
167. Huse, J.T.; Brennan, C.; Hambardzumyan, D.; Wee, B.; Pena, J.; Rouhanifard, S.H.; Sohn-Lee, C.; le Sage, C.; Agami, R.; Tuschl, T.; *et al.* The PTEN-regulating microRNA miR-26a is amplified in high-grade glioma and facilitates gliomagenesis *in vivo*. *Genes Dev.* **2009**, *23*, 1327–1337.
168. Liu, Z.L.; Wang, H.; Liu, J.; Wang, Z.X. MicroRNA-21 (miR-21) expression promotes growth, metastasis, and chemo- or radioresistance in non-small cell lung cancer cells by targeting PTEN. *Mol. Cell. Biochem.* **2013**, *372*, 35–45.
169. Vanin, E.F. Processed pseudogenes: Characteristics and evolution. *Annu. Rev. Genet.* **1985**, *19*, 253–272.
170. Fujii, G.H.; Morimoto, A.M.; Berson, A.E.; Bolen, J.B. Transcriptional analysis of the PTEN/MMAC1 pseudogene, psiPTEN. *Oncogene* **1999**, *18*, 1765–1769.
171. Poliseno, L.; Salmena, L.; Zhang, J.; Carver, B.; Haveman, W.J.; Pandolfi, P.P. A coding-independent function of gene and pseudogene mRNAs regulates tumour biology. *Nature* **2010**, *465*, 1033–1038.
172. Lund, A.H.; van Lohuizen, M. Epigenetics and cancer. *Genes Dev.* **2004**, *18*, 2315–2335.
173. Karreth, F.A.; Tay, Y.; Perna, D.; Ala, U.; Tan, S.M.; Rust, A.G.; DeNicola, G.; Webster, K.A.; Weiss, D.; Perez-Mancera, P.A.; *et al. In vivo* identification of tumor- suppressive PTEN ceRNAs in an oncogenic BRAF-induced mouse model of melanoma. *Cell* **2011**, *147*, 382–395.
174. Faghihi, M.A.; Modarresi, F.; Khalil, A.M.; Wood, D.E.; Sahagan, B.G.; Morgan, T.E.; Finch, C.E.; St Laurent, G., 3rd.; Kenny, P.J.; Wahlestedt, C. Expression of a noncoding RNA is elevated in Alzheimer's disease and drives rapid feed-forward regulation of beta-secretase. *Nat. Med.* **2008**, *14*, 723–730.
175. Cesana, M.; Cacchiarelli, D.; Legnini, I.; Santini, T.; Sthandier, O.; Chinappi, M.; Tramontano, A.; Bozzoni, I. A long noncoding RNA controls muscle differentiation by functioning as a competing endogenous RNA. *Cell* **2011**, *147*, 358–369.
176. Nakagawa, S.; Ip, J.Y.; Shioi, G.; Tripathi, V.; Zong, X.; Hirose, T.; Prasanth, K.V. Malat1 is not an essential component of nuclear speckles in mice. *RNA* **2012**, *18*, 1487–1499.
177. Zhang, A.; Zhou, N.; Huang, J.; Liu, Q.; Fukuda, K.; Ma, D.; Lu, Z.; Bai, C.; Watabe, K.; Mo, Y.Y. The human long non-coding RNA-RoR is a p53 repressor in response to DNA damage. *Cell Res.* **2013**, *23*, 340–350.
178. Yoon, J.H.; Abdelmohsen, K.; Srikantan, S.; Yang, X.; Martindale, J.L.; De, S.; Huarte, M.; Zhan, M.; Becker, K.G.; Gorospe, M. LincRNA-p21 suppresses target mRNA translation. *Mol. Cell* **2012**, *47*, 648–655.
179. Azzalin, C.M.; Reichenbach, P.; Khoriauli, L.; Giulotto, E.; Lingner, J. Telomeric repeat containing RNA and RNA surveillance factors at mammalian chromosome ends. *Science* **2007**, *318*, 798–801.

180. Redon, S.; Reichenbach, P.; Lingner, J. The non-coding RNA TERRA is a natural ligand and direct inhibitor of human telomerase. *Nucleic Acids Res.* **2010**, *38*, 5797–5806.

181. Wang, X.; Arai, S.; Song, X.; Reichart, D.; Du, K.; Pascual, G.; Tempst, P.; Rosenfeld, M.G.; Glass, C.K.; Kurokawa, R. Induced ncRNAs allosterically modify RNA-binding proteins in cis to inhibit transcription. *Nature* **2008**, *454*, 126–130.

182. Jauliac, S.; Lopez-Rodriguez, C.; Shaw, L.M.; Brown, L.F.; Rao, A.; Toker, A. The role of NFAT transcription factors in integrin-mediated carcinoma invasion. *Nat. Cell Biol.* **2002**, *4*, 540–544.

183. Yoeli-Lerner, M.; Yiu, G.K.; Rabinovitz, I.; Erhardt, P.; Jauliac, S.; Toker, A. Akt blocks breast cancer cell motility and invasion through the transcription factor NFAT. *Mol. Cell* **2005**, *20*, 539–550.

184. Willingham, A.T.; Orth, A.P.; Batalov, S.; Peters, E.C.; Wen, B.G.; Aza-Blanc, P.; Hogenesch, J.B.; Schultz, P.G. A strategy for probing the function of noncoding RNAs finds a repressor of NFAT. *Science* **2005**, *309*, 1570–1573.

185. Sharma, S.; Findlay, G.M.; Bandukwala, H.S.; Oberdoerffer, S.; Baust, B.; Li, Z.; Schmidt, V.; Hogan, P.G.; Sacks, D.B.; Rao, A. Dephosphorylation of the nuclear factor of activated T cells (NFAT) transcription factor is regulated by an RNA-protein scaffold complex. *Proc. Natl. Acad. Sci. USA* **2011**, *108*, 11381–11386.

186. Sanger, H.L.; Klotz, G.; Riesner, D.; Gross, H.J.; Kleinschmidt, A.K. Viroids are single-stranded covalently closed circular RNA molecules existing as highly base-paired rod-like structures. *Proc. Natl. Acad. Sci. USA* **1976**, *73*, 3852–3856.

187. Jeck, W.R.; Sorrentino, J.A.; Wang, K.; Slevin, M.K.; Burd, C.E.; Liu, J.; Marzluff, W.F.; Sharpless, N.E. Circular RNAs are abundant, conserved, and associated with ALU repeats. *RNA* **2013**, *19*, 141–157.

188. Salzman, J.; Gawad, C.; Wang, P.L.; Lacayo, N.; Brown, P.O. Circular RNAs are the predominant transcript isoform from hundreds of human genes in diverse cell types. *PLoS One* **2012**, *7*, e30733.

189. Capel, B.; Swain, A.; Nicolis, S.; Hacker, A.; Walter, M.; Koopman, P.; Goodfellow, P.; Lovell-Badge, R. Circular transcripts of the testis-determining gene Sry in adult mouse testis. *Cell* **1993**, *73*, 1019–1030.

190. Hansen, T.B.; Jensen, T.I.; Clausen, B.H.; Bramsen, J.B.; Finsen, B.; Damgaard, C.K.; Kjems, J. Natural RNA circles function as efficient microRNA sponges. *Nature* **2013**, *495*, 384–388.

191. Hansen, T.B.; Wiklund, E.D.; Bramsen, J.B.; Villadsen, S.B.; Statham, A.L.; Clark, S.J.; Kjems, J. miRNA-dependent gene silencing involving Ago2-mediated cleavage of a circular antisense RNA. *EMBO J.* **2011**, *30*, 4414–4422.

9

Dynamic Nature of Non-Coding RNA Regulation of Adaptive Immune Response

Graziella Curtale [1,2,*] and Franca Citarella [3]

[1] Department of Medical Biotechnologies and Translational Medicine, University of Milan, Rozzano 20089, Italy

[2] Humanitas Clinical and Research Center, Rozzano 20089, Italy

[3] Department of Cellular Biotechnology and Hematology, Sapienza University of Rome, Rome 00185, Italy; E-Mail: citarella@bce.uniroma1.it

* Author to whom correspondence should be addressed;
E-Mail: graziella.curtale@humanitasresearch.it;

Abstract: Immune response plays a fundamental role in protecting the organism from infections; however, dysregulation often occurs and can be detrimental for the organism, leading to a variety of immune-mediated diseases. Recently our understanding of the molecular and cellular networks regulating the immune response, and, in particular, adaptive immunity, has improved dramatically. For many years, much of the focus has been on the study of protein regulators; nevertheless, recent evidence points to a fundamental role for specific classes of noncoding RNAs (ncRNAs) in regulating development, activation and homeostasis of the immune system. Although microRNAs (miRNAs) are the most comprehensive and well-studied, a number of reports suggest the exciting possibility that long ncRNAs (lncRNAs) could mediate host response and immune function. Finally, evidence is also accumulating that suggests a role for miRNAs and other small ncRNAs in autocrine, paracrine and exocrine signaling events, thus highlighting an elaborate network of regulatory interactions mediated by different classes of ncRNAs during immune response. This review will explore the multifaceted roles of ncRNAs in the adaptive immune response. In particular, we will focus on the well-established role of miRNAs and on the emerging role of lncRNAs and circulating ncRNAs, which all make indispensable contributions to the understanding of the multilayered modulation of the adaptive immune response.

Keywords: noncoding RNAs; adaptive immunity; lymphocytes; epigenetic

1. Introduction

Adaptive immune response, generated by clonal selection of antigen-specific lymphocytes, provides an extremely versatile mechanism of host organism defense and increases its protection against subsequent exposure to the same antigen and/or reinfection with the same pathogen. Cells involved in the adaptive immune response are characterized by temporal development in primary lymphoid organs and functional differentiation in secondary lymphoid organs, phenomena whose successful regulation relies on a multilayered system of control [1,2]. In fact, it is now clear that the process of transcriptional regulation, carried out by general and specific transcription factors, must be framed in a picture in which the epigenetic control and the post-transcriptional control, which are also the result of non-protein coding RNA (ncRNA) activities, contribute in a fundamental manner. A growing number of observations indicate that different genome-encoded functional RNA species orchestrate the development and the differentiation state of complex tissues and specific cell types [3,4]. With regard to the adaptive immune response, while the regulative roles of several microRNAs (miRNAs) are well-established [5], recent, yet emerging, evidence demonstrates the relevance of a new heterogeneous repertoire of functional noncoding RNAs [6], which can be collectively referred to as long noncoding RNAs (lncRNAs).

The first and most generic association of miRNAs with the immune system was due to the detection of different miRNA expression profiles either in normal lymphocytes or lymphocytes derived from patients affected by immunological malignancies [7]. As a logical consequence of the identification of specific miRNA expression signatures, the next step has been the investigation of a possible causal relationship between miRNAs and the disease. Despite several *in vitro* studies identifying specific miRNA target genes involved in different aspects of cell biology, the first strong genetic evidence of the pivotal role of miRNAs in immune biology came from the analysis of mice knock-out for genes that are involved in the biogenesis of mature miRNAs, namely Dicer and Drosha. These studies established that miRNAs, as a whole, control the development and the homeostasis of the immune system [8–10]. Successively, it was established that the expression of several miRNAs is restricted to hematopoietic cells and is associated with specific developmental stages or functional differentiation of lymphocytes [11]. Moreover, it has been shown that miRNA concentration is dynamically modulated during the immune response, due to the existence of feed-back loops in which miRNAs control their own expression and, in so doing, contribute to the homeostasis of the immune system and the maintenance of cell identity [12,13]. In this context, uncovering the regulation of miRNA expression in the immune system became a major issue.

Genes coding for miRNAs are mostly transcribed by RNA polymerase II and, therefore, are subjected to the regulation of transcription factors, whose repertoire is different in different cells and whose activation can be regulated by extracellular stimuli. This, then, makes way for the transcriptional regulation of miRNAs by co-activators and co-repressors, including chromatin remodeling factors and other epigenetic factors. In addition, recent studies indicate that the expression of mature miRNAs in the cell can be controlled at a post-transcriptional level through several mechanisms, such as processing, subcellular localization and decoy. The latter mechanisms call into question lncRNA functions.

Concerning lncRNAs, global genome profiling studies uncovered that the expression of many of them is lymphocyte-specific and change dynamically during T-cell differentiation, suggesting that, also, these functional RNAs may contribute to the development and homeostasis of the immune system [14,15]. This new evidence, together with the observation that cells are able to exchange information not only via hormones, but also via genetic material, is dictating the need to build a model of multilayered control of the development and the differentiation of the immune system cells, as well as of the regulation of the immune response. Perturbations in any component of this regulatory system will be associated with the onset of immune diseases.

This review will explore the proposed multifaceted roles of these classes of ncRNAs in adaptive immunity.

2. MicroRNAs

Over the past decades, it has been estimated that only 2% of the mammalian genome encodes for proteins, whereas the vast majority of it is extensively transcribed to give rise to a large transcriptome of ncRNA species [16]. This pervasive transcription, which has been recently unraveled, thanks to the new improved RNA-sequencing technologies, challenges our traditional understanding of RNAs as simple intermediates between genes and proteins. Extensive studies demonstrated that ncRNAs are functional RNA species, involved in the modulation of primary biological functions, and miRNAs, that are, to date, the most characterized class of small ncRNAs, have been described as fundamental to drive the development and differentiation of immune cells [17–20] .

miRNAs are small, single-stranded endogenous ncRNAs, of about 22 nucleotides (nt), many of which are highly conserved through evolution. They are encoded by genes that produce transcripts for either single or multiple miRNAs or by intronic sequences of protein coding genes. Their expression is highly regulated and can be tissue- and time-specific. Following transcription, the primary RNA transcripts (pri-miRNAs), which form a distinctive hairpin structure, are processed by a complex formed by ribonuclease III-type Drosha and DiGeorge syndrome critical region gene 8 (Drosha/DGCR8) and, then, exported from the nucleus into the cytoplasm as pre-miRNA [21]. In the cytoplasm, pre-miRNAs are recognized and cleaved by another endoribonuclease, Dicer, which generates a 22 nt double-stranded miRNA. The double stranded miRNA is unwound, and one strand, the guide strand, is loaded into the RNA-induced silencing complex (RISC). Within the RISC, the mature miRNA interacts with Argonaute (Ago) proteins, driving Ago and other associated factors to partially complementary target sites, mainly located in the 3' UTR (3' untranslated region) of the target messenger RNAs (mRNAs) [21–23]. The resulting effect of miRNA-target mRNA interaction inside the RISC complex is the translational inhibition, degradation or deadenylation of the target mRNAs [24,25] (Figure 1).

Figure 1. Key steps in microRNA (miRNA) biogenesis and activity. miRNAs originate from the nucleus as pri-miRNA precursor molecules, organized as single transcriptional units or as a cluster of miRNAs, co-transcribed as a polycistronic transcript. They are processed by the RNAse III-type enzyme, Drosha, in association with the RNA-binding protein, DGCR8, into smaller precursor miRNAs (pre-miRNAs), then exported to cytoplasm, where they are cleaved by Dicer to their mature form of 22 nt double stranded miRNA. The guide strand of the mature miRNA is incorporated into the miRNA-induced silencing complex (miRISC), where it binds to target mRNA by partial complementarity with its 3'UTR. This results in translational inhibition, mRNA degradation or mRNA deadenylation of the recognized miRNA target. Ago, Argonaute.

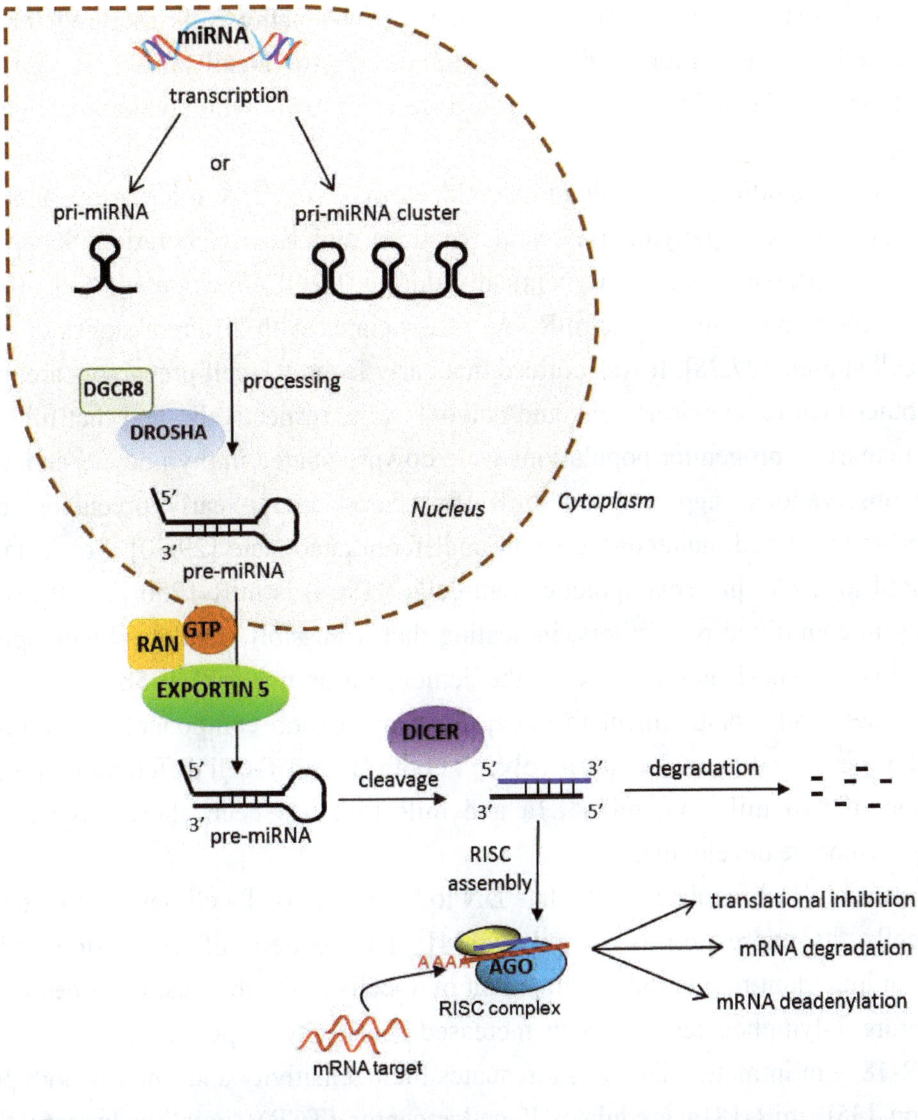

MiRNAs have been demonstrated to affect the expression of numerous mRNAs within a cell, as well mRNAs can be targets of different miRNAs that act cooperatively. Although the effect of miRNAs on single targets can be relatively modest, the coordinated effect on multiple components of the same pathway adds strength and dynamicity to miRNA-mediated regulation of a biological process.

From the perspective of adaptive immunity, miRNAs have been extensively studied both in physiological and pathological contexts; they have been shown to regulate cell development and differentiation decisions playing a role in lineage commitment and in maintaining lineage stability. These effects are achieved by regulating either, positively or negatively, several signaling pathways.

2.1. miRNAs in T-Cell Development

Lymphoid progenitors developed from hematopoietic stem cells in the bone marrow migrate to the thymus to complete their antigen-independent maturation into functional T-cells. Thymic microenvironment directs T-lymphocyte maturation, characterized by the expression of well-defined cell-surface markers: CD4 and CD8, with thymocytes first starting as $CD4^{-}CD8^{-}$ double negative (DN), then becoming $CD4^{+}CD8^{+}$ double positive (DP) and, lastly, maturing into single positive (SP) $CD4^{+}$ or $CD8^{+}$ T-cells [26].

T-cell-specific deletions of Dicer revealed that the miRNA pathway is required for the normal development of T-cells, as well as for the differentiation of effector T-cell subsets. Indeed, conditional deletions of Dicer in the T-cell lineage resulted in a decrease of T-lymphocytes in thymus and the periphery [8,9,13].

Expression profiling studies, using oligonucleotide arrays (miRNA microarray analysis), small RNA cloning, real-time PCR (polymerase chain reaction) and Next-Generation Sequencing, have shown that miRNA expression changes dynamically during T-cell development and differentiation, implying that the expression of individual miRNAs is associated with different stages of development and different T-cell subsets [27,28]. It was noticed that early B- and T-cell precursors are more closely related to each other than to splenic B-cells and naive T-cells, respectively, and that miRNAs highly expressed in bone marrow progenitor populations were downregulated in thymocytes and other mature lineages. These observations suggested that miRNAs shared among early precursors conceivably regulate their self-renewal and maintenance of an undifferentiated state [29,30]. For instance, one of the most expressed miRNAs in hematopoietic stem cells (HSCs) is miR-125b [29,30], whose levels drop significantly in committed progenitors, indicating that it possibly regulates hematopoiesis at the stem-cell level. This hypothesis is supported by the demonstration that miR-125b increases the size of the hematopoietic stem cell compartment in mice, targeting several component of the pro-apoptotic pathway, as well as genes coding molecules involved in both B- and T-cell differentiation [29,31–33].

The expression of two miRNAs, miR-181a and miR-150, has been shown to be dynamically regulated during thymocyte development.

MiR-181a is transiently upregulated at the late DN to DP stages of T-cell development, while in SP and mature T-cells, its expression is very low [34]. This pattern of expression indicates that miR-181a plays an important role in the development of T-cells in the thymus. Enforced expression of miR-181a in mature T-lymphocytes results in increased sensitivity to peptide antigens, whereas the inhibition of miR-181a in immature DP cells attenuates their sensitivity and impairs both positive and negative selection [35]. miR-181a modulates T-cell receptor (TCR) signaling by targeting several phosphatases that restrain ERK activity. In light of this evidence, it has been proposed that miR-181a may function as an intrinsic antigen-sensitivity "rheostat", operating during T-cell development, able to quantitatively modulate T-cell sensitivity to antigen, setting a TCR signaling threshold for proper agonist selection [35]. Along this line, expression of miR-181a is altered during autoimmune disease and significantly downregulated in pediatric Systemic Lupus Erythematosus (SLE) patients [36].

Concerning miR-150, it is expressed at high levels in mature naive B- and T-cells and strongly downregulated in their precursors and upon activation. This leads to the conclusion that a common set of miRNAs could be employed in B- and T lineages to regulate similar effector functions, such as tissue homing and cytokine production [37–40]. In particular, miR-150 is strongly upregulated during human T-cell maturation; its expression is low in DN T-cells, increases in DP and $CD8^+$ cells and is higher in $CD4^+$ cells [20]. miR-150 upregulation in this context could be important to downregulate genes associated with an immature phenotype, such as NOTCH 3, which has been identified as a target of miR-150 [20]. The observation that downregulation of miR-150 confers a growth advantage to transformed lymphocytes corroborated the hypothesis that one of the functions of miR-150 in lymphocyte development is to hamper proliferation during specific stages of maturation to allow for differentiation. Importantly, a proper immune response also requires downregulation of miR-150 after activation of mature B- and T-cells in order to release them from growth arrest and permit rapid proliferation to occur [37].

It is worth mentioning that, besides changes in the expression of specific miRNAs, the dynamic regulation of miRNA expression relies also on precursor processing, which appears to vary during T-cell development. Kirigin *et al.* have reported a significant degree of polymorphisms in mature miRNA ends at the late DN to DP stages, suggesting that a stage-specific processing of miRNA precursors may concurrently affect target recognition, thus contributing to proteome changes associated with T-cell development [29].

2.2. miRNAs in T-Cell Function

Antigen-specific naive T-cells, upon activation, proliferate and differentiate into effector T-cells, specialized in either cytotoxic function or cytokine production. Upon antigen recognition, naive $CD4^+$ T-cells proliferate and differentiate into distinct types of memory and effector T-cell subsets, that home to different tissue and produce specific types of cytokines. Among $CD4^+$ effector T-cells, T-helper 1 (Th1) cells are critical for host defense against intracellular pathogens and secrete interferon-γ (IFN-γ), whereas T-helper 2 (Th2) cells control parasitic infections and secrete interleukin 4 (IL-4), IL-5 and IL-13, important for mucosal barrier function, as well as for the induction of B-cell proliferation. T-helper 17 (Th-17) cells promote the response against extracellular pathogens by producing IL-17 and are also involved in autoimmunity. An additional effector population, more recently described, includes Th9 and Th22 cells, respectively, producing IL-9 and IL-22, whose *in vivo* role has not been clearly elucidated, yet [41]. It has been noted that during T-cell differentiation, the global level of miRNA expression inversely correlates with the activation status of the cells being highest in quiescent naive and memory T-cells and sharply decreasing in proliferating and functionally differentiated effector cells. This occurrence suggests that miRNAs may play a general role in stabilizing the expression of genes involved in the maintenance of a naive state. Accordingly, it has been recently observed that Ago proteins are post-transcriptionally downregulated by ubiquitination upon T-cell activation [42]. Ago proteins' downregulation, together with miRISC-accelerated turnover, could contribute to a rapid reprogramming of miRNA repertoire in favor of miRNAs whose expression is induced during T-cell activation, such as miR-182 [43].

The human $CD4^+$ T lymphocytes-specific miRNA signature changes during differentiation from naive to memory cells. In particular, miR-125b, which is highly expressed in human naive $CD4^+$ T-cells and is downregulated during differentiation toward effector T-cells [44], may contribute to maintenance of the naive state of primary human T-cells. In human, miR-125b directly regulates the expression of genes encoding molecules required for the differentiation of different effector T-lymphocytes, such as IFN-γ, peculiar to the Th1 subset, IL-2 receptor-β (IL2RB), which contributes to memory cell identity, IL-10 receptor-α (IL10RA) and the transcriptional repressor, Blimp-1 (PRDM19), which promote T-cell terminal differentiation. Thereby, activation-induced miR-125b down-modulation is associated with increased expression of these genes and acquisition of an effector memory phenotype by $CD4^+$ T-cells [44].

A key feature that characterizes the transition from a naive to an activated status of lymphocytes is the active proliferation, triggered by antigen recognition, which leads to the activation of the TCR signaling pathway. Once T-helper (Th) lymphocytes are induced to proliferate, their clonal expansion becomes independent of further antigen stimulation and is fostered by autocrine or paracrine IL-2 production. In this scenario, miR-182, whose expression is induced by IL-2, has been described as an important modulator of Th cell population, by promoting clonal expansion through post-transcriptional inhibition of forkhead box O1 (Foxo1), a suppressor of proliferation expressed in resting Th lymphocytes [42]. This finding is consistent with the present assumption that miRNAs induced during T-cell activation may contribute to downregulating the expression of genes involved in the maintenance of the naive state to allow for effector cell proliferation and differentiation. Moreover, it has been shown that activated proliferating T-cells express mRNAs with shortened 3' UTR regions and, consequently, fewer miRNA target sites, thus exploiting another mechanism to regulate immune system homeostasis [45].

Among the other miRNAs whose expression is enhanced upon T-cell activation, we will discuss miR-155, miR-146a and cluster miR-17-92 (Table 1).

Table 1. miRNAs in adaptive immunity. HSC, hematopoietic stem cell; DN, double negative; DP, double positive; SP, single positive.

	MicroRNA	Expression	Targets	Function	Related alterations	References
T-cell development	miR-125b	↑ in HSC	Bak	Regulation of HSC compartment size	HSC exhaustion	[29–33]
	miR-181	↑ in DN and DP cells; ↓ in SP and mature T-cells	Dusp5, Dusp6, Shp2, Ptpn22	B-cell lineage differentiation	↑ in SLE	[34–36]
	miR-150	↑ DP CD8+ cells ↓ DN CD8+ cells	Notch3	T-cell development	↓in T-CLL	[20]

Table 1. *Cont.*

	MicroRNA	Expression	Targets	Function	Related alterations	References
B-cell development	miR-17-92	↑ in progenitor B-cells; ↓ in mature B-cells	Pten, Bim	Defective central memory development	↑ B-cell lymphomas	[46,47]
	miR-150	↑ resting B-cells	cMyb	Impaired B1 cell maturation and Ab response	↓ B-cell lymphomas	[37,38]
	miR-34a	↓ pro-B lymphocytes	FoxP1	Required for pro-B to pre-B-cell transition	↑ B-cell lymphomas	[48]
	miR-155	↑ activated mature B-cells	Ship, C/EbpB, Hdac4	Impaired germinal center B-cell response	↑ diffuse large B-cell lymphoma	[13,49,50]
	miR15a-16	↑ CD5+ B-cells	Bcl2	Defect in apoptosis	↓ in B-CLL	[51]
T-cell function	miR-125b	↑ naïve CD4+ T-cells	Ifng, IL10Ra, IL2Rb, Prdm1	Differentiation of effector T-cells	↓ in SLE	[44]
	miR-182	↑ activated T-cells	Foxo1	Altered T-cell induced inflammation	Altered Treg mediated control of Th2 response	[43]
	miR155	↑ activated CD4+ T-cells	cMaf	Th lineage decisions: KO mice shows ↑ Th2 cells and ↓ Th1 and Th17 cells	↑ in SLE and RA	[13,52]
	miR-146a	↑ CD4+ and CD8+ memory cells	Irak1, Traf6, Fadd	Modulation of IL-2 production and AICD	↑ in RA and ↓ in SLE	[18,52–55]
	miR-17-92	↑ CD4+ and CD8+ memory cells	Pten, Bim	Defect in apoptosis	Lymphoproliferation	[46,56]
Treg cells	miR-10a	↑ Treg	Bcl6, Ncor2	Altered Th17 differentiation	Breakdown of peripheral tolerance	[57]
	miR-155	↑ Treg	Socs1	Altered Treg proliferation and homeostasis	Defect in Treg cell-mediated tolerance	[58–60]
	miR-146a	↑ Treg	Stat1	Altered Treg number, compromised suppressor activity	Autoimmunity due to altered Treg-mediated control of Th1 response	[61]

miR-155, which maps within an exon of the noncoding RNA, Bic [13], has been proven to regulate several aspects of the immune system [62]. miR-155 expression increases upon T-cell activation and controls the differentiation of $CD4^+$ T-cells into different Th cell subsets [11,13], as well as the development of regulatory T (Treg)-cells [58,59]. In miR-155 deficient mice, $CD4^+$ T-cells preferentially differentiate toward the Th2 subtype. Accordingly, *in vivo* expression data from miR-155-deficient mice and *in vitro* reporter assays indicate that musculoaponeurotic fibrosarcoma oncogene homolog (c-Maf), a potent trans-activator of the IL-4 gene promoter, is a target of miR-155 [13]. Therefore, miR-155 may promote differentiation toward a Th1 phenotype by limiting the expression of IL-4, a cytokine whose expression characterizes the Th2 phenotype.

In addition to its function in innate immunity [12,17,63,64], miR-146a is also known to be implicated in the adaptive immune response. In mice, miR-146a expression is higher in Th1 cells and lower in Th2 cells when compared to naive T-cells, suggesting that miR-146a may be involved in the lineage differentiation of T-lymphocytes [11]. In human T-cells, miR-146a is induced upon TCR stimulation and is highly expressed in central memory T-cells [18]. Consistent with this expression profile, miR-146a has been shown to modulate activation-induced cell death (AICD) by targeting the Fas-associated death domain (FADD) and to impair both activator protein 1 (AP-1) activity and interleukin-2 (IL-2) production induced by TCR engagement [18]. Moreover, the observation of increased miR-146a levels in both synovial tissues and in peripheral blood mononuclear cells (PBMCs) of rheumatoid arthritis patients suggest that miR-146a may promote the survival of self-reactive T-cells in autoimmune diseases [52–55]. Finally, in miR-146a-deficient mice, it has been demonstrated that miR-146a contributes to proper resolution of T-cell response, being a part of the negative feedback loop that modulates TCR signaling to the nuclear factor κ-light chain enhancer of activated B-cells (NF-κB). Indeed, NF-κB, a transcription factor activated upon TCR stimulation, induces the expression of miR-146a, which, in turn, downregulates NF-κB activation through repression of NF-κB activators, TNF receptor associated factor 6 (TRAF6) and interleukin-1 receptor-associated kinase 1 (IRAK1) [12,64]. NF-κB binds to the human miR-17-92 promoter, indicating that, also, the miR-17-92 cluster may be one of the intracellular signaling components that promotes proliferation and effector differentiation in response to antigen stimulation. miR-17-92 expression is induced upon T-cell activation during viral infection, while it is downregulated after clonal expansion and during memory cell differentiation [56]. Notably, failure in miR-17-92 downregulation leads to defective central memory cell development [56]. In mice, ectopic expression of the miR-17-92 cluster in the lymphocyte compartment induces a severe lymphoproliferative disease characterized by an expansion of almost all lymphocyte populations, especially the $CD4^+$ T subset [46]. Moreover, B- and T-lymphocytes from transgenic mice show increased proliferation and survival after activation *in vitro* [46]. The phenotype associated with miR17-92 cluster overexpression can be, at least in part, the result of the downregulation of two direct targets of the miR17-92 cluster: bcl-2 interacting protein (Bim), a pro-apoptotic protein belonging to the Bcl2 family, and the tumor suppressor phosphatase and tensin homolog (PTEN).

Studies comparing naive, effector and memory $CD8^+$ T-cells show that a small set of miRNAs is downregulated in effector T-cells compared to naive cells, but also that their expression tends to come back in memory T-cells [52]. Activated T-cells committed to a central memory fate activate a program that enables the acquisition of a central memory phenotype through downregulation of miR-155 and upregulation of miR-150, whose expression is suppressed upon T-cell activation (Table 1). This opposite regulation of miR-150 and miR-155 with respect to the activation state appears coherent, as an effort of the emerging central memory T-cells to distance themselves from the activated state and acquire a maintenance state [39].

2.3. miRNAs in Treg Cells

Treg cells produce inhibitory cytokines, such as IL-10 and transforming growth factor-β (TGF-β), which help to limit immune responses and prevent autoimmunity by suppressing T-lymphocyte activity. Treg cells are critical for the maintenance of immune cell homeostasis, as evidenced by the lethal condition consequent to their ablation in mice [65]. miRNAs have been established to be essential for Treg cell function: indeed, conditional deletion of Drosha or Dicer in forkhead box protein $P3^+$ ($Foxp3^+$) Treg cells induces a fatal, early-onset autoimmune pathology indistinguishable from that observed in Foxp3-knockout mice devoid of Treg cells [10,66,67]. The loss of suppressor function observed in Dicer-deficient Treg cells is most likely due to the deficiency of miRNAs that are over-represented in these cells. Among them, miR-10a, miR-155 and miR-146a have been shown to contribute to distinct aspects of Treg cells homeostasis and function (Table 1). miRNAs confer robustness to differentiation processes by tuning gene expression networks; therefore, they may help to preserve certain phenotypes by targeting transcriptional factor pathways that promote alternative fates. An interesting recent study [57] has shown that miR-10a contributes to the stability of the Treg phenotype and limits the conversion of Treg cells into T-follicular helper (TFH) by targeting the transcriptional repressor B-cell leukemia/lymphoma 6 (Bcl-6). miR-10a is expressed at high levels in naturally occurring Treg (nTreg) cells and is induced by TGF-β and retinoic acid (RA) during inducible Treg (iTreg) cell differentiation from $CD4^+$ T-cells. Actually, TGF-β induces retinoic acid receptor α (RARα) expression, which, upon stimulation by RA, activates miR-10a expression. miR-10a directly targets not only Bcl-6, a master positive regulator of TFH differentiation, but also nuclear receptor corepressor 2 (Ncor2), a corepressor of RARα and, in so doing, activates a positive feedback loop that amplifies its own induction by RA. On the other hand, the miR-10a-induced downregulation of Bcl-6 [57], a negative regulator of T-cell-specific T-box transcription factor (Tbet), elicits higher levels of Tbet, which is known to inhibit Th17 differentiation. Therefore, miR-10a modulates T helper cell stability by restraining, either directly or indirectly, multiple differentiation pathways.

miR-155 not only favors T-cell differentiation toward a Th1 phenotype, but also influences Treg cell homeostasis. miR-155-deficient mice have reduced numbers of Treg cells, both in the thymus and periphery; however, miR-155-deficient Treg cells maintain their suppressor activity [60]. Remarkably, Foxp3, the key transcription factor controlling Treg cell development and function, upregulates the expression of miR-155, which, in turn, positively regulates Treg cell differentiation by targeting the suppressor of cytokine signaling 1 (Socs1) [60]. Accordingly, T-cell-specific deletion of Socs1 results in an increase in the proportion and absolute numbers of Treg cells in the thymus without affecting their suppressive function [59,60,68]. Therefore, miR-155 stabilizes the transcriptional program induced by Foxp3 by repressing genes whose expression counteracts Treg cell differentiation. In contrast to miR-155-deficient mice that exhibited a decreased number of Treg cells in thymus and in the periphery, miR-146a-deficient mice showed an increased number of Treg cells in the periphery; however, miR-146a-deficient Treg cell suppressor activity was seriously compromised [61]. Fatal immune-mediated lesions observed in mice in the presence of miR-146a-deficient Treg cells were accompanied by sharply augmented Th1 responses and were dependent upon increased amounts of IFNγ. Li-Fan Lu *et al.* demonstrated that miR-146a ensures Treg cell-mediated regulation of Th1 responses, in a significant way, through targeting signal transducer and activator of transcription (STAT1), a transcription factor that regulates IFNγ expression [61].

2.4. miRNAs in B-Cells

In the adaptive immune response, B-cells are responsible for antibody-mediated response. B-cell development begins in primary lymphoid tissue, with subsequent functional maturation in secondary lymphoid tissue. It is a continuum of stages defined by the expression and re-arrangement of functional B-cell receptor (BCR)/immunoglobulin (Ig) genes [2]. The first developmental stage exhibiting commitment to the B-cell lineage is called pro-B and is characterized by rearrangement of the Ig heavy chain. In the pro-B stage, Iga (CD79a) and Igβ (CD79b) are expressed at the cell surface. Rearrangements at the immunoglobulin locus result in the generation and surface expression of the pre-B-cell-receptor (pre-BCR), composed of an Igμ heavy chain and surrogate light chains and, finally, a mature BCR, capable of binding antigen [2]. Mature B-cells that move into the periphery can be activated by antigen, undergo clonal expansion and become an antibody-secreting plasma cell or a memory B-cell, which will respond more quickly to a second exposure to antigen. Unlike conventional B-cells (B-2 cells), which originate from adult bone marrow progenitors, another population of B-lymphocytes, called B-1, appear during fetal life and express surface IgM, but little or no IgD. In mouse, B1 cells can be further subdivided into B-1a ($CD5^+$) and B-1b ($CD5^-$) subtypes [69]. B1 lymphocytes do not undergo much class switch, thus producing antibodies with low affinity and having been implicated in the pathogenesis of autoimmune diseases and many chronic leukemias.

Conditional ablation of Dicer or Ago2 in early B-cell progenitors results in an almost complete block of B-cell development at the pro-B to pre-B transition, which underscores the relevance of miRNA function in B-cell development [70]. Moreover, Dicer-deficient B-cells show increased expression of terminal deoxynucleotidyl transferase (TDT) and increased antibody diversification [47,70,71]. Several specific miRNAs have been proven to control B-cell development, which is impaired in Dicer-deficient mice (Table 1). The miR-17-92 cluster is highly expressed in progenitor B-cells, and its expression decreases during B-cell maturation, suggesting that it is a positive regulator of B-cell differentiation [47]. In miR-17-92-deficient mice, transition from pro-B-cells to pre-B-cells is compromised, and increased apoptosis occurs, which correlates with higher expression of the pro-apoptotic protein, Bim, a target of the miR-17-92 cluster [46]. Accordingly, overexpression of the miR-17-92 cluster, often detected in human B-cell lymphomas, possibly facilitates transformation by inappropriate repression of the same pro-apoptotic gene, Bim, and the tumor suppressor, PTEN [56]. Another positive regulator of B-cell differentiation is miR-181a, whose ectopic expression in hematopoietic stem cells leads to an increased production of B-lineage cells [72]. Conversely, the overexpression of miR-150 in hematopoietic progenitor/stem cells significantly reduces the mature B-cell population by targeting c-Myb, a transcription factor critically important in B-cell development. MiR-150 is selectively expressed in mature, resting B- and T-cells, but not in their progenitors [9,38]. The expression pattern of c-Myb in B-cells inversely correlates with that of miR-150: it is highly expressed in progenitor cells and downregulated in mature B-lymphocytes. Premature downregulation of c-Myb by ectopic miR-150 expression triggers apoptosis during the pro-B stage, determining the observed phenotype. These findings substantiate the view that higher levels of miR-150 are required during the pre-B to mature B-cell transition to downregulate c-Myb expression and guarantee normal B-cell development [37,38].

The constitutive expression of miR-34a, whose expression is downregulated at the pro-B to pre-B transition, determines a block of the early B-cell development with an accumulation of pro-B-cells and a reduction in pre-B-cells and mature B-cells [48]. These effects have been proven to be the result of targeting Foxp1, a transcription factor required for early B-cell development. Interestingly, the investigators highlight that the effect of miR-34a on the B-cell developmental pathway [48] is consistent with previously reported abnormalities associated with a deficiency of p53, namely, an increased number of pre-B-cells, as well as mature B-cells, with the latter finding being also a consequence of the loss of miR-34a function. This evidence implies a connection between p53 and Foxp1 through the action of miR-34a, which has been demonstrated to be a direct transcriptional target of p53 and to participate in the control of inappropriate cell proliferation [73,74].

miRNAs also regulate B-cell development in the periphery (Table 1). Indeed, Dicer ablation in mature B-cells results in augmented B-cells in the marginal zone and diminished follicular B-cells. miR-155, whose levels are low in progenitor and mature B-cells and are induced upon B-cell receptor ligation, has been implicated in the differentiation of normal activated B-cells. miR-155-deficient mice show multiple defects in the germinal center (GC) response, such as a decreased number of GC B-cells, decreased IgG production, and decreased affinity maturation. Using mouse models, miR-155 has been demonstrated to affect the regulation of the GC response through modulation of cytokine production [13,49] and by direct post-transcriptional regulation of the transcription factor, Pu.1 [75], and activation-induced cytidine deaminase (AID) [50,76].

2.5. miRNAs and the Anti-Viral Immune Response

The ability of small noncoding RNAs to act as immune sentinels was initially described in plants, nematodes, fungi and insects [22,55,64,77] in which the RNAi-mediated recognition of viral RNAs has evolved as a defense mechanism against RNA viruses. In mammals, another class of small RNA, the piwi RNAs (piRNAs), has been shown to suppress transposable element (TE) expression and mobilization through their silencing [78]. Recently, human cellular miRNAs have also been described as intracellular immune mediators, adopted by the organism to fight viruses. The short seed sequence and the tolerance for mismatches represent peculiar properties of miRNAs, which confer them the capability to recognize and downregulate viral mRNAs, required for successful replication and host infection. In mammals, the repression of Dicer and, hence, miRNAs production, confers increased susceptibility to human immunodeficiency virus type 1 (HIV-1), influenza A virus and vesicular stomatitis virus (VSV) [79–81]. In particular, the anti-HIV effects of Dicer are mediated by two cellular miRNAs, miR-17-5p and miR-20a, which target histone acetylase p300/CBP-associated factor (PCAF), an important cofactor of the HIV protein, Tat [79]. Two other miRNAs, miR-93 and miR-24, instead, contribute to the host defense against the rhabdovirus vesicular stomatitis virus (VSV), by targeting complementary sites in the VSV genome [81]]. On the other hand, viruses evolved several mechanisms to evade host defense, including quenching of the miRNA machinery [79], and, also, co-opted miRNA genes to suppress the host antiviral response; therefore, further studies are required to fully understand the role of miRNAs in the complex interaction between viruses and their mammalian hosts.

2.6. miRNAs and Autoimmunity

Given the pivotal role of miRNAs in the modulation of development, maturation and function of lymphocytes in normal immune response, it is not surprising that they have been also involved in the pathogenesis of several immune disorders. An integral aspect of miRNA-mediated regulation is their ability to function over a narrow range of concentration during normal immune responses; they induce subtle variations in the expression of key proteins that, if inappropriate, may lead to an overt pathological phenotype. Many immune diseases arise from the altered balance of gene expression patterns rather than from heavy changes at the levels of protein-coding regions of the genome. As in the case of multiple polymorphisms, which individually have a small effect [82], miRNAs exert their regulatory role through the subtle individual modulation of multiple targets involved in a common signaling pathway rather than through a strong repression of unique targets, thus contributing to lowering the threshold for the onset of an autoimmune response. Notably, up to one half of immune genes are predicted to be under the direct regulation of miRNAs; therefore, mutations that directly affect miRNA expression or alter the miRNA biogenesis pathway may have a causative effect during disease initiation and progression, owing to inappropriate repression or de-repression of key protein targets.

The tight regulation of immune cell activation and the proper selection of a functional, non-self-reactive repertoire of specific antigen receptors is at the base of the fine balance between a normal immune response and the development of autoimmunity. Because of the miRNA ability to regulate the survival and death of lymphocytes [18,51], control over their expression is essential to prevent adaptive immune cells from dysregulated proliferation and activity leading to leukemia and autoimmunity [83,84]. Several miRNAs exhibit a different expression profile in patients affected by autoimmune diseases compared to normal subjects [84–86], suggesting a causative role for the pathogenesis of human autoimmune diseases, including rheumatoid arthritis (RA), psoriasis and SLE. The expression of some miRNAs, like miR-146a and miR-155, appears to be constantly dysregulated in the aforementioned autoimmune diseases.

SLE is an autoimmune disease characterized by the loss of immune tolerance, resulting in activation and expansion of autoreactive $CD4^+$ T-helper cells [87]. In MRL/lpr mice, a murine model of SLE characterized by the accumulation of $CD4^+CD25^+Foxp3^+$ Treg cells in lymphoid organs [88], the onset of the autoimmune disease correlates with a reduced functional capacity of Treg cells that show a distinct phenotype (*i.e.*, increased CD69 and reduced CD62L expression). CD62L, an L-selectin, which determines the homing properties of T-lymphocytes, is a target of miR-155, one of the miRNAs upregulated in SLE. These data indicate that miR-155 dysregulated expression may directly contribute to the accumulation of Treg cells in the lymphoid organs of MRL/lpr mice [89]. On the contrary, miR-146a expression is reduced in the PBMCs of SLE patients. MiR-146a downregulation inversely correlates with clinical disease activity and with interferon (IFN) scores and may be responsible for the IFN overproduction observed in SLE. This phenotype is consistent with miR-146a-mediated repression of the type I IFN pathway, through the direct targeting of STAT-1 and IFN regulatory factor 5 (IRF5) [90].

In RA patients, upregulation of miR-146a and miR-155 has been found in PBMCs, as well as in the synovial tissue [52–54]. miR-146a also is highly expressed in CD4$^+$ T lymphocytes from RA patients and exhibits a strong correlation with increased levels of TNFα [86]. Interestingly, T-cells isolated from joint tissue and synovial fluid of RA patients show an activated and memory phenotype and are resistant to apoptosis, although the high levels of pro-apoptotic factors, like Fas-ligand (FasL), TNFα and TNF-related apoptosis-inducing ligand (TRAIL). Given that our study [18] and others [54,91] demonstrated an anti-apoptotic role of miR-146a in T-lymphocytes and other cell types, the increased survival of T-lymphocytes in RA patients could be in part explained by the enhanced expression of miR-146a. Another anti-apoptotic miRNA that has been related to the pathogenesis of autoimmune disorders is miR-17-92. Its overexpression results in a decreased activation-induced cell death, sustained lymphoproliferation and a marked presence of serum autoantibodies and lymphoid infiltrates [46]. Although the development of T-lymphocytes does appear normal, the relevant increase in the number of mature CD4$^+$ T-cells and their activated profile strongly suggest a failure of peripheral tolerance. The molecular mechanisms contributing to this immune disease-related phenotype of miR-17-92 transgenic mice are based on the downregulation of the tumor suppressor, PTEN, and the pro-apoptotic, Bim [46]. An interesting study demonstrating the involvement of miRNAs in the possible alteration of autoimmune-related signaling pathways was performed in 2007 in sanroque mice, a murine model resembling a lupus-like autoimmune syndrome [92]. These mice are homozygotes for an M199R mutation in the ROQ domain of Roquin, a protein that binds to mRNAs and targets gene expression post-transcriptionally. The evident accumulation of autoimmune lymphocytes in this murine model depends, at least in part, on the failure to repress the expression of the co-stimulatory molecule inducible co-stimulatory molecule (ICOS), particularly in naive T-cells. Roquin normally limits ICOS expression by promoting the degradation of ICOS mRNA, by a mechanism that involves the recognition of a conserved miRNA binding sequence within the 3' UTR. It has been reported that miR-101 is required for the Roquin-mediated degradation of ICOS mRNA [92], and the alteration of the seed site for miR-101 in the 3' UTR of ICOS disrupts the inhibitory activity by Roquin. These results demonstrate a critical miRNA-mediated regulatory pathway that prevents lymphocyte accumulation and autoimmunity. Further evidence supporting a role of miRNAs in autoimmunity has been provided by Ebert *et al.*, who demonstrated the relevance of miR-181a expression for the elimination of self-reactive thymocytes; indeed, the inhibition of miR-181a expression during thymic development converts endogenous positively selecting peptides into autoantigens [93].

Along with the miRNA-mediated deregulation of disease susceptibility genes, few studies also detected autoantibodies targeting key components of the RNAi/miRNA machinery, including Argonaute proteins (referred to as anti-Su antibodies) and Dicer, in patient blood. The anti-Su autoimmune sera recognize several members of the Ago protein family, which are known to associate into RISC and which have a high degree of sequence identity. Despite that the first characterization of autoantigens recognized by anti-Su autoantibodies [94] had been provided in 1984, the disease specificity of anti-Su antibodies has been addressed in only a few studies, so far. They were initially reported to be specifically associated with SLE [94], but later studies showed that anti-Su antibodies are, instead, found in a variety of systemic rheumatic diseases [95]. The existence of an autoimmune response directed against macromolecular complexes involved in the post-transcriptional regulation of gene expression may indicate the involvement of the miRNA biogenesis pathway in the production of autoantibodies, further supporting the intriguing hypothesis that dysfunction of the RNAi/miRNA machinery lies at the core of the pathogenesis of autoimmune diseases.

3. Long Noncoding RNAs

At least 80% of mammalian genome transcription results in the generation of lncRNAs. It has been estimated that 10,000 to 200,000 different types of lncRNAs are transcribed in the human genome, but functional and molecular mechanisms have been elucidated only for a few [96]. Given their unexpected abundance, lncRNAs were initially considered as spurious transcriptional noise, without biological function, and this assumption was corroborated by the observation that many of the transcripts are expressed at very low levels and exhibit no sequence conservation [97,98]. Nevertheless, it is now apparent that lncRNAs are functional, and many of them are differentially expressed in particular developmental stages.

Figure 2. Long non coding RNAS (lncRNAs): mechanism of action and genomic organization. Illustrative representation of lncRNAs classification relative to their genomic position or mechanism of action. (**A**) lncRNAs are generally classified according to their proximity to protein coding genes in the genome: intergenic lncRNAs are distant from protein coding regions; divergent lncRNAs are located on the opposite strand closed by the transcription starting site of a protein coding gene; intragenic lncRNAs overlap protein coding genes and can be sense, antisense, intronic or exonic; (**B**) lncRNAs can recruit chromatin remodeling complexes to specific genomic loci, to epigenetically mark the region for gene silencing; (**C**) lncRNAs can permit the recruitment of transcription factors activating gene expression or can interfere with the transcription machinery, occluding the access of transcription factors and, thereby, silencing the gene; (**D**) lncRNAs that are antisense to protein coding genes may regulate the splicing or induce the degradation of their corresponding mRNA transcripts.

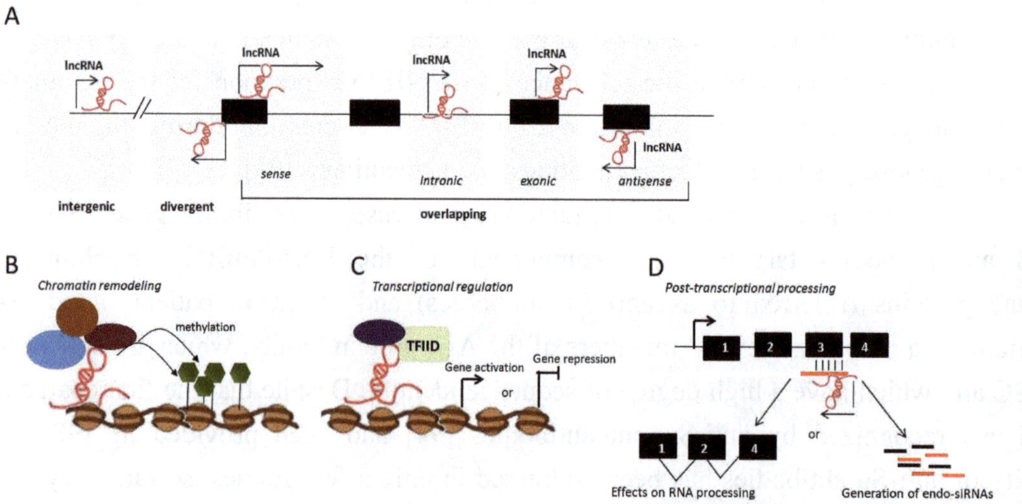

LncRNAs comprise many different transcripts, ranging from several hundreds to ten-thousand nucleotides and can be classified according to their genomic position as intergenic or intragenic lncRNAs (Figure 2A). Intergenic lncRNAs are located distant from protein coding regions, whereas divergent lncRNAs, often referred to as bidirectional lncRNAs, are located on the opposite strand from a coding gene whose transcription is initiated in close proximity. Intragenic (overlapping) lncRNAs are transcribed along one or multiple exons (exonic lncRNA) or introns (intronic lncRNA) of another transcript on the same or opposite strand and can then be spliced into a mature RNA (Figure2A).

Unlike miRNAs, lncRNAs possess heterogeneous molecular mechanisms of action, making them master regulators of intracellular networks and pathways in both physiology and diseases [96]. Some mammalian lncRNAs have two unique relevant properties. The first one is the ability to target a single location through the use of a large sequence space. While transcription factors are effectively recruited to the DNA promoter region by recognizing short DNA motifs, which typically occur thousands of times in the genome, lncRNAs, like Tsix and RepA/Xist, are able to deliver epigenetic complexes to a unique site, thus providing a spatio-temporal regulatory specificity not achievable with proteins and small RNAs [63,99]. A second characteristic of some lncRNA-mediated regulation is the "allelic memory". Proteins loose the memory of their transcriptional origin when the mRNA is shuttled to the cytoplasm to be translated; lncRNAs, instead, remain tethered to the site of transcription and can, therefore, have allele-specific action. Differently from miRNAs, which are mainly post-transcriptional repressors, lncRNAs have a full range of functions, being able to act as scaffold RNAs, co-activators or co-repressors of trans-acting RNAs and epigenetic regulators of protein-coding gene expression. In particular, they can mediate epigenetic changes by recruiting chromatin remodeling complexes to specific genomic loci, thus altering the chromatin status at gene promoters to affect gene expression [100] (Figure 2B). Another major mechanism of lncRNA action is transcriptional regulation, exerted through specific interaction with transcription factors or other protein components of the transcriptional machinery. LncRNAs may sequester transcription factors in the cytoplasm, preventing their translocation into the nucleus [101] or occlude a transcription factor binding site, thereby promoting gene repression [102]. LncRNAs can also promote gene expression, through interactions with components of the core transcriptional machinery, thus enhancing the specific binding of transcription factors to promoter regions (Figure 2C). Moreover, lncRNAs have been implicated in posttranscriptional regulation, either by the direct regulation of alternative splicing through base-pairing interactions or by gene silencing via the generation of endo-siRNAs and the consequent degradation of the targeted transcript (Figure 2D).

3.1. lncRNAs and the Adaptive Immune Response

In contrast to the wealth of recent evidence demonstrating lncRNA involvement in cancer, at present, a role of lncRNAs in immune system regulation is attested by only a handful of mechanistically distinct examples. The importance of lncRNAs in the immune system has been recently highlighted by a report that identified and characterized a set of lncRNAs preferentially expressed in naive and memory mammalian $CD8^+$ T-lymphocytes [14,92]. Over 1000 lncRNAs have been detected and characterized in humans and mice, many of which expressed in a stage- or cell-specific manner. In particular, the authors identified 96 lymphoid-specific lncRNAs. Interestingly,

29 of these transcripts were specific for CD8$^+$ T lymphocytes, 21 lncRNAs were significantly modulated during the T-cell differentiation process and 81 were regulated during effector T-cell activation [14]. These expression patterns strongly suggest a possible role of lncRNA in the modulation of the key cellular processes of T-lymphocyte biology. Remarkably, lncRNAs, as well as the recently discovered large intergenic noncoding RNAs (lincRNAs), have been shown to regulate the expression of adjacent protein-coding genes. Actually, several of the lncRNAs identified in the mentioned study are associated with protein-coding genes that exert key functional roles in T-cells. For example, an lncRNA, AK009498, overlaps the transcription start site of caspase-8 (FLICE)-like inhibitory protein (Flip), a gene fundamental for the survival and development of T-lymphocytes. Flip is upregulated during the transition to an activated or a memory phenotype, whereas the host lncRNA is downregulated, thus suggesting a potential negative function of AK009498 [14] (Table 2).

Table 2. lncRNAs in adaptive immunity.

	lncRNA	Genomic coordinates	Lenght	Expression	Characteristics	Proposed function	References
has lncRNAs	Tmevpg1 (NeST)	chr12: 68, 383, 225-68, 415, 107	32 kb	Activated CD8$^+$ T cells CD4$^+$ Th1 and Th17 cells	Its expression is regulated by Tbet and STAT4	Epigenetic control of Ifnγ locus	[103–105]
	NRON	chr9: 129, 170, 054-129, 172, 783	2.7 kb	CD4$^+$ and CD8$^+$ T cells	Sequence omology conserved back to chicken	Modulator of NFAT nuclear trafficking	[101]
mmu lncRNAs	AK009498	chr1: 58, 711, 508-58, 713, 886	710 bp	CD4$^+$ and CD8$^+$ effector and memory cells	Overlaps flip transcription start site	Potential negative regulator of Flip	[14]
	Lef1as (AK029296)	chr3: 131, 109, 026-131, 112, 090	3 kb	naive CD8$^+$ T cells	Located antisense to Lef1 mRNA	Possible role in suppressionLef1	[14]
	AK053349	chr14: 115, 044, 497-115, 046, 726	2.2 kb	effector CD8$^+$ T cells	Partially overlaps miR-17-92 cluster	unknown	[14]
	AK020764	chr11: 87, 755, 577-87, 757, 267	1.6 kb	effector CD8$^+$ T cells	It hosts in its first intron miR-142	Possible functional link with miR-142	[14]

Nettoie Theiler's Pas Salmonella (NeST), also called Theiler's Murine Encephalitis Virus Possible Gene1 (TMEVPG1), a nuclear RNA initially described as involved in the control of viral load during persistent infection, is the first lncRNA of the immune system that has been proven to regulate the expression of a master cytokine, such as IFN-γ [103]. This lncRNA is expressed selectively in Th1 cells relative to Th2 and Th17 cells, and its expression depends on Th1-specific transcription factors,

STAT4 and Tbet [104]. Moreover, NeST is located adjacent to the IFNγ locus in both humans and mice and is encoded in antisense respect to the IFNγ gene (Table 2). Recent evidence demonstrated that NeST acts in trans and positively regulates the IFNγ gene at the transcriptional level, by interacting with chromatin-modifying complexes. More precisely, it binds WDR5, a member of the histone H3 lysine 4 methyltransferase complex, and promotes permissive methylation marks on histone H3 at the IFNγ locus. The effect of such epigenetic control is the modulation of the host response to viral and bacterial pathogens, thus suggesting a role for lncRNAs as functional links between immune regulation and susceptibility to infections [105]. Indeed, alteration of NeST expression could contribute to differences in T-lymphocyte response and disease susceptibility. The discovery of functional intergenic lncRNAs, whose expression is critical for proper gene regulation, may explain the relevance of some intergenic regions as heritable causes of human diseases.

Another interesting feature of lncRNAs is that a small number of lncRNA genes host in their sequence small RNAs and may be processed into, and exert their effects via, smaller functional ncRNAs. Pang *et al.* identified 18 lncRNAs expressed in murine $CD8^+$ T-lymphocytes overlapping with annotated miRNAs and 21 that harbor internally small nucleolar RNAs (snoRNAs) [14]. In this context, AK053349, one of the most highly upregulated lncRNA during effector T-cell activation, partially overlaps miR-17-92, a cluster of miRNAs, often amplified in lymphomas, whose enforced expression in mouse lymphocytes results in lymphoproliferative disease and autoimmunity [46]. Furthermore, miR-142-5p and miR-142-3p, which have been reported as highly expressed in naive T-cells and are involved in the modulation of Foxp3 in Treg lymphocytes, are hosted in the first intron of a lncRNA (AK020764), strongly upregulated in $CD8^+$ T-cells [14] (Table 2). The colocalization could imply a functional link, but it has not yet been verified.

Finally, lncRNAs may influence cell regulatory networks not only at the transcriptional level, but also by controlling protein trafficking, as it has been shown for the noncoding repressor of NFAT (NRON), a negative regulator of the transcription nuclear factor of activated T-cells (NFAT). NRON is an lncRNA enriched in lymphoid tissues, such as the thymus, the spleen and the lymph nodes. This lncRNA has a cytosolic localization and has been proven to interfere with NFAT function by interacting with several proteins (Table 2). One of these proteins is importin beta1 (KPNB1), which, in the presence of NRON, sequesters NFAT in the cytosol. In addition, the interactions of NRON with two other proteins, the serine/threonine protein, phosphatase 2A (PP2A), and the IQ motif containing GTPase activating protein 1 (IQGAP1), may also be significant [101]. These examples underscore the possibility of limitless mechanisms of gene regulation based on cooperation between ncRNAs and proteins.

3.2. lncRNAs as Biomarkers in Autoimmunity and Leukemia

Genome-wide analysis identified a number of highly conserved transcripts, some of which do not encode for proteins and are, therefore, considered non-genic. Among these, the ultra-conserved regions (UCRs) are noncoding genomic sequences of over 200 bases, located in both intra- and inter-genic regions, and are strongly conserved between human, mouse and rat genomes. The unexpected degree of conservation of UCRs among distant species suggests that they may have essential functional importance for the ontogeny and phylogeny of mammals and vertebrates, in general. A large proportion of transcripts derived from these UCRs have significant RNA secondary structures and are components of clusters containing other sequences with functional noncoding significance [106]. Notably, the highest fraction of transcribed UCRs was found in B-lymphocytes, and distinct UCR signatures have been described as associated with human leukemias and carcinomas. In particular, Calin *et al.* reported a differential expression of 19 UCRs (eight up- and 11 down-regulated) in chronic lymphatic leukemia (CLL) with respect to normal hematopoietic tissues [106,107]. The authors also identified a cluster of seven UCRs (uc. 347 to uc. 353) located within the cancer-associated genomic region (CAGR), and two of them—uc. 349A(P) and uc. 352(N)—are among the transcribed UCRs (T-UCRs) differentially expressed between normal and malignant B-CLL $CD5^+$ cells [106]. Furthermore, the authors correlated a signature of five T-UCRs, namely, uc. 269A(N), uc. 160(N), uc. 215(N), uc. 346A(P) and uc. 348(N), with different CLL prognosis groups, consistent with the expression of 70-kDa zeta-associated protein (ZAP-70), an established prognostic marker for CLL. These data suggest that T-UCRs could be candidate genes for cancer susceptibility. Another interesting aspect emerging from this study is the finding that the expression levels of the five T-UCRs described above negatively correlated with a miRNA expression signature reported in CLL [107]. This evidence represents just an example of how lncRNAs and small RNA biology are strictly connected: a significant fraction of lncRNAs seem to act as precursors for small RNA species, in particular for miRNAs. These findings raise the possibility of complex regulatory mechanisms between lncRNAs and small RNAs, suggesting a possible interplay between the different classes of noncoding RNAs.

Given the emerging role of lncRNAs in biomolecular regulatory interactions within cells and their relevance in the etiology of human disease and cancer, it becomes highly pertinent and imperative to perform an exhaustive evaluation of their functionalities in order to shed light on the specific pathways and regulatory circuits in which they are involved.

4. Secreted Noncoding RNAs

The introduction of Next-Generation Sequencing (NGS) techniques led to the identification of a large number of unexpected ncRNAs, not only at the level of different human tissues, but also in extracellular fluids (e.g., serum, plasma, urine, milk and saliva) [108–112]. The study of circulating RNA populations has almost exclusively focused on miRNAs, most likely due to the availability of array hybridization techniques to detect these small RNAs and, also, because of their recognized role in several biological processes. Alterations in the level and composition of these extracellular miRNA populations are strictly related to different pathologies, including cancer [20,108], diabetes [108] and tissue injury [113]. Extracellular ncRNAs circulate in human plasma within non-vesicular Ago2 ribonucleoprotein complexes, which confer them stability [114]. Alternatively, they can be stably carried in body fluids, as packaged within membranous vesicles (including exosomes, shedding vesicles and apoptotic bodies) [115–117] and spread signals that affect neighboring or distant cells. These different types of microvesicles protect ncRNAs from degradation during systemic transport and, also, are responsible for the specific delivery of them to recipient cells. Exosomes are small vesicles ranging from 30 to 100 nm in size, originated from endosomes. They consist of a lipid bilayer membrane surrounding a small amount of cytosol. The formation and release of exosomes are tightly regulated by multiple signaling mechanisms. Different stimuli can influence exosome secretion, including bacterial lipopolysaccharide on macrophages and dendritic cells [118]. Shedding vesicles are much larger than exosomes and are heterogeneous in size, ranging from 100 nm to 1 μm [119]. The presence of a ncRNA pool has been reported in exosomes and shedding vesicles derived from a variety of sources, including mast cells [115] blood [120], stem cells [121] and adipocytes [122]. The existence of different secretion mechanisms suggests that specific ncRNA populations may be delivered from different cell types and, therefore, have different fates and functions. It has been speculated that differences between vesicle-enclosed and Ago2-associated ncRNAs may reflect cell type-specific ncRNA release mechanisms. Additional studies are needed to unravel the pathways responsible for ncRNA loading and secretion and to better define the biological function of the different circulating ncRNA species.

Role of Secreted ncRNAs in Immunological Processes

The first indication of the horizontal transfer of nuclei acids in mammalian cells reported the release of miRNAs and mRNAs by mast cells and their functional transfer to vesicle-targeted cells [115]. This evidence, together with intensive studies describing the delivery of ncRNA-loaded exosomes by immune cells [123,124], strongly suggests that secreted RNAs could represent a new intricate level of cellular communication and regulation of immunological processes. To mount an effective immune response, different immune cells need to exchange information and develop highly specialized structures, called "cell synapses". During the formation of immunological synapse (IS) between T-lymphocytes and antigen presenting cells (APCs), membrane and transmembrane-associated molecules are rearranged into a highly organized structure at the T-cell-APC contact site [125]. The formation of IS is consequent to antigen recognition and allows the initiation and tuning of T-lymphocyte activation. Interestingly, it has been reported that miRNAs are exchanged during cognate immune interactions, and this delivery is strictly dependent on the formation of IS [124]. These specialized structures promote the unidirectional transfer of miRNA-loaded exosomes from T-lymphocytes to APCs, and this kind of genetic communication is antigen-driven and CD63-dependent. Synaptically transferred miRNAs are functional in recipient cells, as demonstrated by Mittelburn *et al.* [124], showing that antigen-dependent transfer of miR-335 from T-lymphocytes to APC resulted in the targeting of specific genes, involved in the modulation of immune response. Notably, not all miRNAs can be incorporated into exosomes, and their presence into delivered microvesicles is a selective process. The miRNA composition in circulating RNAs does not merely reflect the cellular miRNA pool: certain miRNAs are equally abundant between cellular and extracellular environment; others (like miR-760, miR-362, miR-654-5p and miR-671-5p) are significantly more abundant in exosomal fractions derived from all cell types; and finally, another set (for example, miR-21-3p, miR-101 and miR-32) are, instead, more represented in cells than in exosomes [124]. Moreover, it has been also observed that there is a higher similarity between miRNAs composition in exosomes of different cellular origin than between shuttle RNA and their corresponding cellular miRNAs [123,124]. These data demonstrate that specific miRNA populations are selectively sorted into shuttle RNA, strongly suggesting that cells specifically release these RNA molecules to affect the function of target cells. However, to date, the precise mechanism controlling this selection is still unclear. Together with synaptic-dependent shuttling of miRNA from T-cells to

APC, another important source of RNA-loaded exosomes is represented by dendritic cells (DCs), which have a role in the activation of T-lymphocytes. In that context, among the most representative miRNAs vehiculated by exosomes, miR-223 and miR-93 have validated target genes that play important roles in immunity. In particular, miR-223 downregulates myocyte enhancer factor 2C (MEF2C), involved in the transcription of interleukins, whereas miR-93 targets STAT3, thus affecting T-cell activation processes [123]. More interestingly, profiling of shuttle RNA isolated from DC-derived exosomes also revealed the presence of many small noncoding RNA species, other than miRNAs, that could act as regulatory RNAs, in particular, snoRNAs. Altogether, this first piece of evidence suggests a wide range of biological effects that could be mediated by shuttle RNA, and the recent findings that the shuttle RNA population is not restricted to miRNAs, but includes other ncRNAs reciprocally exchanged between immune cells, provide a further level of cellular inter-communication that contributes to affecting the activation and the efficacy of immunological response.

5. Concluding Remarks

The proper functioning of the immune system is accomplished through several steps of development and differentiation that must be strictly regulated to respond, at the right time and with the appropriate effector cells, to different insults. Thanks to the advancement of technologies that allow us to have a global picture of which proteins and functional RNAs are expressed in a cell at a defined time, we have now the instruments to dissect the complexity of this regulation, which is time- and cell-specific, allowing the immune system to dynamically adapt to the many challenges it faces. The crosstalk between signaling pathways and epigenetic processes during development and cell differentiation is still poorly understood. Changes in these networks might lead to deregulation of gene expression, which ultimately results in diseases, e.g., autoimmunity and cancer.

It is conceivable that transcriptional control, mediated by epigenetic modifications, is preferentially responsible for the developmental decision, while posttranscriptional control plays a major role during activation and differentiation. This concept is particularly relevant in the immunology field, where recent advances in the "-omics" techniques have revealed a multilayered network composed of protein-coding RNAs, as well as of ncRNAs.

Although the miRNA and lncRNA pathways were initially believed to be independent and distinct, the lines distinguishing them continue to fade. There is now compelling evidence that, despite their differences, these distinct ncRNA pathways are interconnected, interact, compete and rely on each other at several levels as they regulate genes and protect the genome from external and internal threats. However, there are still major gaps in our understanding of their function at the molecular level. It is becoming evident that miRNAs, more than being molecules that act at the post-transcriptional level to switch on and off the expression of specific genes at a precise stage, have a more intricate role. On the one hand, being part of regulatory loops, miRNAs may contribute to the maintenance of a certain state of the cell; on the other hand, because of their transient rising, induced by external stimuli and activation of specific signaling pathways, miRNAs help to rapidly remodel the proteome of a cell, rendering it functional for the needs of the organism. Furthermore, we have now to consider the primary contribution of lncRNAs, regulatory molecules that have the prerogative to join some functional peculiarity of proteins and the capability to recognize unique loci in the genome. LncRNAs are responsible for transcriptional control of gene expression, both by site-specific recruitment of chromatin remodeling complexes (epigenetic changes) and by favoring or inhibiting the recruitment of transcription factors on defined promoters. These mechanisms may affect the expression timing of proteins, as well as of miRNAs, thus contributing to shape both the proteome and the miRNome of a developing cell. However, lncRNAs may exert their control also post-transcriptionally by driving molecular complexes responsible for editing, splicing, transport and degradation towards mRNA molecules. At the post-transcriptional level, lncRNAs will work alongside with miRNAs, thus contributing to establish a more flexible and fine-tuned system of regulation. This functional link, evident in adaptive immunity system, reveals a more general mechanism for the control of patterning and lineage commitment. Understanding how ncRNAs are regulated and exert their function in the context of the immune system will encourage further advancements on studies in this field and the exploitation of ncRNAs as possible candidates for future therapies.

Acknowledgments

We thank C. Cicchini and A. Marchetti for critical reading of this manuscript.

This study was supported by Ministero dell'Istruzione dell'Università e della Ricerca PRIN Research Grants 2002061255.

References

1. Lim, P.S.; Li, J.; Holloway, A.F.; Rao, S. Epigenetic regulation of inducible gene expression in the immune system. *Immunology* **2013**, *139*, 285–293.
2. Blom, B.; Spits, H. Development of human lymphoid cells. *Annu. Rev. Immunol.* **2006**, *24*, 287–320.
3. Sheik Mohamed, J.; Gaughwin, P.M.; Lim, B.; Robson, P.; Lipovich, L. Conserved long noncoding RNAs transcriptionally regulated by Oct4 and Nanog modulate pluripotency in mouse embryonic stem cells. *RNA* **2010**, *16*, 324–337.
4. Orom, U.A.; Derrien, T.; Beringer, M.; GumiReddy, K.; Gardini, A.; Bussotti, G.; Lai, F.; Zytnicki, M.; Notredame, C.; Huang, Q.; *et al.* Long noncoding RNAs with enhancer-like function in human cells. *Cell* **2010**, *143*, 46–58.
5. O'Connell, R.M.; Rao, D.S.; Chaudhuri, A.A.; Baltimore, D. Physiological and pathological roles for microRNAs in the immune system. *Nat. Rev. Immunol.* **2010**, *10*, 111–122.
6. Mercer, T.R.; Dinger, M.E.; Mattick, J.S. Long non-coding RNAs: Insights into functions. *Nat. Rev. Genet.* **2009**, *10*, 155–159.
7. Calin, G.A.; Croce, C.M. microRNA signatures in human cancers. *Nat. Rev. Cancer* **2006**, *6*, 857–866.
8. Cobb, B.S.; Nesterova, T.B.; Thompson, E.; Hertweck, A.; O'Connor, E.; Godwin, J.; Wilson, C.B.; Brockdorff, N.; Fisher, A.G.; Smale, S.T.; *et al.* T-cell lineage choice and differentiation in the absence of the RNase III enzyme Dicer. *J. Exp. Med.* **2005**, *201*, 1367–1373.
9. Muljo, S.A.; Ansel, K.M.; Kanellopoulou, C.; Livingston, D.M.; Rao, A.; Rajewsky, K. Aberrant T-cell differentiation in the absence of Dicer. *J. Exp. Med.* **2005**, *202*, 261–269.

10. Chong, M.M.; Rasmussen, J.P.; Rudensky, A.Y.; Littman, D.R. The RNAseIII enzyme Drosha is critical in T-cells for preventing lethal inflammatory disease. *J. Exp. Med.* **2008**, *205*, 2005–2017.
11. Monticelli, S.; Ansel, K.M.; Xiao, C.; Socci, N.D.; Krichevsky, A.M.; Thai, T.H.; Rajewsky, N.; Marks, D.S.; Sander, C.; Rajewsky, K.; *et al.* microRNA profiling of the murine hematopoietic system. *Genome Biol.* **2005**, *6*, R71.
12. Taganov, K.D.; Boldin, M.P.; Chang, K.J.; Baltimore, D. NF-kappaB-dependent induction of microRNA miR-146, an inhibitor targeted to signaling proteins of innate immune responses. *Proc. Natl. Acad. Sci. USA* **2006**, *103*, 12481–12486.
13. Rodriguez, A.; Vigorito, E.; Clare, S.; Warren, M.V.; Couttet, P.; Soond, D.R.; van Dongen, S.; Grocock, R.J.; Das, P.P.; Miska, E.A.; *et al.* Requirement of bic/microRNA-155 for normal immune function. *Science* **2007**, *316*, 608–611.
14. Pang, K.C.; Dinger, M.E.; Mercer, T.R.; Malquori, L.; Grimmond, S.M.; Chen, W.; Mattick, J.S. Genome-wide identification of long noncoding RNAs in CD8+ T-cells. *J. Immunol.* **2009**, *182*, 7738–7748.
15. Pagani, M.; Rossetti, G.; Panzeri, I.; de Candia, P.; Bonnal, R.J.; Rossi, R.L.; Geginat, J.; Abrignani, S. Role of microRNAs and long-non-coding RNAs in CD4(+) T-cell differentiation. *Immunol. Rev.* **2013**, *253*, 82–96.
16. Hangauer, M.J.; Vaughn, I.W.; McManus, M.T. Pervasive transcription of the human genome produces thousands of previously unidentified long intergenic noncoding RNAs. *PLoS Genet.* **2013**, *9*, e1003569.
17. Curtale, G.; MiRolo, M.; Renzi, T.A.; Rossato, M.; Bazzoni, F.; Locati, M. Negative regulation of Toll-like receptor 4 signaling by IL-10-dependent microRNA-146b. *Proc. Natl. Acad. Sci. USA* **2013**, *110*, 1499–11504.
18. Curtale, G.; Citarella, F.; Carissimi, C.; Goldoni, M.; Carucci, N.; Fulci, V.; Franceschini, D.; Meloni, F.; Barnaba, V.; Macino, G. An emerging player in the adaptive immune response: microRNA-146a is a modulator of IL-2 expression and activation-induced cell death in T lymphocytes. *Blood* **2010**, *115*, 265–273.
19. Rossato, M.; Curtale, G.; Tamassia, N.; Castellucci, M.; Mori, L.; Gasperini, S.; Mariotti, B.; de Luca, M.; MiRolo, M.; Cassatella, M.A.; *et al.* IL-10-induced microRNA-187 negatively regulates TNF-alpha, IL-6, and IL-12p40 production in TLR4-stimulated monocytes. *Proc. Natl. Acad. Sci. USA* **2012**, *109*, E3101–E3110.
20. Ghisi, M.; Corradin, A.; Basso, K.; Frasson, C.; Serafin, V.; Mukherjee, S.; Mussolin, L.; Ruggero, K.; Bonanno, L.; Guffanti, A.; *et al.* Modulation of microRNA expression in human T-cell development: targeting of NOTCH3 by miR-150. *Blood* **2011**, *117*, 7053–7062.
21. Lee, Y.; Ahn, C.; Han, J.; Choi, H.; Kim, J.; Yim, J.; Lee, J.; Provost, P.; Radmark, O.; Kim, S.; *et al.* The nuclear RNase III Drosha initiates microRNA processing. *Nature* **2003**, *425*, 415–419.
22. Hutvagner, G.; McLachlan, J.; Pasquinelli, A.E.; Balint, E.; Tuschl, T.; Zamore, P.D. A cellular function for the RNA-interference enzyme Dicer in the maturation of the let-7 small temporal RNA. *Science* **2001**, *293*, 834–838.

23. Matranga, C.; Tomari, Y.; Shin, C.; Bartel, D.P.; Zamore, P.D. Passenger-strand cleavage facilitates assembly of siRNA into Ago2-containing RNAi enzyme complexes. *Cell* **2005**, *123*, 607–620.
24. Lewis, B.P.; Shih, I.H.; Jones-Rhoades, M.W.; Bartel, D.P.; Burge, C.B. Prediction of mammalian microRNA targets. *Cell* **2003**, *115*, 787–798.
25. Nilsen, T.W. Mechanisms of microRNA-mediated gene regulation in animal cells. *Trends Genet.* **2007**, *23*, 243–249.
26. Kisielow, P.; von Boehmer, H. Development and selection of T-cells: Facts and puzzles. *Adv. Immunol.* **1995**, *58*, 87–209.
27. Wu, H.; Neilson, J.R.; Kumar, P.; Manocha, M.; Shankar, P.; Sharp, P.A.; Manjunath, N. miRNA profiling of naive, effector and memory CD8 T-cells. *PLoS One* **2007**, *2*, e1020.
28. Neilson, J.R.; Zheng, G.X.; Burge, C.B.; Sharp, P.A. Dynamic regulation of miRNA expression in ordered stages of cellular development. *Genes Dev.* **2007**, *21*, 578–589.
29. Kirigin, F.F.; Lindstedt, K.; Sellars, M.; Ciofani, M.; Low, S.L.; Jones, L.; Bell, F.; Pauli, F.; Bonneau, R.; Myers, R.M.; *et al.* Dynamic microRNA gene transcription and processing during T-cell development. *J. Immunol.* **2012**, *188*, 3257–3267.
30. O'Connell, R.M.; Chaudhuri, A.A.; Rao, D.S.; Gibson, W.S.; Balazs, A.B.; Baltimore, D. MicroRNAs enriched in hematopoietic stem cells differentially regulate long-term hematopoietic output. *Proc. Natl. Acad. Sci. USA* **2010**, *107*, 14235–14240.
31. Ooi, A.G.; Sahoo, D.; Adorno, M.; Wang, Y.; Weissman, I.L.; Park, C.Y. microRNA-125b expands hematopoietic stem cells and enriches for the lymphoid-balanced and lymphoid-biased subsets. *Proc. Natl. Acad. Sci. USA* **2010**, *107*, 21505–21510.
32. Shi, X.B.; Xue, L.; Yang, J.; Ma, A.H.; Zhao, J.; Xu, M.; Tepper, C.G.; Evans, C.P.; Kung, H.J.; deVere White, R.W. An androgen-regulated miRNA suppresses Bak1 expression and induces androgen-independent growth of prostate cancer cells. *Proc. Natl. Acad. Sci. USA* **2007**, *104*, 19983–19988.
33. Surdziel, E.; Cabanski, M.; Dallmann, I.; Lyszkiewicz, M.; Krueger, A.; Ganser, A.; Scherr, M.; Eder, M. Enforced expression of miR-125b affects myelopoiesis by targeting multiple signaling pathways. *Blood* **2011**, *117*, 4338–4348.
34. Henao-Mejia, J.; Williams, A.; Goff, L.A.; Staron, M.; Licona-Limon, P.; Kaech, S.M.; Nakayama, M.; Rinn, J.L.; Flavell, R.A. The microRNA miR-181 is a critical cellular metabolic rheostat essential for NKT-cell ontogenesis and lymphocyte development and homeostasis. *Immunity* **2013**, *38*, 984–997.
35. Li, Q.J.; Chau, J.; Ebert, P.J.; Sylvester, G.; Min, H.; Liu, G.; Braich, R.; Manoharan, M.; Soutschek, J.; Skare, P.; *et al.* miR-181a is an intrinsic modulator of T-cell sensitivity and selection. *Cell* **2007**, *129*, 147–161.
36. Lashine, Y.A.; Seoudi, A.M.; Salah, S.; Abdelaziz, A.I. Expression signature of microRNA-181-a reveals its crucial role in the pathogenesis of paediatric systemic lupus erythematosus. *Clin. Exp. Rheumatol.* **2011**, *29*, 351–357.
37. Zhou, B.; Wang, S.; Mayr, C.; Bartel, D.P.; Lodish, H.F. miR-150, a microRNA expressed in mature B and T-cells, blocks early B-cell development when expressed prematurely. *Proc. Natl. Acad. Sci. USA* **2007**, *104*, 7080–7085.

38. Xiao, C.; Calado, D.P.; Galler, G.; Thai, T.H.; Patterson, H.C.; Wang, J.; Rajewsky, N.; Bender, T.P.; Rajewsky, K. MiR-150 controls B-cell differentiation by targeting the transcription factor c-Myb. *Cell* **2007**, *131*, 146–159.
39. Almanza, G.; Fernandez, A.; Volinia, S.; Cortez-Gonzalez, X.; Croce, C.M.; Zanetti, M. Selected microRNAs define cell fate determination of murine central memory CD8 T-cells. *PLoS One* **2010**, *5*, e11243.
40. Lin, Y.C.; Kuo, M.W.; Yu, J.; Kuo, H.H.; Lin, R.J.; Lo, W.L.; Yu, A.L. c-Myb is an evolutionary conserved miR-150 target and miR-150/c-Myb interaction is important for embryonic development. *Mol. Biol. Evol.* **2008**, *25*, 2189–2198.
41. Zhu, J.; Paul, W.E. Peripheral CD4+ T-cell differentiation regulated by networks of cytokines and transcription factors. *Immunol. Rev.* **2010**, *238*, 247–262.
42. Bronevetsky, Y.; Villarino, A.V.; Eisley, C.J.; Barbeau, R.; Barczak, A.J.; Heinz, G.A.; Kremmer, E.; Heissmeyer, V.; McManus, M.T.; Erle, D.J.; *et al.* T-cell activation induces proteasomal degradation of Argonaute and rapid remodeling of the microRNA repertoire. *J. Exp. Med.* **2013**, *210*, 417–432.
43. Stittrich, A.B.; Haftmann, C.; Sgouroudis, E.; Kuhl, A.A.; Hegazy, A.N.; Panse, I.; Riedel, R.; Flossdorf, M.; Dong, J.; Fuhrmann, F.; *et al.* The microRNA miR-182 is induced by IL-2 and promotes clonal expansion of activated helper T lymphocytes. *Nat. Immunol.* **2010**, *11*, 1057–1062.
44. Rossi, R.L.; Rossetti, G.; Wenandy, L.; Curti, S.; Ripamonti, A.; Bonnal, R.J.; Birolo, R.S.; Moro, M.; Crosti, M.C.; Gruarin, P.; *et al.* Distinct microRNA signatures in human lymphocyte subsets and enforcement of the naive state in CD4+ T-cells by the microRNA miR-125b. *Nat. Immunol.* **2011**, *12*, 796–803.
45. Sandberg, R.; Neilson, J.R.; Sarma, A.; Sharp, P.A.; Burge, C.B. Proliferating cells express mRNAs with shortened 3' untranslated regions and fewer microRNA target sites. *Science* **2008**, *320*, 1643–1647.
46. Xiao, C.; Srinivasan, L.; Calado, D.P.; Patterson, H.C.; Zhang, B.; Wang, J.; Henderson, J.M.; Kutok, J.L.; Rajewsky, K. Lymphoproliferative disease and autoimmunity in mice with increased miR-17–92 expression in lymphocytes. *Nat. Immunol.* **2008**, *9*, 405–414.
47. Mendell, J.T. miRiad roles for the miR-17–92 cluster in development and disease. *Cell* **2008**, *133*, 217–222.
48. Rao, D.S.; O'Connell, R.M.; Chaudhuri, A.A.; Garcia-Flores, Y.; Geiger, T.L.; Baltimore, D. microRNA-34a perturbs B lymphocyte development by repressing the forkhead box transcription factor Foxp1. *Immunity* **2010**, *33*, 48–59.
49. Thai, T.H.; Calado, D.P.; Casola, S.; Ansel, K.M.; Xiao, C.; Xue, Y.; Murphy, A.; Frendewey, D.; Valenzuela, D.; Kutok, J.L.; *et al.* Regulation of the germinal center response by microRNA-155. *Science* **2007**, *316*, 604–608.
50. Teng, G.; Hakimpour, P.; Landgraf, P.; Rice, A.; Tuschl, T.; Casellas, R.; Papavasiliou, F.N. microRNA-155 is a negative regulator of activation-induced cytidine deaminase. *Immunity* **2008**, *28*, 621–629.
51. Cimmino, A.; Calin, G.A.; Fabbri, M.; Iorio, M.V.; Ferracin, M.; Shimizu, M.; Wojcik, S.E.; Aqeilan, R.I.; Zupo, S.; Dono, M.; *et al.* miR-15 and miR-16 induce apoptosis by targeting BCL2. *Proc. Natl. Acad. Sci. USA* **2005**, *102*, 13944–13949.

52. Stanczyk, J.; Pedrioli, D.M.; Brentano, F.; Sanchez-Pernaute, O.; Kolling, C.; Gay, R.E.; Detmar, M.; Gay, S.; Kyburz, D. Altered expression of microRNA in synovial fibroblasts and synovial tissue in rheumatoid arthritis. *Arthritis Rheum.* **2008**, *58*, 1001–1009.
53. Nakasa, T.; Miyaki, S.; Okubo, A.; Hashimoto, M.; Nishida, K.; Ochi, M.; Asahara, H. Expression of microRNA-146 in rheumatoid arthritis synovial tissue. *Arthritis Rheum.* **2008**, *58*, 1284–1292.
54. Pauley, K.M.; Satoh, M.; Chan, A.L.; Bubb, M.R.; Reeves, W.H.; Chan, E.K. Upregulated miR-146a expression in peripheral blood mononuclear cells from rheumatoid arthritis patients. *Arthritis Res. Ther.* **2008**, *10*, R101.
55. Nagata, S. Human autoimmune lymphoproliferative syndrome, a defect in the apoptosis-inducing Fas receptor: A lesson from the mouse model. *J. Hum. Genet.* **1998**, *43*, 2–8.
56. Wu, T.; Wieland, A.; Araki, K.; Davis, C.W.; Ye, L.; Hale, J.S.; Ahmed, R. Temporal expression of microRNA cluster miR-17–92 regulates effector and memory CD8+ T-cell differentiation. *Proc. Natl. Acad. Sci. USA* **2012**, *109*, 9965–9970.
57. Takahashi, H.; Kanno, T.; Nakayamada, S.; Hirahara, K.; Sciume, G.; Muljo, S.A.; Kuchen, S.; Casellas, R.; Wei, L.; Kanno, Y.; *et al.* TGF-beta and retinoic acid induce the microRNA miR-10a, which targets Bcl-6 and constrains the plasticity of helper T-cells. *Nat. Immunol.* **2012**, *13*, 587–595.
58. Kohlhaas, S.; Garden, O.A.; Scudamore, C.; Turner, M.; Okkenhaug, K.; Vigorito, E. Cutting edge: The Foxp3 target miR-155 contributes to the development of regulatory T-cells. *J. Immunol.* **2009**, *182*, 2578–2582.
59. Lu, L.F.; Thai, T.H.; Calado, D.P.; Chaudhry, A.; Kubo, M.; Tanaka, K.; Loeb, G.B.; Lee, H.; Yoshimura, A.; Rajewsky, K.; *et al.* Foxp3-dependent microRNA155 confers competitive fitness to regulatory T-cells by targeting SOCS1 protein. *Immunity* **2009**, *30*, 80–91.
60. Yao, R.; Ma, Y.L.; Liang, W.; Li, H.H.; Ma, Z.J.; Yu, X.; Liao, Y.H. microRNA-155 modulates Treg and Th17 cells differentiation and Th17 cell function by targeting SOCS1. *PLoS One* **2012**, *7*, e46082.
61. Lu, L.F.; Boldin, M.P.; Chaudhry, A.; Lin, L.L.; Taganov, K.D.; Hanada, T.; Yoshimura, A.; Baltimore, D.; Rudensky, A.Y. Function of miR-146a in controlling Treg cell-mediated regulation of Th1 responses. *Cell* **2010**, *142*, 914–929.
62. Tsai, C.Y.; Allie, S.R.; Zhang, W.J.; Usherwood, E.J. microRNA miR-155 affects antiviral effector and effector memory CD8 T-cell differentiation. *J. Virol.* **2013**, *87*, 2348–2351.
63. Brown, C.J.; Hendrich, B.D.; Rupert, J.L.; Lafreniere, R.G.; Xing, Y.; Lawrence, J.; Willard, H.F. The human XIST gene: analysis of a 17 kb inactive X-specific RNA that contains conserved repeats and is highly localized within the nucleus. *Cell* **1992**, *71*, 527–542.
64. Yang, L.; Boldin, M.P.; Yu, Y.; Liu, C.S.; Ea, C.K.; Ramakrishnan, P.; Taganov, K.D.; Zhao, J.L.; Baltimore, D. miR-146a controls the resolution of T-cell responses in mice. *J. Exp. Med.* **2012**, *209*, 1655–1670.
65. Hobeika, A.C.; Morse, M.A.; Osada, T.; Peplinski, S.; Lyerly, H.K.; Clay, T.M. Depletion of human regulatory T-cells. *Methods Mol. Biol.* **2011**, *707*, 219–231.
66. Zhou, X.; Jeker, L.T.; Fife, B.T.; Zhu, S.; Anderson, M.S.; McManus, M.T.; Bluestone, J.A. Selective miRNA disruption in T reg cells leads to uncontrolled autoimmunity. *J. Exp. Med.***2008**, *205*, 1983–1991.

67. Liston, A.; Lu, L.F.; O'Carroll, D.; Tarakhovsky, A.; Rudensky, A.Y. Dicer-dependent microRNA pathway safeguards regulatory T-cell function. *J. Exp. Med.* **2008**, *205*, 1993–2004.
68. Zhan, Y.; Davey, G.M.; Graham, K.L.; Kiu, H.; Dudek, N.L.; Kay, T.; Lew, A.M. SOCS1 negatively regulates the production of Foxp3+ CD4+ T-cells in the thymus. *Immunol. Cell Biol.* **2009**, *87*, 473–480.
69. Berland, R.; Wortis, H.H. Origins and functions of B-1 cells with notes on the role of CD5. *Annu. Rev. Immunol.* **2002**, *20*, 253–300.
70. Xu, S.; Guo, K.; Zeng, Q.; Huo, J.; Lam, K.P. The RNase III enzyme Dicer is essential for germinal center B-cell formation. *Blood* **2012**, *119*, 767–776.
71. Koralov, S.B.; Muljo, S.A.; Galler, G.R.; Krek, A.; Chakraborty, T.; Kanellopoulou, C.; Jensen, K.; Cobb, B.S.; Merkenschlager, M.; Rajewsky, N.; *et al.* Dicer ablation affects antibody diversity and cell survival in the B lymphocyte lineage. *Cell* **2008**, *132*, 860–874.
72. Chen, C.Z.; Li, L.; Lodish, H.F.; Bartel, D.P. MicroRNAs modulate hematopoietic lineage differentiation. *Science* **2004**, *303*, 83–86.
73. Chang, T.C.; Wentzel, E.A.; Kent, O.A.; Ramachandran, K.; Mullendore, M.; Lee, K.H.; Feldmann, G.; Yamakuchi, M.; Ferlito, M.; Lowenstein, C.J.; *et al.* Transactivation of miR-34a by p53 broadly influences gene expression and promotes apoptosis. *Mol. Cell* **2007**, *26*, 745–752.
74. Bommer, G.T.; Gerin, I.; Feng, Y.; Kaczorowski, A.J.; Kuick, R.; Love, R.E.; Zhai, Y.; Giordano, T.J.; Qin, Z.S.; Moore, B.B.; *et al.* p53-mediated activation of miRNA34 candidate tumor-suppressor genes. *Curr. Biol.* **2007**, *17*, 1298–1307.
75. Vigorito, E.; Perks, K.L.; Abreu-Goodger, C.; Bunting, S.; Xiang, Z.; Kohlhaas, S.; Das, P.P.; Miska, E.A.; Rodriguez, A.; Bradley, A.; *et al.* microRNA-155 regulates the generation of immunoglobulin class-switched plasma cells. *Immunity* **2007**, *27*, 847–859.
76. Dorsett, Y.; McBride, K.M.; Jankovic, M.; Gazumyan, A.; Thai, T.H.; Robbiani, D.F.; di Virgilio, M.; Reina San-Martin, B.; Heidkamp, G.; Schwickert, T.A.; *et al.* MicroRNA-155 suppresses activation-induced cytidine deaminase-mediated Myc-Igh translocation. *Immunity* **2008**, *28*, 630–638.
77. Li, H.; Li, W.X.; Ding, S.W. Induction and suppression of RNA silencing by an animal virus. *Science* **2002**, *296*, 1319–1321.
78. Watanabe, T.; Tomizawa, S.; Mitsuya, K.; Totoki, Y.; Yamamoto, Y.; Kuramochi-Miyagawa, S.; Iida, N.; Hoki, Y.; Murphy, P.J.; Toyoda, A.; *et al.* Role for piRNAs and noncoding RNA in *de novo* DNA methylation of the imprinted mouse Rasgrf1 locus. *Science* **2011**, *332*, 848–852.
79. Triboulet, R.; Mari, B.; Lin, Y.L.; Chable-Bessia, C.; Bennasser, Y.; Lebrigand, K.; Cardinaud, B.; Maurin, T.; Barbry, P.; Baillat, V.; *et al.* Suppression of microRNA-silencing pathway by HIV-1 during virus replication. *Science* **2007**, *315*, 1579–1582.
80. Matskevich, A.A.; Moelling, K. Dicer is involved in protection against influenza A virus infection. *J. Gen. Virol.* **2007**, *88*, 2627–2635.
81. Otsuka, M.; Jing, Q.; Georgel, P.; New, L.; Chen, J.; Mols, J.; Kang, Y.J.; Jiang, Z.; Du, X.; Cook, R.; *et al.* Hypersusceptibility to vesicular stomatitis virus infection in Dicer1-deficient mice is due to impaired miR24 and miR93 expression. *Immunity* **2007**, *27*, 123–134.
82. Cookson, W.; Liang, L.; Abecasis, G.; Moffatt, M.; Lathrop, M. Mapping complex disease traits with global gene expression. *Nat. Rev. Genet.* **2009**, *10*, 184–194.

83. Pauley, K.M.; Cha, S.; Chan, E.K. microRNA in autoimmunity and autoimmune diseases. *J. Autoimmun.* **2009**, *32*, 189–194.
84. Furer, V.; Greenberg, J.D.; Attur, M.; Abramson, S.B.; Pillinger, M.H. The role of microRNA in rheumatoid arthritis and other autoimmune diseases. *Clin. Immunol.* **2010**, *136*, 1–15.
85. Iborra, M.; Bernuzzi, F.; Invernizzi, P.; Danese, S. microRNAs in autoimmunity and inflammatory bowel disease: crucial regulators in immune response. *Autoimmun. Rev.* **2012**, *11*, 305–314.
86. Li, J.; Wan, Y.; Guo, Q.; Zou, L.; Zhang, J.; Fang, Y.; Fu, X.; Liu, H.; Lu, L.; Wu, Y. Altered microRNA expression profile with miR-146a upregulation in CD4+ T-cells from patients with rheumatoid arthritis. *Arthritis Res. Ther.* **2010**, *12*, R81.
87. Nambiar, M.P.; Krishnan, S.; Tsokos, G.C. T-cell signaling abnormalities in human systemic lupus erythematosus. *Methods Mol. Med.* **2004**, *102*, 31–47.
88. Zhang, D.; Yang, W.; Degauque, N.; Tian, Y.; Mikita, A.; Zheng, X.X. New differentiation pathway for double-negative regulatory T-cells that regulates the magnitude of immune responses. *Blood* **2007**, *109*, 4071–4079.
89. Divekar, A.A.; Dubey, S.; Gangalum, P.R.; Singh, R.R. Dicer insufficiency and microRNA-155 overexpression in lupus regulatory T-cells: An apparent paradox in the setting of an inflammatory milieu. *J. Immunol.* **2011**, *186*, 924–930.
90. Tang, Y.; Luo, X.; Cui, H.; Ni, X.; Yuan, M.; Guo, Y.; Huang, X.; Zhou, H.; de Vries, N.; Tak, P.P.; *et al.* microRNA-146A contributes to abnormal activation of the type I interferon pathway in human lupus by targeting the key signaling proteins. *Arthritis Rheumat.* **2009**, *60*, 1065–1075.
91. Suzuki, Y.; Kim, H.W.; Ashraf, M.; Haider, H. Diazoxide potentiates mesenchymal stem cell survival via NF-kappaB-dependent miR-146a expression by targeting Fas. *Am. J. Physiol. Heart Circ. Physiol.* **2010**, *299*, H1077–H1082.
92. Yu, D.; Tan, A.H.; Hu, X.; Athanasopoulos, V.; Simpson, N.; Silva, D.G.; Hutloff, A.; Giles, K.M.; Leedman, P.J.; Lam, K.P.; *et al.* Roquin represses autoimmunity by limiting inducible T-cell co-stimulator messenger RNA. *Nature* **2007**, *450*, 299–303.
93. Ebert, P.J.; Jiang, S.; Xie, J.; Li, Q.J.; Davis, M.M. An endogenous positively selecting peptide enhances mature T-cell responses and becomes an autoantigen in the absence of microRNA miR-181a. *Nat. Immunol.* **2009**, *10*, 1162–1169.
94. Treadwell, E.L.; Alspaugh, M.A.; Sharp, G.C. Characterization of a new antigen-antibody system (Su) in patients with systemic lupus erythematosus. *Arthritis Rheumat.* **1984**, *27*, 1263–1271.
95. Satoh, M.; Langdon, J.J.; Chou, C.H.; McCauliffe, D.P.; Treadwell, E.L.; Ogasawara, T.; Hirakata, M.; Suwa, A.; Cohen, P.L.; Eisenberg, R.A.; *et al.* Characterization of the Su antigen, a macromolecular complex of 100/102 and 200-kDa proteins recognized by autoantibodies in systemic rheumatic diseases. *Clin. Immunol. Immunopathol.* **1994**, *73*, 132–141.
96. Lipovich, L.; Johnson, R.; Lin, C.Y. MacroRNA underdogs in a microRNA world: Evolutionary, regulatory, and biomedical significance of mammalian long non-protein-coding RNA. *Biochim. Biophys. Acta* **2010**, *1799*, 597–615.

97. Cabili, M.N.; Trapnell, C.; Goff, L.; Koziol, M.; Tazon-Vega, B.; Regev, A.; Rinn, J.L. Integrative annotation of human large intergenic noncoding RNAs reveals global properties and specific subclasses. *Genes Dev.* **2011**, *25*, 1915–1927.
98. Clark, M.B.; Johnston, R.L.; Inostroza-Ponta, M.; Fox, A.H.; Fortini, E.; Moscato, P.; Dinger, M.E.; Mattick, J.S. Genome-wide analysis of long noncoding RNA stability. *Genome Res.* **2012**, *22*, 885–898.
99. Zhao, J.; Sun, B.K.; Erwin, J.A.; Song, J.J.; Lee, J.T. Polycomb proteins targeted by a short repeat RNA to the mouse X chromosome. *Science* **2008**, *322*, 750–756.
100. Lee, J.T. Epigenetic regulation by long noncoding RNAs. *Science* **2012**, *338*, 1435–1439.
101. Willingham, A.T.; Orth, A.P.; Batalov, S.; Peters, E.C.; Wen, B.G.; Aza-Blanc, P.; Hogenesch, J.B.; Schultz, P.G. A strategy for probing the function of noncoding RNAs finds a repressor of NFAT. *Science* **2005**, *309*, 1570–1573.
102. Martianov, I.; Ramadass, A.; Serra Barros, A.; Chow, N.; Akoulitchev, A. Repression of the human dihydrofolate reductase gene by a non-coding interfering transcript. *Nature* **2007**, *445*, 666–670.
103. Vigneau, S.; Rohrlich, P.S.; Brahic, M.; Bureau, J.F. Tmevpg1, a candidate gene for the control of Theiler's virus persistence, could be implicated in the regulation of gamma interferon. *J. Virol.* **2003**, *77*, 5632–5638.
104. Collier, S.P.; Collins, P.L.; Williams, C.L.; Boothby, M.R.; Aune, T.M. Cutting edge: Influence of Tmevpg1, a long intergenic noncoding RNA, on the expression of Ifng by Th1 cells. *J. Immunol.* **2012**, *189*, 2084–2088.
105. Gomez, J.A.; Wapinski, O.L.; Yang, Y.W.; Bureau, J.F.; Gopinath, S.; Monack, D.M.; Chang, H.Y.; Brahic, M.; Kirkegaard, K. The NeST long ncRNA controls microbial susceptibility and epigenetic activation of the interferon-gamma locus. *Cell* **2013**, *152*, 743–754.
106. Calin, G.A.; Liu, C.G.; Ferracin, M.; Hyslop, T.; Spizzo, R.; Sevignani, C.; Fabbri, M.; Cimmino, A.; Lee, E.J.; Wojcik, S.E.; *et al.* Ultraconserved regions encoding ncRNAs are altered in human leukemias and carcinomas. *Cancer Cell* **2007**, *12*, 215–229.
107. Calin, G.A.; Ferracin, M.; Cimmino, A.; di Leva, G.; Shimizu, M.; Wojcik, S.E.; Iorio, M.V.; Visone, R.; Sever, N.I.; Fabbri, M.; *et al.* A MicroRNA signature associated with prognosis and progression in chronic lymphocytic leukemia. *N. Engl. J. Med.* **2005**, *353*, 1793–1801.
108. Chen, X.; Ba, Y.; Ma, L.; Cai, X.; Yin, Y.; Wang, K.; Guo, J.; Zhang, Y.; Chen, J.; Guo, X.; *et al.* Characterization of microRNAs in serum: a novel class of biomarkers for diagnosis of cancer and other diseases. *Cell Res.* **2008**, *18*, 997–1006.
109. Lawrie, C.H.; Gal, S.; Dunlop, H.M.; Pushkaran, B.; Liggins, A.P.; Pulford, K.; Banham, A.H.; Pezzella, F.; Boultwood, J.; Wainscoat, J.S.; *et al.* Detection of elevated levels of tumour-associated microRNAs in serum of patients with diffuse large B-cell lymphoma. *Br. J. Haematol.* **2008**, *141*, 672–675.
110. Park, N.J.; Zhou, H.; Elashoff, D.; Henson, B.S.; Kastratovic, D.A.; Abemayor, E.; Wong, D.T. Salivary microRNA: Discovery, characterization, and clinical utility for oral cancer detection. *Clin. Cancer Res.* **2009**, *15*, 5473–5477.
111. Kosaka, N.; Izumi, H.; Sekine, K.; Ochiya, T. microRNA as a new immune-regulatory agent in breast milk. *Silence* **2010**, *1*, 7.

112. Mitchell, P.S.; Parkin, R.K.; Kroh, E.M.; Fritz, B.R.; Wyman, S.K.; Pogosova-Agadjanyan, E.L.; Peterson, A.; Noteboom, J.; O'Briant, K.C.; Allen, A.; *et al.* Circulating microRNAs as stable blood-based markers for cancer detection. *Proc. Natl. Acad. Sci. USA* **2008**, *105*, 10513–10518.
113. Ji, X.; Takahashi, R.; Hiura, Y.; Hirokawa, G.; Fukushima, Y.; Iwai, N. Plasma miR-208 as a biomarker of myocardial injury. *Clin. Chem.* **2009**, *55*, 1944–1949.
114. Arroyo, J.D.; Chevillet, J.R.; Kroh, E.M.; Ruf, I.K.; Pritchard, C.C.; Gibson, D.F.; Mitchell, P.S.; Bennett, C.F.; Pogosova-Agadjanyan, E.L.; Stirewalt, D.L.; *et al.* Argonaute2 complexes carry a population of circulating microRNAs independent of vesicles in human plasma. *Proc. Natl. Acad. Sci. USA* **2011**, *108*, 5003–5008.
115. Valadi, H.; Ekstrom, K.; Bossios, A.; Sjostrand, M.; Lee, J.J.; Lotvall, J.O. Exosome-mediated transfer of mRNAs and microRNAs is a novel mechanism of genetic exchange between cells. *Nat. Cell Biol.* **2007**, *9*, 654–659.
116. Zernecke, A.; Bidzhekov, K.; Noels, H.; Shagdarsuren, E.; Gan, L.; Denecke, B.; Hristov, M.; Koppel, T.; Jahantigh, M.N.; Lutgens, E.; *et al.* Delivery of microRNA-126 by apoptotic bodies induces CXCL12-dependent vascular protection. *Sci. Signal.* **2009**, *2*, ra81.
117. Zhang, Y.; Liu, D.; Chen, X.; Li, J.; Li, L.; Bian, Z.; Sun, F.; Lu, J.; Yin, Y.; Cai, X.; *et al.* Secreted monocytic miR-150 enhances targeted endothelial cell migration. *Mol. Cell* **2010**, *39*, 133–144.
118. Bhatnagar, S.; Schorey, J.S. Exosomes released from infected macrophages contain *Mycobacterium avium* glycopeptidolipids and are proinflammatory. *J. Biol. Chem.* **2007**, *282*, 25779–25789.
119. Cocucci, E.; Racchetti, G.; Meldolesi, J. Shedding microvesicles: Artefacts no more. *Trends Cell Biol.* **2009**, *19*, 43–51.
120. Hunter, M.P.; Ismail, N.; Zhang, X.; Aguda, B.D.; Lee, E.J.; Yu, L.; Xiao, T.; Schafer, J.; Lee, M.L.; Schmittgen, T.D.; *et al.* Detection of microRNA expression in human peripheral blood microvesicles. *PLoS One* **2008**, *3*, e3694.
121. Collino, F.; Deregibus, M.C.; Bruno, S.; Sterpone, L.; Aghemo, G.; Viltono, L.; Tetta, C.; Camussi, G. Microvesicles derived from adult human bone marrow and tissue specific mesenchymal stem cells shuttle selected pattern of miRNAs. *PLoS One* **2010**, *5*, e11803.
122. Muller, G.; Schneider, M.; Biemer-Daub, G.; Wied, S. Microvesicles released from rat adipocytes and harboring glycosylphosphatidylinositol-anchored proteins transfer RNA stimulating lipid synthesis. *Cell. Signal.* **2011**, *23*, 1207–1223.
123. Nolte-'t Hoen, E.N.; Buermans, H.P.; Waasdorp, M.; Stoorvogel, W.; Wauben, M.H.; 't Hoen, P.A. Deep sequencing of RNA from immune cell-derived vesicles uncovers the selective incorporation of small non-coding RNA biotypes with potential regulatory functions. *Nucleic Acids Res.* **2012**, *40*, 9272–9285.
124. Mittelbrunn, M.; Gutierrez-Vazquez, C.; Villarroya-Beltri, C.; Gonzalez, S.; Sanchez-Cabo, F.; Gonzalez, M.A.; Bernad, A.; Sanchez-Madrid, F. Unidirectional transfer of microRNA-loaded exosomes from T-cells to antigen-presenting cells. *Nat. Commun.* **2011**, *2*, 282.
125. Monks, C.R.; Freiberg, B.A.; Kupfer, H.; Sciaky, N.; Kupfer, A. Three-dimensional segregation of supramolecular activation clusters in T-cells. *Nature* **1998**, *395*, 82–86.

10

Advances in the Techniques for the Prediction of microRNA Targets

Hao Zheng [1,2,†], Rongguo Fu [3,†], Jin-Tao Wang [4], Qinyou Liu [5], Haibin Chen [6,*] and Shi-Wen Jiang [2,*]

[1] School of Electrical and Computer Engineering, Georgia Institute of Technology, 225 North Avenue NW, Atlanta, GA 30301, USA; E-Mails: hzheng7@gatech.edu

[2] Department of Biomedical Science, Mercer University School of Medicine, Savannah Campus, 4700 Waters Ave, Savannah, GA 31404, USA

[3] Department of Nephrology, Second Affiliated Hospital, School of Medicine, Xi'an Jiaotong University, Xi'an 710004, Shannxi, China; E-Mail: furongguo@mail.xjtu.edu.cn

[4] Department of Epidemiology, School of Public Health, Shanxi Medical University, 56 Xin Jian Nan Road, Taiyuan 030001, Shanxi, China; E-Mail: wjtxw@yahoo.com.cn

[5] Animal Reproduction Institute of Guangxi University, 100 Daxue Road, Nanning 530005, Guangxi, China; E-Mail: qyliu2002@126.com

[6] Department of Histology and Embryology, Shantou University Medical College, 22 Xinling Road, Shantou 515041, Guangdong, China

† These authors contributed equally to this work.

* Authors to whom correspondence should be addressed; E-Mails: jiang_s@mercer.edu (S.-W.J.); chenhb@stu.edu.cn (H.C.);

Abstract: MicroRNAs (miRNAs) are small, non-coding, endogenous RNA molecules that play important roles in a variety of normal and diseased biological processes by post-transcriptionally regulating the expression of target genes. They can bind to target messenger RNA (mRNA) transcripts of protein-coding genes and negatively control their translation or cause mRNA degradation. miRNAs have been found to actively regulate a variety of cellular processes, including cell proliferation, death, and metabolism. Therefore, their study is crucial for the better understanding of cellular functions in eukaryotes. To better understand the mechanisms of miRNA: mRNA interaction and their cellular functions, it is important to identify the miRNA targets accurately. In this paper, we provide a brief review for the advances in the animal miRNA target prediction methods and available resources to facilitate further study of miRNAs and their functions.

Keywords: prediction; microRNA; feature selection

1. Introduction

In addition to DNA methylation and histone modification, epigenetic mechanisms have recently been extended to microRNAs (miRNAs), which are important regulators of gene expression in many biological systems. miRNAs are small, non-coding, endogenous RNA molecules, about 19–24 nucleotides in length that can negatively control their target gene expression post-transcriptionally [1]. This is mainly achieved by recognizing and binding to the 3' untranslated region of the target messenger RNA (mRNA) sequences [2]. miRNAs have been found to actively regulate a variety of cellular processes, including cell proliferation, death, and metabolism, and therefore, their study is crucial for the better understanding of cellular functions in eukaryotes [3].

Mature miRNAs are incorporated into the RNA-induced silencing complex (RISC), where miRNAs specifically interact with target mRNAs. Approximately one thousand miRNAs have been discovered in humans and are believed to control more than half of the protein coding genes, where a single miRNAs might regulate hundreds of such genes [4]. This one-to-multiple mapping presents a hurdle in accurately identifying the miRNA targets. Furthermore, miRNAs are only partially complementary to their mRNA target sequences. Such imperfections in base matching (e.g., a mismatch or bulge) make it even more difficult to accurately predict the miRNA targets *in silico* [4].

In this paper, we provide a brief review on the advances in the miRNA target prediction methods and available resources. The readers are referred to the literature cited in this review, and the references therein for further details.

2. Methods for miRNA Target Recognition

A key step in the identification of miRNA target is the selection of features that are potentially of predictive power. Many researchers are devoted to such an effort, and quite a number of predictive features have been discovered. Such features include dinucleotide composition of flanking sequence [5,6], strong base pairing between the 3' UTR of mRNAs and the miRNA seed region [7], thermodynamic stability of binding sites [8], evolutionary conservation of binding sites (particularly the seed region) [5,9], secondary structure accessibility [10,11], and host genes expression profiles [12].

The most commonly used predictive features include characteristics in the seed regions and the phylogenetic conservation of miRNA binding sites, and almost all the existing methods take advantage of such features in the algorithm.

For example, by identifying mRNAs with strong base pairing to the 5' region of the miRNA and evaluating the number and quality of these complementary sites, Lewis *et al.* identified more than 400 regulatory target genes for the conserved vertebrate miRNAs [7]. Likewise, another popular algorithm PicTar [13–17] similarly incorporated seed constraints for the identification of miRNA targets. The new doRiNA database offers computational miRNA target site predictions for human, mouse and worm, and these predictions constitute the most recent update of PicTar predictions [17]. It is notable that some researchers have questioned the universality of the seed assumption, demonstrating that several experimentally confirmed miRNA targets do not seem to meet the seed region criterion. So far, the seed assumption is not unanimously accepted as a method to identify all miRNA targets, and that some relevant miRNA:mRNA interactions might not exhibit the seed region property [18].

With the purpose of enhancing the specificity of prediction for functional target sites, many computational studies also incorporated the evolution conservation [9,14,19–22] or flagged conserved putative targets [8,23]. Particularly, ElMMo [22] incorporated such conservation statistics in a more general, rigorous and miRNA-dependent manner. Also, Friedman *et al.* developed a quantitative method for evaluating evolutionary conservation of binding sites and applied this to the study of vertebrate miRNA targeting With this method, they found three times as many preferentially conserved sites as detected previously, further increasing the known scope and density of conserved miRNA regulatory interactions [9].

Another commonly used feature for target recognition includes the thermodynamic stability of binding sites. It is believed that the formation of a stable miRNA:target binding *in vivo*, to some extent, must be governed by thermodynamic stability. With the rationale that this binding is a process where free energy changes occur through the formation of a miRNA:target duplex, such changes may help detect miRNA targets [24,25]. The computation of energy can vary, but most methods focus only on a particular form of energy (*i.e.*, hybridization) [7,14,23,26,27]. For example, Rehmsmeier *et al.* developed a program, named RNA-hybrid, which predicts multiple potential binding sites of miRNAs in large target RNAs based on the thermodynamic stability of binding sites [8].

However, more recently, combining target accessibility and duplex stability [11,28], integrated thermodynamic features for miRNA target prediction demonstrated more effectiveness. In addition, based on the immuno-precipitation (IP) of the RISC components, AIN-1 and AIN-2, Hammell *et al.* presented that total free energy change and target accessibility yielded enrichments in miRISC-enriched transcripts [25,29]. In addition to incorporating accessibility into an energy parameter [28], methods to calculate target accessibility differ, including A/U nucleotides [5,10] and larger nucleotide window to the 5' of the binding site [29]. More specifically, for example, the Sfold method was used to fold whole 3' UTR sequences plus 300 nucleotides of adjacent coding sequence for all predicted *C. elegans* transcripts. The output of Sfold was then used to calculate the average accessibility over 25 nucleotide windows flanking each potential microRNA binding site [29].

Expression-based approaches are also becoming popular to elucidate miRNA-mRNA associations. Based on expression profiles of host genes, Radfar *et al.* introduced a new computational method InMiR, which uses a linear-Gaussian model for the prediction of targets of intronic miRNAs [12]. They separated intronic miRNAs into three classes: those that are tightly regulated with their host gene; those that are likely to be expressed from the same promoter but whose host gene is highly regulated by miRNAs; and those likely to have independent promoters. Compared to a method considering only correlation, this method recovered nearly twice as many true positives as the same fixed false positive rate [12]. Engelmann *et al.* recently also showed that entire mRNA expression profiles or large groups of them can be reconstructed only from miRNA expression, and *vice versa.*

This introduced a regression model for the prediction of canonical and non-canonical miRNA-mRNA interactions [30].

Furthermore, machine learning algorithms can also be used to intelligently search for the parameters with most predictive power of genuine miRNA binding sites. An example of a method for miRNA target prediction is TargetBoost, which uses machine learning based on a set of validated miRNA targets in lower organisms to create weighted sequence motifs that capture binding characteristics between miRNAs and their targets [31]. Combining genetic programming with boosting, TargetBoost generates a metric that represents the likelihood of a site being targeted by the miRNA.

3. Resources for miRNA Target Prediction

Various popular resources for miRNA target predictions are summarized in Table 1. Different miRNA target prediction algorithms can provide differing results, and often researchers need to cross check multiple algorithms to get an additional layer of confidence for the true positive targets. For example, Ryland *et al.* incorporated miRanda [32], microCOSM Targets [33], DIANA-MicroT [27,34] and TargetScan [9] to determine whether the variants detected in mRNA 3' UTRs occurred within miRNA binding sites [35]. To facilitate that end, starBase was developed to provide a comprehensive exploration of miRNA-target interaction maps from CLIP-Seq and Degradome-Seq data [36]. This allows for a search of commonly agreed upon targets predicted by different algorithms, including TargetScan, PicTar, PITA, miRanda and RNA22 [37]. For example, when TargetScan and PicTar are selected, the database will output target sites predicted by both TargetScan and PicTar programs. This resource greatly facilitates inter-method and inter-database consensus comparison of miRNA targets. In addition, miRTar, an integrated system for miRNA target prediction, enables biologists to easily identify biological functions and regulatory relationships between a group of known/putative miRNAs and protein coding genes. Furthermore, this database delivers perspective information on miRNA targets and their alternatively spliced transcripts [38].

Table 1. Summary of prediction techniques for miRNA target recognition.

Method	Feature	References	Availability
TargetScan(S)	Database of microRNA targets conserved in 5 vertebrates.	[7,19]	http://genes.mit.edu/tscan/targetscanS2005.html
miRanda	Optimizes sequence complementarity based on position-specific rules and interspecies conservation.	[23,32,39]	http://www.microrna.org
RNA-hybrid	Determines the most favourable hybridization site between two sequences.	[8,40]	http://bibiserv.techfak.uni-bielefeld.de/rnahybrid
PicTar (including doRiNA)	Provides details about 3' UTR alignments with predicted sites, and links to various public databases.	[13–17]	http://pictar.mdc-berlin.de

Table 1. *Cont.*

Method	Feature	References	Availability
TargetBoost	Learns the hidden rules of miRNA-target site hybridization based on machine learning.	[31]	http://www.interagon.com/demo
PITA	Investigates the role of target-site accessibility, as determined by base-pairing interactions within the mRNA.	[11]	http://genie.weizmann.ac.il/pubs/mir07/index.html
ElMMo	Infers miRNA targets using evolutionary conservation and pathway analysis.	[22]	http://www.mirz.unibas.ch/ElMMo2/
Singh's	Predicts and characterizes 45 miRNAs by genome-wide homology search against all the reported miRNAs.	[41]	http://www.cdfd.org.in/lmg/PDF/imb816.pdf
mirWIP	Employs structural accessibility of target sequences, the total free energy of microRNA:target hybridization, and the topology of base-pairing to the 5 seed region of the microRNA.	[29]	http://ambroslab.org
microCOSM Targets	Web resource containing computationally predicted targets for microRNAs across many species.	[33]	http://www.ebi.ac.uk/enright-srv/microcosm/htdocs/targets/v5/
DIANA-microT 3.0	Individually calculate several parameters for each microRNA and combines conserved and non-conserved microRNA recognition elements into a final prediction score.	[27,34]	http://www.microrna.gr/microT
starBase	Database with intersections among targets by five predictive softwares.	[36]	http://starbase.sysu.edu.cn/clipSeqIntersection.php
InMiR	Uses a linear-Gaussian model, and provides a dataset of 1,935 predicted mRNA targets for 22 intronic miRNAs.	[12]	http://www.plosone.org
miRTar	Identifies the biological functions and regulatory relationships between a group of known/putative miRNAs and protein coding genes.	[38]	http://mirtar.mbc.nctu.edu.tw/human/

4. Next-Generation Sequencing for miRNA Target Identification

With the advances of next-generation sequencing, high-throughput, systematic identification of specific miRNAs targets in a relatively short time became realistic. Several resources using CLIP-seq data to identify miRNA targets were developed, including Piranha [42], CLIPZ [43] and starBase [36].

Piranha [42] provides a utility for peak-calling based on a zero-truncated negative binomial regression model, which is able to incorporate external information to help guide the target identification process. CLIPZ provides a database and analysis environment for experimentally determined binding sites of RNA-binding proteins [43].

5. Future Work

Although quite a number of methods and databases have been developed for the identification of miRNA targets, most methods have a false positive rate (FPR) greater than 0.3, which means that the specificity is often lower than 70%. FPR is evaluated as (1-specificity), where specificity is defined as the ratio of the number of true negatives and true negatives plus false positives. Filtering for true positive targets from the large predicted target lists is challenging and time consuming. Although conservation and functional similarities have been taken advantage of to reduce false positives, there is still much room for improvement. Since different miRNA target prediction algorithms still provide varying results, this indicates that such methods also suffer from higher rates of false negatives. As a result, highly accurate prediction algorithms with small false positive and false negative rates need to be further developed. Such algorithms are crucial to studying the exact role of miRNA in signaling pathways, as well as associations with various disease pathways.

To better perform the comparative study of different methods, it is imperative to have some "gold standard" data sets, and quantitatively evaluate different methods based on a fixed set of metrics. The establishment of a gold standard requires strong experimental evidence (reporter assay or western blot analysis) as well as consensus across independent experiments.

Acknowledgments

The authors would like to thank Charlie Bodine for his efforts to edit the manuscript.

Shi-Wen Jiang is a Distinguished Cancer Scholar supported by Georgia Cancer Coalition (GCC). This project is partially supported by Mercer University School of Medicine Seed Research Funding (S-W Jiang), The Memorial Health Hospital and Anderson Cancer Institute Pancreatic Cancer Program (S-W Jiang), and the research grant of the Natural Science Foundation of China (81070590; 81100530; R Fu).

References

1. Chuang, J.; Jones, P. Epigenetics and MicroRNAs. *Pediatr. Res.* **2007**, *61*, 24R–29R.
2. Yue, D.; Liu, H.; Huang, Y. Survey of computational algorithms for microRNA target prediction. *Curr. Genomics* **2009**, *10*, 478–492.
3. Bartel, D. MicroRNAs: Genomics, biogenesis, mechanism, and function. *Cell* **2004**, *116*, 281–297.
4. Yue, D.; Meng, J.; Lu, M.; Chen, C.; Guo, M.; Huang, Y. Understanding microRNA regulation: A computational perspective. *Signal Proc. Mag. IEEE* **2012**, *29*, 77–88.
5. Nielsen, C.; Shomron, N.; Sandberg, R.; Hornstein, E.; Kitzman, J.; Burge, C. Determinants of targeting by endogenous and exogenous microRNAs and siRNAs. *RNA* **2007**, *13*, 1894–1910.
6. Ohler, U.; Yekta, S.; Lim, L.; Bartel, D.; Burge, C. Patterns of flanking sequence conservation and a characteristic upstream motif for microRNA gene identification. *RNA* **2004**, *10*, 1309–1322.

7. Lewis, B.; Shih, I.; Jones-Rhoades, M.; Bartel, D.; Burge, C. Prediction of mammalian microRNA targets. *Cell* **2003**, *115*, 787–798.
8. Rehmsmeier, M.; Steffen, P.; Hochsmann, M.; Giegerich, R. Fast and effective prediction of microRNA/target duplexes. *RNA* **2004**, *10*, 1507–1517.
9. Friedman, R.; Farh, K.; Burge, C.; Bartel, D. Most mammalian mRNAs are conserved targets of microRNAs. *Genome Res.* **2009**, *19*, 92–105.
10. Grimson, A.; Farh, K.; Johnston, W.; Garrett-Engele, P.; Lim, L.; Bartel, D. MicroRNA targeting specificity in mammals: Determinants beyond seed pairing. *Mol. Cell* **2007**, *27*, 91–105.
11. Kertesz, M.; Iovino, N.; Unnerstall, U.; Gaul, U.; Segal, E. The role of site accessibility in microrna target recognition. *Nat. Genet.* **2007**, *39*, 1278–1284.
12. Radfar, M.; Wong, W.; Morris, Q. Computational prediction of intronic microRNA targets using host gene expression reveals novel regulatory mechanisms. *PLoS One* **2011**, *6*, e19312.
13. Grun, D.; Wang, Y.; Langenberger, D.; Gunsalus, K.; Rajewsky, N. microRNA target predictions across seven drosophila species and comparison to mammalian targets. *PLoS Comput. Biol.* **2005**, *1*, e13.
14. Krek, A.; Grun, D.; Poy, M.; Wolf, R.; Rosenberg, L.; Epstein, E.; MacMenamin, P.; Piedade, I.; Gunsalus, K.; Stoffel, M.; *et al.* Combinatorial microrna target predictions. *Nat. Genet.* **2005**, *37*, 495–500.
15. Lall, S.; Grun, D.; Krek, A.; Chen, K.; Wang, Y.; Dewey, C.; Sood, P.; Colombo, T.; Bray, N.; Macmenamin, P.; *et al.* A genome-wide map of conserved microrna targets in *C. elegans*. *Curr. Biol.* **2006**, *16*, 460–471.
16. Chen, K.; Rajewsky, N. Natural selection on human microRNA binding sites inferred from SNP data. *Nat. Genet.* **2006**, *38*, 1452–1456.
17. Anders, G.; Mackowiak, S.; Jens, M.; Maaskola, J.; Kuntzagk, A.; Rajewsky, N.; Landthaler, M.; Dieterich, C. doRiNa: A database of RNA interactions in post-transcriptional regulation. *Nucleic Acids Res.* **2012**, *40*, D180–D186.
18. Liu, B.; Li, J.; Cairns, M.J. Identifying miRNAs, targets and functions. *Brief Bioinforma.* **2012**, doi:10.1093/bib/bbs075.
19. Lewis, B.; Burge, C.; Bartel, D. Conserved seed pairing, often flanked by adenosines, indicates that thousands of human genes are microRNA targets. *Cell* **2005**, *120*, 15–20.
20. Stark, A.; Brennecke, J.; Bushati, N.; Russell, R.; Cohen, S. Animal microRNAs confer robustness to gene expression and have a significant impact on 3'UTR evolution. *Cell* **2005**, *123*, 1133–1146.
21. Johnson, S.M.; Grosshans, H.; Shingara, J.; Byrom, M.; Jarvis, R.; Cheng, A.; Labourier, E.; Reinert, K.L.; Brown, D.; Slack, F.J. RAS is regulated by the let-7 microRNA family. *Cell* **2005**, *120*, 635–647.
22. Gaidatzis, D.; Nimwegen, E.; Hausser, J.; Zavolan, M. Inference of miRNA targets using evolutionary conservation and pathway analysis. *BMC Bioinforma.* **2007**, *8*, 69:1–69:22.
23. John, B.; Enright, A.; Aravin, A.; Tuschl, T.; Sander, C.; Marks, D. Human microRNA targets. *PLoS Biol.* **2004**, *2*, e363.
24. Hammell, M. Computational methods to identify miRNA targets. *Semin. Cell Dev. Biol.* **2010**, *21*, 738–744.

25. Lekprasert, P.; Mayhew, M.; Ohler, U. Assessing the utility of thermodynamic features for microRNA target prediction under relaxed seed and no conservation requirements. *PLoS One* **2011**, *6*, e20622.
26. Stark, A.; Brennecke, J.; Russell, R.; Cohen, S. Identification of drosophila microRNA targets. *PLoS Biol.* **2003**, *1*, e60.
27. Maragkakis, M.; Alexiou, P.; Papadopoulos, G.L.; Reczko, M.; Dalamagas, T.; Giannopoulos, G.; Goumas, G.; Koukis, E.; Kourtis, K.; Simossis, V.A.; *et al.* Accurate microRNA target prediction correlates with protein repression levels. *BMC Bioinforma.* **2009**, *10*, 295:1–295:10.
28. Long, D.; Lee, R.; Williams, P.; Chan, C.; Ambros, V. Potent effect of target structure on microRNA function. *Nat. Struct. Mol. Biol.* **2007**, *14*, 287–294.
29. Hammell, M.; Long, D.; Zhang, L.; Lee, A.; Carmack, C. mir-WIP: MicroRNA target prediction based on microrna-containing ribonucleoprotein-enriched transcripts. *Nat. Methods* **2008**, *5*, 813–819.
30. Engelmann, J.; Spang, R. A least angle regression model for the prediction of canonical and non-canonical miRNA-mRNA interactions. *PLoS One* **2012**, *7*, e40634.
31. Saetrom, O.; Snove, O.; Saetrom, P. Weighted sequence motifs as an improved seeding step in microRNA target prediction algorithms. *Cell* **2005**, *1*, 995–1003.
32. Betel, D.; Wilson, M.; Gabow, A.; Marks, D.; Sander, C. The microrna.org resource: Targets and expression. *Nucleic Acids Res.* **2008**, *36*, D149–D153.
33. Griffiths-Jones, S.; Saini, H.; Dongen, S.; Enright, A. Mirbase: Tools for microRNA genomics. *Nucleic Acids Res.* **2008**, *36*, D154–D158.
34. Maragkakis, M.; Reczko, M.; Simossis, V.; Alexiou, P.; Papadopoulos, G. Diana-microt web server: Elucidating microRNA functions through target prediction. *Nucleic Acids Res.* **2009**, *37*, W273–W276.
35. Ryland, G.; Bearfoot, J.; Doyle, M.; Boyle, S.; Choong, D.; Rowley, S. MicroRNA genes and their target 3' untranslated regions are infrequently somatically mutated in ovarian cancers. *PLoS One* **2012**, *7*, e35805.
36. Yang, J.; Li, J.; Shao, P.; Zhou, H.; Chen, Y.; Qu, L. starBase: A database for exploring microRNAcmRNA interaction maps from Argonaute CLIP-seq and Degradome-Seq data. *Nucleic Acids Res.* **2011**, *39*, D202–D209.
37. Miranda, K.; Huynh, T.; Tay, Y.; Ang, Y.; Tam, W.; Thomson, A.; Lim, B.; Rigoutsos, I. A pattern-based method for the identification of MicroRNA binding sites and their corresponding heteroduplexes. *Cell* **2006**, *126*, 1203–1217.
38. Hsu, J.B.; Chiu, C.M.; Hsu, S.D.; Huang, W.Y.; Chien, C.H.; Lee, T.Y.; Huang, H.D. miRTar: An integrated system for identifying miRNA-target interactions in human. *BMC Bioinforma.* **2011**, *12*, 300:1–300:12.
39. Enright, A.; John, B.; Gaul, U.; Tuschl, T.; Sander, C.; Marks, D. MicroRNA targets in drosophila. *Genome Biol.* **2003**, *5*, R1–R14.
40. Kruger, J.; Rehmsmeier, M. RNAhybrid: MicroRNA target prediction easy, fast and flexible. *Nat. Genet.* **2005**, *37*, 495–500.
41. Singh, J.; Nagaraju, J. *In silico* prediction and characterization of microRNAs from red flour beetle (*Tribolium castaneum*). *Insect Mol. Biol.* **2008**, *17*, 427–436.

42. Uren, P.J.; Bahrami-Samani, E.; Burns, S.C.; Qiao, M.; Karginov, F.V.; Hodges, E.; Hannon, G.J.; Sanford, J.R.; Penalva, L.O.; Smith, A.D. Site identification in high-throughput RNA–protein interaction data. *Bioinformatics* **2012**, *1*, 3013–3020.
43. Khorshid, M.; Rodak, C.; Zavolan, M.; CLIPZ: A database and analysis environment for experimentally determined binding sites of RNA-binding proteins. *Nucleic Acids Res.* **2011**, *39*, D245–D252.

11

Intercellular Communication by Exosome-Derived microRNAs in Cancer

Bethany N. Hannafon * and Wei-Qun Ding *

Department of Pathology, University of Oklahoma Health Sciences Center, Oklahoma City, OK 73104, USA; E-Mails: bethany-hannafon@ouhsc.edu (B.N.H.); weiqun-ding@ouhsc.edu (W-Q.D.);

Abstract: The development of human cancers is a multistep process in which normal cells acquire characteristics that ultimately lead to their conversion into cancer cells. Many obstacles must be overcome for this process to occur; of these obstacles, is the ability to survive an inhospitable microenvironment. It is recognized that the intercommunication between tumor cells and their surrounding microenvironment is essential to overcoming this obstacle and for the tumor to progress, metastasize and establish itself at distant sites. Exosomes are membrane-derived vesicles that have recently been recognized as important mediators of intercellular communication, as they carry lipids, proteins, mRNAs and microRNAs that can be transferred to a recipient cell via fusion of the exosome with the target cell membrane. In the context of cancer cells, this process entails the transfer of cancer-promoting cellular contents to surrounding cells within the tumor microenvironment or into the circulation to act at distant sites, thereby enabling cancer progression. In this process, the transfer of exosomal microRNAs to a recipient cell where they can regulate target gene expression is of particular interest, both in understanding the basic biology of cancer progression and for the development of therapeutic approaches. This review discusses the exosome-mediated intercellular communication via microRNAs within the tumor microenvironment in human cancers, with a particular focus on breast cancer exosomes.

Keywords: exosomes; microvesicles; microRNAs; cancer; breast cancer; microenvironment

---

1. Introduction

1.1. Historical Perspective

Most cell types are known to continually release soluble factors and to exfoliate membrane derived vesicles into the extracellular space, including mast cells, dendritic cells, B-lymphocytes, platelets, neurons, adipocytes, endothelial cells and epithelial cells [1]. These membrane-derived vesicles are generally discriminated by size with two major classes; the larger class is called microvesicles (200–1000 nm) and the smaller class of nanometer size vesicles is called exosomes (30–200 nm). It is important to note that these are distinctly different from apoptotic bodies (0.5–3 μm), which are released from cells undergoing apoptosis or mechanical stress and are of a different cellular origin and molecular composition (see, for review, [2]). Exosomes were first observed three decades ago in differentiating reticulocytes. It was shown that during reticulocyte maturation, the transferrin receptor and many membrane-associated proteins were shed in small membrane vesicles via an unknown secretory process [3,4]. This process was considered as a way for cells to eliminate unwanted proteins and molecules, with exosomes functioning as cellular garbage disposals. However, in recent years, exosomes have emerged as important mediators of cellular communication that are involved in both normal physiological processes, such as lactation [4], immune response [5] and neuronal function [4], and also in the development and progression of diseases, such as liver disease [6], neurodegenerative diseases [7] and cancer. Exosomes have been identified in most bodily fluids, including urine and amniotic fluid [8], serum [9], saliva [10], breast-milk [5], cerebrospinal fluid [11], and nasal secretions [12]. Importantly, cancer cells have been shown to secrete exosomes in greater amounts than normal cells [13], indicating their potential use as biomarkers for diagnosis of disease.

1.2. Exosome Biogenesis and Secretion

Although the detailed mechanism for exosome biogenesis remains incompletely defined, current models suggest that exosomes are formed within the endocytic pathway and released from the plasma membrane via multivesicular bodies (MVBs) [14]. MVBs are formed during the maturation of early endosomes into late endosomes with the accumulation of intraluminal vesicles (ILVs), which correspond to exosomes [15]. Upon maturation, MVBs are either destined for fusion with the lysosome, where their contents will undergo lysosomal degradation, or with the plasma membrane, where their contents are released into the extracellular space. How these vesicles are sorted for either destination is not well understood. However, central players in this process are thought to be the endosomal sorting complexes required for transport (ESCRT). The ESCRT machinery is made up of five distinct complexes (ESCRT-0, -I, -II, -III and Vsp4; reviewed in [16]). This process is best characterized in yeast [17], where it was shown that the ESCRT machinery is responsible for generating MVB vesicles by initiating the budding of the endosome away from the cytoplasm and scission of the membrane to release of the mature MVB vesicles into the lumen of the lysosome (reviewed in [18]). ESCRT-0, -I and -II complexes recognize ubiquitinated proteins in the endosomal membrane [19], whereas the ESCRT-III complex may be responsible for membrane budding and vesicle scission [20]. However, there are additional pathways for MVB formation, sorting and exosome secretion. Most recently, an ESCRT-independent mechanism was described involving the sphingolipid, ceramide. Ceramide is generated during cellular stress and apoptosis either by *de novo* synthesis or by sphingomyelinase, the enzyme that hydrolyzes sphingomyelin into ceramide. Ceramide contributes to cellular signaling by playing a role in membrane microdomain coalescence, receptor clustering, vesicle formation, membrane fusion/fission and vesicular trafficking [21]. Additionally, ceramide is enriched in exosome membranes (see, for review, [21,22]). Further validating the ESCRT-independent process, inhibition of neutral sphingomyelinase (nSMase) decreased exosome formation and release, whereas depletion of different ESCRT components did not reduce exosome secretion or the formation of MVBs [23]. Interestingly, exosomes produced by the ESCRT-independent/sphingomyelinase pathway are enriched in tetraspanins, which are transmembrane proteins that may also be involved in endosomal sorting pathways [24]. Based on these observations, ESCRT-dependent sorting mechanisms may target proteins loaded into ILVs for lysosomal degradation, whereas ESCRT-independent sorting mechanisms may target ILVs for secretion. However, it is likely that this process is much more complex and may depend on the cell type, cargo or other stimulus. Furthermore, the signals that may control the switch between the two mechanisms remain unknown.

Exosome secretion is not considered a random event, but rather, a highly controlled process. Control of exosome secretion or "exocytosis", although largely still under investigation, is thought to be coordinated through the transport and fusion of MVBs with the plasma membrane by the microtubule and actin cytoskeleton, t- and v-SNAREs and Rab GTPases [25]. Rab GTPases are ubiquitously expressed proteins that are responsible for the coordination of various vesicle trafficking events [26]. For example, overexpression of Rab11 has been shown to stimulate exocytosis [27], and Rab27a and Rab27b control different steps of the exosome secretion pathway [28]. Exosome secretion can be initiated by many different mechanical, chemical and biological stimuli. For example, DNA damage due to γ-irradiation activates the p53 tumor suppressor gene and induces the release of exosomes [29]. When breast cancer cells were cultured under hypoxic conditions (1% to 0.1% O_2), their exosome secretion was significantly enhanced, whereas siRNA knockdown of HIF-1α prior to hypoxic exposure prevented this increase in exosome secretion [30]. Heparanase, an enzyme that cleaves heparan sulfate, which is upregulated in many cancers and is associated with enhanced tumor growth, was shown to dramatically increase exosome secretion in several human cancer cell lines [31]. Interestingly, heparanase could also alter the protein cargo carried by these exosomes, with increases in levels of syndecan-1, VEGF and HGF. Mechanical changes can also affect exosome secretion. For example, it was recently shown that detachment of adherent breast cancer cells from various surfaces could induce rapid exosome secretion [32]. Treatment with chemicals, such as calcium ionophores or statins to reduce membrane cholesterol levels and cholesterol biosynthesis, can also stimulate the release of exosomes in many cell types [23,33,34]. Finally, low pH level, a common hallmark of malignancy, in melanoma cells has been shown to increase exosome release and uptake, and pre-treatment with a proton pump inhibitor led to an inhibition of exosome uptake [35].

In order for secreted exosomes to exert any biological function, they must be absorbed by and deliver their contents to a recipient target cell. However, the specific targeting of exosomes to target cells and how this process unfolds in normal physiology or in the diseased state is not well understood. This process must critically depend on the specific adhesion molecules, integrins and antigenic factors expressed on the exosome, as well as the receptors or other docking molecules found on the surface of target cells. Presumably, any cell capable of endocytosis or phagocytosis may participate in the uptake of exosomes. Many studies have documented the uptake of exosomes by target cells; however, to date, only a handful of examples have described specific exosome and target-cell interactions. For example, exosomes from T-, B- and dendritic immune cells were shown to communicate with antigen presenting cells by transferring their contents in a unidirectional manner and modulating gene expression in the recipient cell [36]. Uptake of ovarian cancer secreted exosomes by NK cells has been demonstrated and was found to require the surface expression of phosphatidylserine (PS) as an uptake signal [37]. Lastly, the expression of galectin-5 on the surface of rat reticulocyte exosomes was also found to modulate their uptake by macrophages [38]. These studies have demonstrated that there are distinct signals that mediate exosome and target-cell interactions; however, more work is required to fully understand the distinct mechanisms controlling this process.

1.3. Exosome Components

According to the current version of the exosome content database, Exocarta (Version 4; http://www.exocarta.org), 4,563 proteins, 194 lipids, 1,639 mRNAs and 764 microRNAs have been identified in exosomes of many different cell types and from multiple organisms [39,40], thus demonstrating their complexity. The most frequently identified proteins in exosomes (as compiled by Exocarta) include membrane transport and fusion proteins, such as tetraspanins (CD9, CD63, CD81), heat-shock proteins (Hspa8, Hsp90), GTPases (EEF1A1, EEF2), and MVB biogenesis proteins (Alix). Other identified proteins include cytoskeletal proteins (actin, syntenin, moesin), metabolic enzymes (GAPDH, LDHA, PGK1, aldolase, PKM), signal transduction proteins (annexin, 14-3-3ε, 14-3-3ξ) and the carrier protein, albumin. The specific protein composition will depend on the cell type or tissue source from which the exosome originates and may fluctuate according to physiological changes. In addition, many of these proteins may function as specific exosome markers, particularly the tetraspanins, CD63 and CD81.

Beyond proteins, exosomes are also enriched in lipids and may act as cell-to-cell lipid mediators. Exosomes predominantly contain lipids, such as cholesterol, diglycerides, sphingolipids (including sphingomyelin and ceramide), phospholipids, glycerophospholipids (including phosphatidylcholine (PC), phosphatidylserine (PS), phosphatidylethanolamine (PE) and phosphatidylinositol (PI)) and polyglycerophospholipids (*i.e.*, bisphosphate). The ratio is increased for certain exosomal lipids when compared to parental cell lipids; these include sphingomyelin, PS, PC, PI and cholesterol, which can be present at as much as four times greater amounts and may account for the increased membrane rigidity of exosomes. Exosomes have also been reported to contain bioactive lipids, such as prostaglandins and leukotrienes, and active enzymes of lipid metabolism that may generate these lipids [13,41]. The presence of certain lipids, such as PS, on the outer membrane of exosomes can function in exosome recognition and internalization [42]. In this way, exosomes function as lipid carriers, allowing the transport of the bioactive lipids they carry to a recipient cell. This process of exosomal trafficking, particularly in the context of the tumor microenvironment, could lead to an enrichment of certain tumor progressive/immunosuppressive lipids, such as prostaglandins [43]. On the other hand, it may also lead to a replacement of harmful exosome lipid contents with beneficial ones. For example, docosahexaenoic acid (DHA), an omega-3 polyunsaturated fatty acid with many health and anticancer benefits, could be supplied by exosomes throughout the tumor microenvironment to affect cell-to-cell communication, reduce tumor cell growth and increase sensitivity to therapeutic interventions, particularly in breast tumors [44]. However, further studies are required to determine, which lipids participate in exosomal cell-to-cell communication and whether *ex vivo* manipulation is a plausible and/or effective therapeutic approach.

Exosomal transport of mRNAs and other non-coding RNAs, including microRNAs, was discovered only recently [45], and due to this exciting discovery, the interest in exosomes as carriers of genetic information is burgeoning, particularly in cancer research. Current reports have shown that the majority of the RNA present in exosomes is somewhat degraded and less than 200 nucleotides in length; however, recognizable proteins could be generated from *in vitro* translation of exosomal RNA extracts, thus demonstrating that full-length mRNAs are present [45]. Several studies have demonstrated that the RNA present in exosomes is very different from the parental cell RNA content, with the apparent lack of ribosomal RNA [36,45–47]. In contrast, the exosomal microRNA content is similar to that in the original tumor, thus peaking researchers' interests in the use of exosomal microRNA profiles for cancer diagnostics [9,48]. However, an abundance of certain microRNAs that are not present or present at very low levels in the parental cells has recently been observed [49,50]. These results suggest that certain microRNAs may be preferentially secreted. However, the mechanisms for selective packaging and release of exosomal microRNAs are currently unknown, and whether these microRNAs may serve as reliable markers of disease is yet to be determined.

The first in-depth screening study was recently conducted to examine the entire transcriptome, miRNome and proteome of exosomes derived from melanoma cells and normal melanocytes [51]. Thousands of mRNAs that are associated with melanoma progression and metastasis, as well as several microRNAs (miR-31, miR-185 and miR-34b) that are involved in melanoma invasion were identified. In addition, several differentially expressed proteins, such as HAPLN1, GRP78, syntenin-1, annexin A1 and annexinA2, were identified, which were specific to the melanoma exosomes and may be involved in the malignant conversion of melanocytes. This study demonstrates the need for more in-depth explorations of exosome contents, so that specific targets may be identified and translated into clinical applications for disease biomarkers or potential therapeutic targets for cancer patients.

1.4. Exosome Isolation and Examination

Because exosome membranes are enriched in cholesterol, sphingomyelin, ceramide and lipid raft-associated proteins [52,53], they are highly stable and can be collected from various bodily fluids or from cell culture mediums [54]. Due to their small size and low density, exosome isolation usually involves multiple centrifugation and ultracentrifugation steps with a rotational force up to 100,000 × *g* for sedimentation. Centrifugation is also sometimes combined with 0.1 μm to 0.22 μm filtration in order to separate the nano-sized particles and to exclude larger particles and cellular debris, (see, for review of methods, [54]). For reduction of protein aggregate contamination and for obtaining a purer exosome preparation, sucrose, iodixanol [55], deuterium oxide density gradients (also called cushions) or proprietary reagents, such as ExoQuick (System Biosciences), have also be utilized [56,57]. Immunoaffinity capture methods can be used to isolate exosomes from cancer cells or patient serum using beads coated with antibodies against presumably any exosome-specific surface marker, such as the tetraspanins, CD63 or CD82, as a way to forego any ultracentrifugation. Epithelial cell adhesion molecule (Ep-CAM)-positive exosomes have been collected from the serum from lung [38] and ovarian cancer [48,58] patients, and HER2-positive exosomes have been isolated from HER2 overexpressing breast cancer cells [59] using this method.

Electron microscopy (EM) combined with negative staining is the standard method for visualization of whole-mount exosome preparations. Typical EM results show rounded vesicles with lipid bilayers, and sometimes, a "cup-shaped" morphology is observed, which may depend on the preparation process used. A review of these various isolation procedures and a comparison of the images obtained by EM analysis has been recently published [60].

Once the exosomes are collected, there are many downstream analysis options available. Exosomal proteins may be extracted utilizing standard cell lysis buffers or the TRIzol® reagent (Life Technologies-Invitrogen) [61]. Proteins may be detected and analyzed by immunoblot procedures [61] or mass spectrometry (see, for review, [62]). Because the RNA content of exosomes is mostly small RNAs, the selection of RNA isolation technique is an important consideration. Various RNA extraction techniques, including phenol-based techniques (TRIzol®), silica column (e.g., RNeasy® (Qiagen) or miRCURY™ (Exiqon)) and combined phenol and silica column approaches (e.g., TRIzol® followed by RNeasy (Qiagen), miRNeasy (Qiagen) or mirVana™ (Ambion)) have been utilized and compared [63–65]. The RNA yield can be determined by spectrophotometric analysis at 260 nm, and a profile of the exosomal RNAs can be determined using the Agilent 2100 Bioanalyzer Lab-on-a-Chip instrument system (Agilent Technologies). Typical profiles of RNA extracted from exosomes contain a size distribution of 25–2000 nucleotides and are characteristically absent of ribosomal RNAs [63]. Detection of specific small RNA or microRNA species can be determined by real-time reverse-transcription PCR assay and oligonucleotide microarray analysis [51], or more in-depth analysis next-generation RNA sequencing can be applied [47,51,66].

Due to their nanometer size, the process of quantifying exosomes is somewhat of a challenge. There are two primary approaches currently used to quantify the amount of exosomes isolated from a preparation: quantification of the amount of exosomal protein using enzyme-linked immunosorbent assays (ELISA) or by immunoblotting. However, new approaches to quantify exosome secretion have been demonstrated using cell lines stably expressing GFP tagged CD63 (a specific marker of exosomes), thus generating exosomes with a traceable marker that can be easily measured by fluorescent spectrometry [32]. New nanoparticle/exosome tracking analysis technologies have recently been developed by Nanosight Ltd. [67]. These systems are equipped with a blue laser and camera that can visualize and measure nanoparticles within the 30 nm to 1,000 nm range (as demonstrated in [30,68]). Because the field of exosome research is in a phase of rapid growth, the refinement of isolation, imaging and visualization methods are expected to improve, along with the identification of specific molecular markers for isolation and the development of new technological approaches.

2. Exosomes and Intercellular Communication in Tumor Progression

2.1. Exosome Mediation of Intercellular Communication

It is well recognized that tumor development and progression is dependent on the reciprocal relationship between cancerous cells and their surrounding microenvironment. While the cancerous cells, which harbor many pro-tumorigenic genetic mutations, are the main driving force of tumor development, the surrounding stroma, which includes fibroblasts, endothelial and infiltrating immune cells, play a supportive and enabling role (reviewed in [69]). This reciprocal relationship requires not only a particular spatial interaction, but also the ability for the cancerous cells to communicate with the microenvironment by exchange of certain soluble proteins and genetic factors. Cancer cells are known to secrete factors that can promote the formation of new blood vessels, known as angiogenesis, to obtain oxygen to feed the tumor and to modify their adhesive properties in order to promote migration and invasion into the newly formed vasculature. Tumor cells of many different cancer types have been shown to secrete exosomes in greater amounts than normal cells [13], thus allowing the transfer of tumor-associated signaling molecules, including microRNAs, via fusion of the exosome with the target cell membrane [70].

Tumor-derived exosomes (TD-exosomes) are generally considered pro-tumorigenic. However, some anti-tumorigenic abilities have also been described. For example, exosome-like nanoparticles isolated from pancreatic cancer cells were shown to induce apoptosis in tumor cells [71]. Other studies have focused on the use of TD-exosomes as a source for tumor antigens for the development of exosome-based immunotherapies [72–74]. One report demonstrated that modified cell lines expressing interleukin-2 (IL-2) produced TD-exosomes containing IL-2 with increased antitumor effects [72]. A Chinese phase I clinical trial demonstrated that ascites-derived exosomes combined with granulocyte-macrophage colony stimulating factor could modulate the immune response and induce an antitumor cytotoxic T-lymphocyte (CTLs) response in 40 patients with colorectal cancer [73]. Another study showed that exosomes derived from IL-2 GPI-anchored renal cancer cells could induce CTLs and significant cytotoxic and antitumor effects *in vitro*, suggesting a novel strategy for an exosome-based vaccine for renal cell carcinoma [75]. Heat-stressed tumor cells were shown to produce exosomes that could attract and activate dendritic and T-cells, induce specific antitumor immune responses and inhibit tumor growth *in vivo* [76]. However, despite the above-described antitumor characteristics of exosomes, it is still unclear whether the immunomodulatory effects of exosomes secreted from tumor cells are either cancer-promoting or cancer-inhibiting, as no studies have demonstrated any immune stimulatory effects of TD-exosomes. Rather, many immune-evasion characteristics have been described. In this context, TD-exosomes from the ascites of ovarian cancer patients were recently shown to express the death ligands, FasL and TRAIL, which could trigger apoptosis in immune system cells, thereby inhibiting a tumor growth inhibitory immune response [77].

As described above, most initial studies on TD-exosomes were focused on their interaction with the immune system, while the effects of TD-exosomes on the tumor microenvironment has been less characterized. However, recent studies have shown that TD-exosomes have many pro-tumorigenic functions and are able to transfer their phenotypic traits (such as onco-proteins or onco-microRNAs) to a recipient cell and promote cancer stimulatory activities, such as proliferation, extracellular matrix remodeling, migration and invasion and angiogenesis, and contribute to the pre-metastatic niche formation for the promotion of metastasis. The following sections will discuss these topics in greater detail.

2.2. Exosome Modulation of Extracellular Matrix, Stromal Cells and Invasion

Remodeling of the extracellular matrix and alterations in cell-cell and cell-extracellular matrix interactions are the first barriers that must be overcome for a cancer cell to migrate and travel to distant sites. Several studies have shown that TD-exosomes can alter the extracellular matrix through secretion of matrix metalloproteinases (MMPs) or activators of MMPs, such as heat shock proteins. MMPs are zinc-dependent plasma membrane endo-peptidases that can degrade extracellular matrix proteins, such as collagen, fibronectin, proteoglycans and laminins. In fibrosarcoma and melanoma cells, it was shown that MT1-MMP was secreted in exosomes and could activate pro-MMP-2 and degrade collagen and gelatin [78]. Other studies have demonstrated that heat shock proteins, such as hsp90, are also secreted via exosomes and can activate MMP-2 to enhance invasion of cancer cells [79]. The role of platelets in tumor progression has been recently investigated, where it was found that platelet-derived microvesicles contribute to cancer progression in lung cancer cell lines by stimulating proliferation, cyclin D2 expression, adhesion to endothelial cells, invasion and angiogenesis through activation of mitogen-activated protein kinase (MAPK) p42/44, MT1-MMP, MMP-9, IL-8 and VEGF and other factors controlling these processes [80]. The effects of TD-exosomes on mesenchymal stem cells (MSCs) have also been explored. For example, TD-exosomes from ovarian cancer cells can induce a tumor-associated myofibroblast-like phenotype on adipose-derived MSCs, suggesting that exosomes contribute to the generation of tumor-associated fibroblasts in the tumor stroma [81]. Recently, Bobrie *et al.* showed that inhibition of Rab27a in breast cancer cells decreases the secretion of exosomes and MMP-9, resulting in a decrease in primary tumor growth and lung dissemination *in vivo*, thus further demonstrating that exosome secretion promotes tumor formation and progression and that inhibition of exosome secretion may impede tumor growth [82].

2.3. Exosome Stimulation of Tumor Angiogenesis and Metastatic Niche Formation

In order for a tumor to sustain its growth and survival, they have to obtain greater amounts of oxygen and nutrients through the formation of new blood vessels or by metastasizing to more hospitable organ sites. TD-exosomes have been shown to transport oncogenic and pro-angiogenic factors to cells within the tumor microenvironment to induce neoangiogenic activity and to promote premetastatic niche formation. For example, a hypoxic tumor microenvironment has been shown to enhance the secretion and transport of exosomes and pro-angiogenic protein factors that could potentially modulate the microenvironment to facilitate angiogenesis and metastasis [83]. Recently, *in vitro* hypoxia experiments in glioma cells and patient samples showed an enrichment of hypoxia-regulated mRNAs and proteins, such as MMPs, IL-8, platelet derived growth factor (PDGF), caveolin-1 and lysyl oxidase, in secreted exosomes. In addition, it was demonstrated that these exosomes are also potent inducers of angiogenesis through possessing these growth factors and cytokines [84]. Colorectal cancer cell-derived microvesicles/exosomes are enriched in cell cycle-related mRNAs that could promote proliferation of vascular endothelial cells [85] and in several metastatic and signal transduction molecules [86]. Microvesicles released from human renal cancer stem cells were shown to stimulate angiogenesis and promote the formation of a premetastatic niche in the lungs *in vivo* [87]. As mentioned above, MSCs can also promote tumor growth; for example, bone marrow MSC-derived exosomes can enhance VEGF expression in tumor cells by activating the ERK1/2 pathway [88]. Likewise, platelet-derived exosomes have also been shown to stimulate mRNA expression of angiogenic factors (such as MMP-9), as described above [80]. Exosomes released by chronic myeloid leukemia cells were shown to promote angiogenesis in a Src-dependent fashion [89]. Melanoma-derived exosomes were found to prepare bone marrow progenitor cells for a pro-metastatic phenotype through the receptor tyrosine kinase, MET [90]. A recent review proposes that the interaction and bidirectional exchange of genetic information, via secreted microvesicles, between macrophages and endothelial cells may work to promote vascular growth in the tumor microenvironment [91]. Overall, these studies clearly demonstrate that exosomes indeed function as pro-tumorigenic factors that can mediate intercellular communication in the tumor microenvironment and contribute to cancer progression.

3. Exosome-Derived microRNAs in Tumor Progression

3.1. Exosomes as Transporters of microRNAs

microRNAs are small (17–21 nt), non-coding RNAs that regulate gene expression at the post-transcriptional level through the RNA interference pathway. microRNAs are transcribed by RNA polymerase II as primary-microRNAs (pri-miRNAs) [92–94] and are processed in the nucleus by the enzyme, Drosha, into shorter hairpin structures of approximately 70 nucleotides in length, called pre-miRNAs. Pre-miRNAs are then transported from the nucleus, to the cytoplasm via Exportin 5 [95], where they are further processed into mature microRNA transcripts by the enzyme, Dicer [93]. The mature microRNA is then loaded into the ribonucleoprotein complex, known as the RNA induced silencing complex (RISC), and can bind in a sequence specific manner to the 3' untranslated region (UTR) of target mRNAs, resulting in either translational inhibition or mRNA degradation [96]. Presently, ~2000 microRNAs have been described in humans [97] and a single microRNA may regulate many mRNAs; likewise, a single mRNA may be targeted by many microRNAs, establishing microRNAs as the largest class of gene regulators [98]. Through this mechanism, microRNAs are an essential component to regulating most cellular and developmental processes, including developmental timing, organ development, differentiation, proliferation, apoptosis and immune regulation (see, for review, [99]). Therefore, it is of no surprise that microRNAs are involved in cancer development and progression, and depending upon their target gene and level of expression, microRNAs may function as either tumor suppressors or oncogenes and assist in the promotion or suppression of cancer growth and progression [100]. Aberrant microRNA expression has been described across many cancer types, with global downregulation of microRNA expression seen as a common trend [101,102]. The transport of mRNAs and microRNAs by exosomes was realized only recently, but has led to an explosion of interest in cancer research. The first study to demonstrate exchange of nucleic acids via exosomes examined secreted exosomes from mouse and human mast cell lines [45]. Using standard RNA and DNA extraction techniques of exosomes isolated by ultracentrifugation, Valadi *et al.* discovered the presence of small RNAs and mRNAs (but not DNA) from approximately 1300 genes present in exosomes that are not present in the parental cell and proposed that these RNAs be referred to as exosomal shuttle RNAs (esRNAs) to distinguish them from circulating microRNAs [45]. In addition, these esRNAs could be *in vitro* translated into functional proteins and transferred to other human and mouse mast cells, where new proteins were generated in the recipient cells [45]. This seminal study has generated much interest in the study of cell-cell communication via delivery of small RNAs by transfer through exosomes.

3.2. Secretion and Uptake of Exosome-Derived microRNAs

The exact mechanism of microRNA loading into MVBs in the endocytic pathway and secretion via exosomes is not well understood. Studies thus far have demonstrated that MVBs are associated with GW-bodies, also known as P-bodies, which are cytoplasmic foci, where post-transcriptional regulation of mRNAs occurs, and are enriched in GW182 and AGO2 proteins, two main components of the RISC [103]. In this study, it was shown that endosomes or MVBs are sites of microRNAs, microRNA-repressible mRNAs and RISC accumulation and action and that exosomes secreted via MVBs are enriched in GW182, suggesting a mechanism for microRNA loading. Depletion of some of the ESCRT components compromised microRNA-mediated gene silencing and led to an over-accumulation of GW182, thus suggesting that GW182 and microRNA-loaded AGO2 are sorted in to MVBs via ESCRT components [103]. However, as previously mentioned, a more recent study has demonstrated that microRNAs are released through a ceramide-dependent secretory mechanism [104]. Furthermore, a tumor-suppressive microRNA, secreted via this mechanism, was taken up by a recipient cell, where it exerted gene silencing and growth inhibition [104]. The GW182 protein may also be important for microRNA stability and secretion via exosomes [105]. In this study, it was shown that knockdown of GW182 by siRNA increased microRNA instability and reduced secretion via exosomes, whereas replenishment of GW182 restored microRNA stability, thereby demonstrating a role of GW182 in protecting AGO2-bound microRNA [105].

Other studies have demonstrated that certain microRNAs are selectively secreted in exosomes. For example, the let-7 microRNA family is selectively secreted via exosomes in metastatic gastric cancer cell lines. Since this family of microRNAs targets oncogenes, such as Ras and HMGA2, they are generally considered a tumor-suppressive group of microRNAs; however, whether their release via exosomes is to promote or inhibit oncogenesis remains unclear [106]. Breast cancer cell lines selectively release the majority of miR-451 and miR-1246 via exosomes as compared to their parental cell, whereas these specific microRNAs are retained in non-malignant mammary epithelial cells and normal fibroblast cells [49]. The authors note that the biogenesis of miR-451 is dicer-independent [107], thus raising the possibility that non-canonical processing of microRNAs may target them for selective exosome release. Overall, these results suggest that there is a specific selection mechanism for microRNA release. However, the exact mechanism remains to be elucidated.

While the exact mechanism of exosome-derived microRNA uptake and processing in recipient cells is largely unknown, many types of cells have been shown to absorb exosomal microRNAs, where they can induce post-translational repression of target mRNAs. For example, T-cells have been shown to transfer microRNAs-loaded exosomes in an antigen-driven unidirectional manner to antigen presenting cells in the immune synapsis, where they modulate gene expression [36]. Additionally, mouse dendritic cells (DCs) release exosomes containing different microRNAs, depending on their level of maturation. These microRNAs were absorbed by recipient DCs, where they were shown to repress target mRNA expression [108]. One of the first studies to show that this unique intercellular method of communication could contribute to the initiation and progression of cancer demonstrated that hepatocellular carcinoma cells produce exosomes that can be internalized by other cells. These exosomes were shown to transmit microRNAs that modulated the expression of transforming growth factor β activated kinase-1 (TAK1), whose loss is implicated in hepatocarcinogenesis [109]. Thus, these studies have begun to shed light on the potential mechanisms of exosome uptake and the functional consequences of microRNA transfer.

3.3. microRNA Profiling of Tumor-Derived Exosomes in Clinical Samples

The discovery of the transport and exchange of microRNAs via exosomes has also generated much interest in the use of circulating TD-exosomes and their resident esRNAs as clinical diagnostic markers for cancer. To this end, a handful of studies have examined the microRNA profile from circulating TD-exosomes and compared the expression levels to the original tumor cells. For example, circulating TD-exosomes isolated from the serum of patients with ovarian cancer, age-matched controls and primary tumor cell cultures and matched sera were examined for microRNA expression changes. This study found that the expression of eight microRNAs, previously demonstrated to be diagnostic in ovarian cancer, were similar between cellular and exosomal microRNA preparations, thus suggesting that circulating TD-exosomes could be used as surrogate diagnostic markers for biopsy profiling, particularly in asymptomatic patient populations [9]. Another study from Taylor *et al.* evaluated the levels of microRNAs from the plasma of patients with lung adenocarcinoma, matched tumor samples and controls; they similarly observed no significant differences in exosome microRNA levels between microRNAs derived from circulating exosomes or from microRNAs from the primary tumors [48]. In another study, circulating exosomal/microvesicle-derived microRNAs were profiled from the plasma of prostate cancer patients with and without metastases [110], and a distinct set of 11 microRNAs was present at significantly greater amounts in patients with metastases compared to those without metastases. The association of two of these 11 microRNAs (miR-141 and miR-375) were confirmed in plasma exosomes from a separate patient cohort with recurrent or non-recurrent disease, thus demonstrating that changes in microRNA concentrations present in circulating exosomes from prostate cancer patients may be used for diagnosis and tumor staging [110]. In serum obtained from esophageal squamous cell carcinoma (ESCC) patients, microRNA expression profiling showed that miR-1246 was consistently elevated in patients *versus* controls and was an independent risk factor for poor survival [111]. The authors also indicate that miR-1246 was not upregulated in ESCC tissue samples; however, this observation is consistent with the previously mentioned report of preferential exosome secretion of miR-1246 from breast cancer cells [49]. Overall, these exosomal microRNA profiling studies, summarized in Table 1, have found that microRNA expression signatures are not significantly different between TD-exosomes and tumor cells, with the exception of miR-1246, suggesting that these circulating TD-exosome microRNAs could be utilized as a surrogate for biopsy microRNA profiling. In addition, a database called miRandola has been created to catalog all extracellular circulating microRNAs and currently contains 2312 entries with 581 unique mature microRNAs identified in circulation from 21 different types of samples [112].

Table 1. Summary of clinical microRNA profiling studies of circulating tumor exosomes/microvesicles.

Cancer type	Clinical samples	Exosome isolation method	Major findings	Potential diagnostic microRNAs	Reference
Ovarian cancer	Sera from patients with serous papillary adenocarcinoma ($n = 50$); sera from age-matched controls with benign ovarian adenoma ($n = 10$); primary ovarian adenocarcinoma tumor cell cultures and matched patient sera ($n = 6$).	Magnetic activated cell sorting using beads coupled with anti-EpCAM.	Exosomal microRNA profiles were similar in ovarian cancer patient samples and distinctly different from benign disease samples. microRNAs were elevated in exosomes *versus* tumor cells (31 out of 467).	miR-21, miR-141, miR-200a, miR-200c, miR-203, miR-205 and miR-214	[9]
Lung adenocarcinoma	Plasma from patients with lung adenocarcinoma ($n = 27$); plasma from controls ($n = 9$); matched plasma and lung tumor tissue ($n = 4$).	Size exclusion chromatography and magnetically activated cell sorting using beads coupled with anti-EpCAM.	No significant differences in exosome microRNA levels between microRNAs derived from circulating exosomes or from microRNAs from the primary tumor were observed.	miR-17-3p, miR-21, miR-106a, miR-146, miR-155, miR-191, miR-192, miR-203, miR-205, miR-210, miR-212, miR-214	[48]
Prostate cancer	Plasma from prostate cancer patients ($n = 78$); plasma from normal controls ($n = 28$); urine samples ($n = 135$); serum from patients with recurrent metastatic prostate cancer ($n = 47$) or non-recurrent disease ($n = 72$).	Filtration of plasma through a 1.2 µm filter, concentrated with a 150 kDa molecular weight cut-off.	The levels of 12 microRNAs were different between plasma exosomes of prostate cancer patients compared to control. Eleven microRNAs were present in significantly greater amounts in patients with metastases *versus* without. The association of exosomal miR-141 and miR-375 with metastases was confirmed in a second patient population.	miR-107, miR-130b, miR-141, miR-181a-2*, miR-2110, miR-301a, miR-326, miR-331-3p, miR-432, miR-438, miR-574-3p, miR-625*	[110]
Esophageal squamous cell carcinoma	Serum from ESCC patients ($n = 101$); Serum from healthy controls ($n = 46$).	Sequential centrifugation, 0.22 µm filtration and ultracentrifugation.	miR-1246 was markedly elevated in serum and exosomes from ESCC patients and was a strong independent risk factor for poor survival. miR-1246 expression was not increased in ESCC tissue samples.	miR-1246	[111]

Abbreviations: EpCAM = epithelial cell adhesion molecule; ESCC = esophageal squamous cell carcinoma.

Figure 1. Biogenesis, secretion and uptake of tumor-derived exosomes in the tumor microenvironment. Exosomes are formed by the inward budding of the multivesicular body (MVB) membrane in the form of intraluminal vesicles (ILVs). Exosome formation and cargo sorting into lysosomes involves the endosomal sorting complex required for transport (ESCRT), which recognizes ubiquitinated proteins. Exosome production and secretion also occurs through an ESCRT-independent process involving the sphingolipid, ceramide, and the enzyme neutral, sphingomyelinase (the enzyme that converts sphingomyelin to ceramide). Exosomes secretion can be stimulated by various chemical, environmental and mechanical stimuli, such as gamma-irradiation, hypoxia (low oxygen), low pH, matrix detachment, *etc.* Exosomes are secreted in exocytic MVBs following fusion of MVBs with the cell membrane, a process that depends on Rab GTPases (Rab27A, Rab27B). Exosomes released from a primary tumor cell will display similar membrane components as their cell of origin, such as receptor ligands or antigens. Endocytosis of exosomes may occur through activation of cell surface receptors or bioactive lipid ligands. Upon endocytosis by a secondary recipient cell, such as fibroblasts or vascular endothelial cells, exosomes can release their microRNA cargo. The transferred microRNAs are functionally active and can regulate gene expression in the recipient cell through post-translational regulation of target mRNA expression, leading to mRNA degradation or de-stabilization. microRNA-dependent gene regulation can activate various processes involved in tumor development and progression. Abbreviations: TAK1, transforming growth factor β activated kinase-1; MMPs, matrix metalloproteinases; MAPK, mitogen activated protein kinase; NFκB, nuclear factor kappa-light-chain-enhancer of activated B-cells; EZH2, enhancer of zeste homolog 2; VEGF/VEGF, vascular endothelial growth factor/receptor; ECM, extracellular matrix.

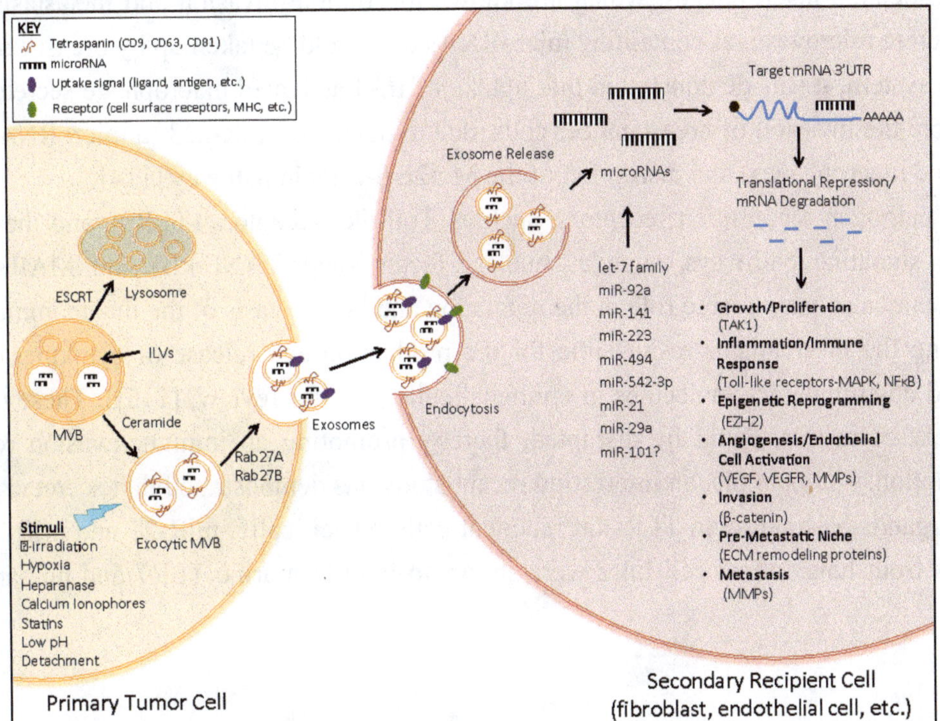

3.4. *Pro-Tumorigenic Effects of Exosome-Derived microRNAs* in Vitro

The pro-tumorigenic effects of exosome-derived microRNAs after uptake by a recipient cell have recently begun to emerge. Thus far, exosome-derived microRNAs, through target gene transcriptional repression, have the demonstrated ability to induce cell migration, inflammation, immune responses, angiogenesis (including endothelial cell migration and tube formation), invasion, pre-metastatic niche formation and metastasis; see Figure 1. Therefore, these studies have implicated cancer cell exosome-derived microRNAs in most aspects of tumor progression. For example, leukemia cell exosomes have been shown to communicate with human umbilical vein endothelial cells (HUVECs), leading to increased cell migration and tube formation [113]. In this study, K562 leukemia cells were transfected with a Cy3-labeled pre-miR-92a and co-cultured with HUVECs. The Cy3-labeled miR-92a, derived from the K562 cells, could be detected in the cytoplasm of the endothelial cells and was co-localized with the exosomal marker, CD63. In addition, the expression of integrin α5, a target of miR-92a, was also greatly reduced in the recipient cells, thus demonstrating that an exosome-derived microRNA can function as an endogenous microRNA in a recipient cell and that exosomal microRNAs play an important role in cancer-to-endothelial cell communication [113]. As previously mentioned, microvesicles released from human renal cancer stem cells containing pro-angiogenic mRNA, and microRNA were shown to greatly stimulate endothelial cell growth and vessel formation and enhance lung metastases after *in vivo* implantation in a severe combined immunodeficient (SCID) mouse model [87]. In this study, molecular characterization of microvesicles, derived from CD105-positive (a mesenchymal stem cell marker) renal cancer stem cells, was conducted and was found to contain a set of pro-angiogenic mRNAs and microRNAs that are implicated in tumor progression and metastases [87]. The previously mentioned study, which profiled the miRNome of melanoma exosomes, identified 228 microRNAs that were differentially regulated in melanoma exosomes *versus* normal melanocyte exosomes, 15 of which are known to be associated with melanoma invasion and metastasis [51].

Tumor-associated macrophages, which are known to promote invasion and metastasis, have been shown to secrete microvesicles containing microRNAs that could be taken up by breast cancer cells. In a co-culture system, it was demonstrated that uptake of IL-4 activated macrophage secreted exosomes could promote the invasion of breast cancer cells, due to uptake of miR-223 (a microRNA specific for IL-4 activated macrophages) and disruption of the Mef2c-β-catenin pathway [114].

Through pathogen recognition receptors, such as Toll-like receptors (TLRs), and their associated downstream signaling pathways, such as nuclear factor kappaB (NF-κB) and MAPK, exosomal microRNAs may also play a large role in the regulation and homeostasis of the innate immune response by fine-tuning the mechanisms responsible for the production and release of cytokines/chemokines, adhesion and co-stimulatory molecules in epithelial cells (see, for review, [115]). These mechanisms, in the context of cancer, could be disrupted, thereby promoting an immune-evasion response and cancer promotion. For example, an interesting recent study has demonstrated that secreted microRNAs may act as ligands by binding to TLRs on recipient cells. Specifically, miR-21 and miR-29a secreted in exosomes from lung cancer cell lines were shown to bind to murine TLR7 and human TLR8 and triggered a TLR-mediated pro-metastatic inflammatory response that could lead to tumor growth and metastasis [116].

The process of malignant transformation may also alter the specific species of microRNAs that are secreted in exosomes or retained in cells. For instance, selective release of certain microRNA populations has been demonstrated in malignant breast cancer cells. Specifically, the microRNAs, miR-451 and miR-1246, produced by malignant breast epithelial cells are released, whereas the majority of these microRNAs are retained in non-malignant mammary epithelial cells [49]. A follow-up study from this same group demonstrated that these selectively exported microRNAs are packaged in exosomes that are larger than conventional exosomes and are enriched in CD44, a protein relevant to breast cancer metastasis. In contrast, they showed that normal cells release microRNAs in a homogenous type of vesicle, suggesting that the process of malignant transformation may alter the pathways by which microRNAs are exported from cells, thus leading to differences in exosome content and morphology [117].

TD-exosomes may also modulate pre-metastatic niche formation via long-range transfer of microRNAs. One study has demonstrated that exosomes from metastatic rat adenocarcinoma cells are preferentially taken up by lymph node stroma cells and lung fibroblasts [118]. The transferred microRNAs significantly affected mRNA translation of many genes, including proteases, adhesion molecules, chemokine ligands, cell cycle- and angiogenesis-promoting genes and oxidative stress response. In particular, miR-494 and miR-542-3p modulated the expression of cadherin-17 with concomitant upregulation of matrix metalloproteinases. Together, these findings demonstrate that TD-exosomes may target non-transformed cells in pre-metastatic tissues, leading to modulation of gene expression in these cells specifically through transfer of microRNA and priming distant tissues for tumor cell hosting [118].

Exosomal transfer of microRNAs could also induce permanent changes in recipient cell phenotypes via transfer of microRNAs that are known to regulate genes involved in epigenetic reprogramming (*i.e.*, miR-101 regulation of the histone methyltransferase EZH2). For example, it has been demonstrated that microvesicles derived from one cell type can deliver mRNAs that could mediate gene expression and alter cell fate in a secondary recipient cell type (reviewed in [119]). While we are unaware of any studies demonstrating changes in epigenetic programming or cell fate through the exchange of exosomal microRNAs, it is certainly a highly plausible occurrence. In conclusion, it is clear from these studies, summarized in Table 2, that transfer of microRNAs via exosomes is indeed a mechanism of intercellular communication that can initiate and promote tumor progression via transfer of genetic information at local and distant cells and tissues.

Table 2. Summary of *in vitro* studies of microRNAs derived from cancer cell exosomes.

Cell Line model	Major findings	Predominant microRNAs	Target genes or pathways	Reference
Human and mouse mast cells	Identified small RNAs, including 121 microRNAs and 1,300 specific mRNAs. Detected mouse exosomal RNA and new mouse proteins in human mast cells after treatment with mouse mast cell exosomes. Coined the term "exosomal shuttle RNA (esRNA)".	let-7, miR-1, miR-15, miR-16, miR-181, miR-375.	None tested	[45]
Metastatic gastric cell line	Profiled microRNA expression by microarray in exosomes isolated from gastric cancer cells. let-7 microRNA family was enriched in exosomes.	let-7 family	None tested	[106]
Co-culture of IL-4-activated macrophages and breast cancer cells	miRNAs can be transferred from macrophages to breast cancer cells. miR-223 released by macrophages was found in MCF7 and MDA-MB-231 cells and promoted invasion.	miR-223	Mef2c-β-catenin pathway	[114]
Mouse dendritic cells	Exosomal microRNA from dendritic cells can be transferred to a recipient dendritic cell and repress microRNA target mRNAs in the acceptor cell.	miR-148a, miR-451	Luciferase reporter containing tandem microRNA target sequences	[108]
Leukemia cells and endothelial cells	Leukemia cells released microRNAs from the miR-17-92 cluster and were taken up by human umbilical vein endothelial cells (HUVECs) and repressed a target mRNA. Did not affect the growth of HUVEC cells, but did enhance cell migration and tube formation.	miR-92a	Integrin α5	[113]
Hepatocellular carcinoma cells	Transmission of exosome microRNAs from hepatocellular carcinoma cells could contribute to the initiation and progression of hepatocellular carcinoma by targeting a tumor suppressor frequently lost in hepatocarcinogenesis.	miR-584, miR-517c, miR-378, miR-520f, miR-142-5p, miR-451, miR-518d, miR-215, miR-376a*, miR-133b, miR-367	Transforming growth factor β activated kinase-1 (TAK1) pathway	[109]
Renal cancer stem cells	Microvesicles were secreted from human renal cell carcinoma that could trigger angiogenesis and premetastatic niche formation in a severe combined immunodeficient (SCID) mouse model.	miR-92, miR-141, miR-29a, miR-650, miR-151, miR-19b, miR-29c	Increase in VEGFR1 and MMP-9 expression	[87]

Table 2. *Cont.*

Cell Line model	Major findings	Predominant microRNAs	Target genes or pathways	Reference
Breast cancer cells	Selective release of certain microRNA populations was demonstrated in malignant breast cancer cells that are retained in non-malignant mammary epithelial cells.	miR-451, miR-1246	None tested	[49]
Metastatic rat adenocarcinoma cells	Exosomes were preferentially taken up by lymph node stroma cells and lung fibroblasts. The transferred microRNAs affected mRNA translation of many genes.	miR-494, miR-542-3p	Cadherin-17 and many proteases, adhesion molecules, chemokine ligands, cell cycle- and angiogenesis-promoting and oxidative stress response genes.	[118]
Lung cancer cell lines	miR-21 and miR-29a were secreted in exosomes and could bind to murine TLR7 and human TLR8 and trigger a Toll-like receptor (TLR)-mediated prometastatic inflammatory response that could lead to tumor growth and metastasis.	miR-21 and miR-29a	Toll-like receptor (TLR) 8 and 9	[116]
Melanoma and normal melanocyte cells	The first in-depth screening to examine the entire exosome transcriptome, miRNome and proteome. Thousands of mRNAs and 15 microRNAs that are associated with melanoma progression and metastasis were identified.	let-7 family, miR-138, miR-125b, miR-130a, miR-34a, miR-196a, miR-199b-3p, miR-25, miR-27a, miR-200b, miR-23b, miR-146a, miR-613, miR-205, miR-149	None tested	[51]

4. Exosomes and Breast Cancer

4.1. Exosomes in Normal Mammary Epithelium

The human mammary gland is comprised of two main compartments: the branching epithelial ductal-lobular system and the supporting stroma. The epithelial component is comprised of two epithelial cell types: the luminal cells, whose function is to maintain the apical-basal polarity within the lumen and to produce and secrete milk into the ducts, and the myoepithelial cells, whose function is to maintain the organization of the mammary gland and to contract and eject the milk in response to hormonal signals. The mammary gland is one of the few organs to undergo significant developmental changes after birth, including growth at the onset of puberty and pregnancy, lactation and regression upon cessation of lactation [120]. Breast milk, which is a complex and nutrient rich liquid that contains proteins, lipids, carbohydrates and trace elements, is known to be the optimal nutrition and an important source of immunoprotective components for infants during their first months of life [121]. Exosomes have been identified in both colostrum and mature human breast milk that have the capacity to potentially influence the immune response in infants [5]. Specifically, breast milk exosomes were found to inhibit anti-CD3-induced IL-2 and interferon (IFN)-γ production and could increase the number of T-regulatory cells from peripheral blood mononuclear cells when incubated with milk vesicle preparations [3]. Furthermore, breast milk exosomes are also known to contain microRNAs [122]. Over 600 unique microRNAs, originating from ~450 microRNA precursors, have been identified in human breast milk exosomes using deep sequencing technology, and greater than 65% of the known immune-related microRNAs were enriched in these exosomes [122]. These results suggest that microRNAs can be transferred from the mother's milk to the infant via the digestive tract, where they could play an important role in the development of the infant's immune system, although further work must be done to confirm these speculations. At this time, however, it is unknown if exosomes are secreted in breast luminal epithelial cells during the growth phase of puberty or pregnancy, during regression or during the mammary resting state. It is also unknown what role the myoepithelial cells play in exosome secretion or in crosstalk between the two epithelial lineages and the breast stroma in the normal mammary gland. Furthermore, the aberrant secretion of exosomes and their specific contribution to breast cancer development and early progression are also unknown at this time.

4.2. Exosomes in Breast Cancer Development and Progression

Breast cancer is one of the most frequently observed cancers in industrialized countries and is the second leading cause of cancer death among women in the US. **Every year, ~200,000 new cases of invasive breast cancer will be diagnosed, and around 40,000 women are expected to die from breast cancer [123].** The general view of the tumorigenic process involves cells that have acquired critical genetic and epigenetic abnormalities that inhibit their responsiveness to normal growth and regulatory signals. In the mammary gland, the majority of these critical genetic changes occurs in the luminal epithelium at the transition from normal, to hyperplastic, to pre-invasive lesions and contains greater measures of genetic and morphological abnormalities, including aneuploidy [124–126], oncogene amplification [127,128] or allele imbalance [129,130]. It is thought that fewer abnormalities exist in precursor lesions and more are acquired as the cancer progresses. In addition, the loss of the normal myoepithelium cell layer and apical-basal polarity are early signs of tumorigenesis. It has also been demonstrated that the microenvironment also undergoes changes and has a dramatic influence on the tumor, even at the pre-invasive stage [131]. It is suggested that a reciprocal relationship between breast cancer cells and their surrounding microenvironment predominantly influences the energetics and growth of the cancer. However, the role of exosomes in mediating this reciprocal relationship is only beginning to emerge. Early studies in breast cancer exosomes demonstrated that exosomes might play a role in the control of tumor growth. For example, mice pretreated with exosomes derived from murine mammary carcinomas had increased rates of tumor growth, due to inhibition of natural killer cells [132]. In co-culture experiments, tumor-associated macrophages were shown to transfer miR-223, a microRNA specific for IL-4 activated macrophages, into breast cancer cells, where it promoted invasion through activation of the Mef2c-β-catenin pathway [114]. Uptake of the epidermal growth factor receptor (EGFR) ligand, amphiregulin, carried by exosomes to breast cancer cells, increased their invasiveness compared to exosomes carrying other EGFR ligands, suggesting a role for exosomes in the cancer "field effect" and metastatic niche priming [133]. Another study examined the effects of exosomes derived from a triple-negative breast cancer (TNBC) cell line (Hs578T) *versus* its more invasive variant ($Hs578T(i)_8$) on three recipient TNBC cell lines. It was shown that exosomes from the more invasive variant increase proliferation, migration, and invasion and stimulate significantly more endothelial tubule formation in all recipient cell lines [134]. The intercellular communication between fibroblasts and breast cancer cells was recently examined, which showed that fibroblast-secreted exosomes could promote breast cancer cell protrusive and motile activity through Wnt-planar cell polarity (PCP) signaling and that co-injection of fibroblasts with breast cancer cells in an orthotopic mouse model could dramatically increase metastasis that was dependent on the PCP pathway [135]. A recent study utilized GFP-tagged CD63 expressing breast cancer cells to examine the fate of cancer-cell derived exosomes in a nude mouse model of breast cancer. They demonstrated that breast cancer cells could transfer their exosomes to other cancer cells and normal lung tissue *in vitro* and into the tumor microenvironment and the circulation of mice with breast cancer metastases *in vivo* [136].

While very few studies thus far have profiled or examined the microRNAs present in breast cancer exosomes, a previously mentioned study has identified a specific set microRNAs that are secreted in exosomes or retained in cells that differ between non-malignant and malignant breast cancer cells [49]. Further studies demonstrated that the selectively exported microRNAs from breast cancer cells are packaged in exosomes that differ from conventional exosomes [117]. However, the mechanism of this selective microRNA transport and altered exosome formation in malignant breast cancer cells is unknown at this time.

In summary, these studies suggest that breast cancer derived-exosomes may contribute significantly to breast tumor growth and development, promotion of angiogenesis, invasion and formation of a pre-metastatic niche to promote tumor growth and metastasis. Moreover, further elucidation of the mechanisms of how exosomes and their residual components mediate the intercellular communication in the breast tumor microenvironment and how this process unfolds during early malignant transformation and cancer progression are of great research interest.

4.3. Potential for Diagnostic, Prognostic and Therapeutic Interference

Much of the excitement surrounding TD-exosome research is due to its high clinical relevance. In particular, because breast cancer exosomes can be easily isolated through minimally invasive procedures, such as from the blood or ductal lavage of breast cancer patients, they have great potential in breast cancer diagnosis. Because exosomal microRNA profiles of circulating tumor exosomes tend to be unique from those in normal controls, breast cancer-specific exosomal-microRNA signatures may also be developed to predict tumor development. It is accepted that breast cancer is a highly heterogeneous disease with phenotypically diverse tumors, which have been categorized by their gene expression profiles as the intrinsic molecular subtypes of breast cancer. These are generally defined as: basal-like breast cancer, which generally corresponds to estrogen receptor (ER) negative, progesterone receptor (PR) and HER-2 negative (*i.e.*, triple negative), luminal A (ER positive, low grade), luminal B (ER positive, high grade) and HER-2 (or ErbB2) positive [137]. These subtypes have distinctly different gene and microRNA expression profiles [138,139]. Therefore, their secreted exosomes may also have distinctly different RNA profiles that could correspond to the molecular subtype of their host tumors. Exosomes may also be important players in chemotherapy and chemoresistance. For example, release of exosomes from HER2-overexpressing breast cancer cell lines (BT474 and SKBR3) or from HER2-positive breast cancer patient serum could bind to Trastuzumab (a monoclonal antibody therapy that interferes with the HER2 receptor) and lead to an inhibition of the anti-proliferative effects of Trastuzumab on SKBR3 cells. These results suggest that HER2-positive exosomes may interfere with anticancer therapy and may promote HER-2 driven tumor aggressiveness [140]. These findings have led to the development of a novel therapeutic strategy for exosome removal as an adjuvant to chemotherapy. In particular, Aethlon Medical, Inc. (San Diego, CA, USA) has introduced the HER2osome^{TM} as a new therapeutic strategy to maximize the effects of anti-HER2 therapies in combating breast cancer [141]. Furthermore, exosomes may also function to shuttle chemotherapies, such as cisplatin, out of the tumor cell, thus reducing their effectiveness [142]. For example, acquired resistance to cisplatin is associated with abnormalities of protein trafficking and secretion. Cisplatin-resistant ovarian cancer cells release more protein via exosomes, including greater levels of the lysosome-associated protein 1 (LAMP1), the putative cisplatin export transporter, MRP2, and the copper export proteins, ATP7A and ATP7B, than those released by cisplatin-sensitive cells [142]. This implicates the exosome secretion pathway in the resistance of breast cancer to cancer therapy. Lastly, because exosomes are naturally produced, cell-derived nucleic acid carriers, they also hold the potential to function as biological therapeutic delivery systems. For example, exosomes engineered to express the transmembrane domain of the PDGF receptor fused to the GE11 peptide were shown to successfully deliver the let-7a microRNA to EGFR-expressing breast cancer cells *in vitro* and EGFR-expressing xenograft breast cancer tissue *in vivo*, thus demonstrating the potential of exosomes to be used therapeutically to specifically target cancer cells with nucleic acid drug targets [143].

5. Future Directions

Current knowledge of TD-exosomes suggests that they can play an important role in the development and progression of cancer through modulation of intercellular communication within the tumor microenvironment by the transfer of protein, lipid and RNA cargo. A further exploration of exosome secretion in the normal physiological state and during cancer development and progression, as well as the specific content exosomes transport under these conditions, will increase our understanding of their role in intercellular communication and tumorigenesis. Identification and modification of cancer cell-derived exosome contents may allow for the development of novel diagnostic, preventive and therapeutic approaches, with potentially minimally invasive procedures. Utilization of exosomes for therapeutic delivery may also prove to be the answer for the field of RNA therapeutics, whose main roadblock has been the development of an effective RNA delivery system. Furthermore, the concept of exploiting these extracellular vesicles or the creation of synthetic exosomes, called "exosome mimetics" for drug delivery may also allow for specific targeting of cancer cells by developing exosomes that harbor cell-specific targeting factors [144]. Although this field is in its infancy, it is easy to imagine all the future possibilities these natural nanoparticles hold.

Acknowledgments

We acknowledge grant support from the American Cancer Society (CNE-117557), the Susan G. Komen for the Cure Foundation (KG081083), the NIH OK-INBRE program (3P20RR016478-09S2), and the Oklahoma Center for the Advancement of Science and Technology (HR09-025).

References

49. Thery, C.; Zitvogel, L.; Amigorena, S. Exosomes: Composition, biogenesis and function. *Nat. Rev. Immunol.* **2002**, *2*, 569–579.
50. Wickman, G.; Julian, L.; Olson, M.F. How apoptotic cells aid in the removal of their own cold dead bodies. *Cell Death Differ.* **2012**, *19*, 735–742.
51. Pan, B.T.; Teng, K.; Wu, C.; Adam, M.; Johnstone, R.M. Electron microscopic evidence for externalization of the transferrin receptor in vesicular form in sheep reticulocytes. *J. Cell Biol.* **1985**, *101*, 942–948.
52. Harding, C.; Heuser, J.; Stahl, P. Receptor-mediated endocytosis of transferrin and recycling of the transferrin receptor in rat reticulocytes. *J. Cell Biol.* **1983**, *97*, 329–339.
53. Admyre, C.; Johansson, S.M.; Qazi, K.R.; Filen, J.J.; Lahesmaa, R.; Norman, M.; Neve, E.P.; Scheynius, A.; Gabrielsson, S. Exosomes with immune modulatory features are present in human breast milk. *J. Immunol.* **2007**, *179*, 1969–1978.
54. Masyuk, A.I.; Masyuk, T.V.; Larusso, N.F. Exosomes in the pathogenesis, diagnostics and therapeutics of liver diseases. *J. Hepatol.* **2013**, doi:10.1016/j.jhep.2013.03.028.
55. Vella, L.J.; Sharples, R.A.; Nisbet, R.M.; Cappai, R.; Hill, A.F. The role of exosomes in the processing of proteins associated with neurodegenerative diseases. *Eur. Biophys. J.* **2008**, *37*, 323–332.
56. Keller, S.; Rupp, C.; Stoeck, A.; Runz, S.; Fogel, M.; Lugert, S.; Hager, H.D.; Abdel-Bakky, M.S.; Gutwein, P.; Altevogt, P. CD24 is a marker of exosomes secreted into urine and amniotic fluid. *Kidney Int.* **2007**, *72*, 1095–1102.
57. Taylor, D.D.; Gercel-Taylor, C. MicroRNA signatures of tumor-derived exosomes as diagnostic biomarkers of ovarian cancer. *Gynecol. Oncol.* **2008**, *110*, 13–21.

58. Gallo, A.; Tandon, M.; Alevizos, I.; Illei, G.G. The majority of microRNAs detectable in serum and saliva is concentrated in exosomes. *PLoS One* **2012**, *7*, e30679.
59. Saman, S.; Kim, W.; Raya, M.; Visnick, Y.; Miro, S.; Saman, S.; Jackson, B.; McKee, A.C.; Alvarez, V.E.; Lee, N.C.; *et al.* Exosome-associated tau is secreted in tauopathy models and is selectively phosphorylated in cerebrospinal fluid in early Alzheimer disease. *J. Biol. Chem.* **2012**, *287*, 3842–3849.
60. Qiu, S.; Duan, X.; Geng, X.; Xie, J.; Gao, H. Antigen-specific activities of CD8+ T cells in the nasal mucosa of patients with nasal allergy. *Asian Pac. J. Allergy Immunol. Launched Allergy Immunol. Soc. Thail.* **2012**, *30*, 107–113.
61. Record, M. Exosomal Lipids in Cell–Cell Communication. In *Emerging Concepts of Tumor Exosome–Mediated Cell-Cell Communication*; Zhang, H.-G., Ed.; Springer: New York, NY, USA, 2013; pp. 47–68.
62. Bobrie, A.; Colombo, M.; Raposo, G.; Thery, C. Exosome secretion: Molecular mechanisms and roles in immune responses. *Traffic* **2011**, *12*, 1659–1668.
63. Hanson, P.I.; Cashikar, A. Multivesicular body morphogenesis. *Annu. Rev. Cell Dev. Biol.* **2012**, *28*, 337–362.
64. Schmidt, O.; Teis, D. The ESCRT machinery. *Curr. Biol.* **2012**, *22*, R116–R120.
65. Raymond, C.K.; Howald-Stevenson, I.; Vater, C.A.; Stevens, T.H. Morphological classification of the yeast vacuolar protein sorting mutants: Evidence for a prevacuolar compartment in class E vps mutants. *Mol. Biol. Cell* **1992**, *3*, 1389–1402.
66. Raiborg, C.; Stenmark, H. The ESCRT machinery in endosomal sorting of ubiquitylated membrane proteins. *Nature* **2009**, *458*, 445–452.
67. Katzmann, D.J.; Babst, M.; Emr, S.D. Ubiquitin-dependent sorting into the multivesicular body pathway requires the function of a conserved endosomal protein sorting complex, ESCRT-I. *Cell* **2001**, *106*, 145–155.
68. Wollert, T.; Wunder, C.; Lippincott-Schwartz, J.; Hurley, J.H. Membrane scission by the ESCRT-III complex. *Nature* **2009**, *458*, 172–177.
69. Van Blitterswijk, W.J.; van der Luit, A.H.; Veldman, R.J.; Verheij, M.; Borst, J. Ceramide: Second messenger or modulator of membrane structure and dynamics? *Biochem. J.* **2003**, *369*, 199–211.
70. Subra, C.; Laulagnier, K.; Perret, B.; Record, M. Exosome lipidomics unravels lipid sorting at the level of multivesicular bodies. *Biochimie* **2007**, *89*, 205–212.
71. Trajkovic, K.; Hsu, C.; Chiantia, S.; Rajendran, L.; Wenzel, D.; Wieland, F.; Schwille, P.; Brugger, B.; Simons, M. Ceramide triggers budding of exosome vesicles into multivesicular endosomes. *Science* **2008**, *319*, 1244–1247.
72. Van Niel, G.; Charrin, S.; Simoes, S.; Romao, M.; Rochin, L.; Saftig, P.; Marks, M.S.; Rubinstein, E.; Raposo, G. The tetraspanin CD63 regulates ESCRT-independent and -dependent endosomal sorting during melanogenesis. *Dev. Cell* **2011**, *21*, 708–721.
73. Rodriguez-Boulan, E.; Kreitzer, G.; Musch, A. Organization of vesicular trafficking in epithelia. *Nat. Rev. Mol. Cell Biol.* **2005**, *6*, 233–247.

74. Hutagalung, A.H.; Novick, P.J. Role of Rab GTPases in membrane traffic and cell physiology. *Physiol. Rev.* **2011**, *91*, 119–149.
75. Savina, A.; Vidal, M.; Colombo, M.I. The exosome pathway in K562 cells is regulated by Rab11. *J. Cell Sci.* **2002**, *115*, 2505–2515.
76. Ostrowski, M.; Carmo, N.B.; Krumeich, S.; Fanget, I.; Raposo, G.; Savina, A.; Moita, C.F.; Schauer, K.; Hume, A.N.; Freitas, R.P.; *et al.* Rab27a and Rab27b control different steps of the exosome secretion pathway. *Nat. Cell Biol.* **2010**, *12*, 19–30.
77. Yu, X.; Harris, S.L.; Levine, A.J. The regulation of exosome secretion: A novel function of the p53 protein. *Cancer Res.* **2006**, *66*, 4795–4801.
78. King, H.W.; Michael, M.Z.; Gleadle, J.M. Hypoxic enhancement of exosome release by breast cancer cells. *BMC Cancer* **2012**, *12*, 421.
79. Thompson, C.A.; Purushothaman, A.; Ramani, V.C.; Vlodavsky, I.; Sanderson, R.D. Heparanase Regulates Secretion, Composition, and Function of Tumor Cell-derived Exosomes. *J. Biol. Chem.* **2013**, *288*, 10093–10099.
80. Koumangoye, R.B.; Sakwe, A.M.; Goodwin, J.S.; Patel, T.; Ochieng, J. Detachment of breast tumor cells induces rapid secretion of exosomes which subsequently mediate cellular adhesion and spreading. *PLoS One* **2011**, *6*, e24234.
81. Llorente, A.; van Deurs, B.; Sandvig, K. Cholesterol regulates prostasome release from secretory lysosomes in PC-3 human prostate cancer cells. *Eur. J. Cell Biol.* **2007**, *86*, 405–415.
82. Faure, J.; Lachenal, G.; Court, M.; Hirrlinger, J.; Chatellard-Causse, C.; Blot, B.; Grange, J.; Schoehn, G.; Goldberg, Y.; Boyer, V.; *et al.* Exosomes are released by cultured cortical neurones. *Mol. Cell. Neurosci.* **2006**, *31*, 642–648.
83. Parolini, I.; Federici, C.; Raggi, C.; Lugini, L.; Palleschi, S.; de Milito, A.; Coscia, C.; Iessi, E.; Logozzi, M.; Molinari, A.; *et al.* Microenvironmental pH is a key factor for exosome traffic in tumor cells. *J. Biol. Chem.* **2009**, *284*, 34211–34222.
84. Mittelbrunn, M.; Gutierrez-Vazquez, C.; Villarroya-Beltri, C.; Gonzalez, S.; Sanchez-Cabo, F.; Gonzalez, M.A.; Bernad, A.; Sanchez-Madrid, F. Unidirectional transfer of microRNA-loaded exosomes from T cells to antigen-presenting cells. *Nat. Commun.* **2011**, *2*, 282.
85. Keller, S.; Konig, A.K.; Marme, F.; Runz, S.; Wolterink, S.; Koensgen, D.; Mustea, A.; Sehouli, J.; Altevogt, P. Systemic presence and tumor-growth promoting effect of ovarian carcinoma released exosomes. *Cancer Lett.* **2009**, *278*, 73–81.
86. Barres, C.; Blanc, L.; Bette-Bobillo, P.; Andre, S.; Mamoun, R.; Gabius, H.J.; Vidal, M. Galectin-5 is bound onto the surface of rat reticulocyte exosomes and modulates vesicle uptake by macrophages. *Blood* **2010**, *115*, 696–705.
87. Mathivanan, S.; Fahner, C.J.; Reid, G.E.; Simpson, R.J. ExoCarta 2012: Database of exosomal proteins, RNA and lipids. *Nucleic Acids Res.* **2012**, *40*, D1241–D1244.
88. Mathivanan, S.; Simpson, R.J. ExoCarta: A compendium of exosomal proteins and RNA. *Proteomics* **2009**, *9*, 4997–5000.
89. Subra, C.; Grand, D.; Laulagnier, K.; Stella, A.; Lambeau, G.; Paillasse, M.; de Medina, P.; Monsarrat, B.; Perret, B.; Silvente-Poirot, S.; *et al.* Exosomes account for vesicle-mediated transcellular transport of activatable phospholipases and prostaglandins. *J. Lipid Res.* **2010**, *51*, 2105–2120.

90. Zakharova, L.; Svetlova, M.; Fomina, A.F. T cell exosomes induce cholesterol accumulation in human monocytes via phosphatidylserine receptor. *J. Cell. Physiol.* **2007**, *212*, 174–181.
91. Xiang, X.; Poliakov, A.; Liu, C.; Liu, Y.; Deng, Z.B.; Wang, J.; Cheng, Z.; Shah, S.V.; Wang, G.J.; Zhang, L.; *et al.* Induction of myeloid-derived suppressor cells by tumor exosomes. *Int. J. Cancer* **2009**, *124*, 2621–2633.
92. Cao, W.; Ma, Z.; Rasenick, M.M.; Yeh, S.; Yu, J. N-3 poly-unsaturated fatty acids shift estrogen signaling to inhibit human breast cancer cell growth. *PLoS One* **2012**, *7*, e52838.
93. Valadi, H.; Ekstrom, K.; Bossios, A.; Sjostrand, M.; Lee, J.J.; Lotvall, J.O. Exosome-mediated transfer of mRNAs and microRNAs is a novel mechanism of genetic exchange between cells. *Nat. Cell Biol.* **2007**, *9*, 654–659.
94. Hessvik, N.P.; Phuyal, S.; Brech, A.; Sandvig, K.; Llorente, A. Profiling of microRNAs in exosomes released from PC-3 prostate cancer cells. *Biochim. Biophys. Acta* **2012**, *1819*, 1154–1163.
95. Bellingham, S.A.; Coleman, B.M.; Hill, A.F. Small RNA deep sequencing reveals a distinct miRNA signature released in exosomes from prion-infected neuronal cells. *Nucleic Acids Res.* **2012**, *40*, 10937–10949.
96. Rabinowits, G.; Gercel-Taylor, C.; Day, J.M.; Taylor, D.D.; Kloecker, G.H. Exosomal microRNA: A diagnostic marker for lung cancer. *Clin. Lung Cancer* **2009**, *10*, 42–46.
97. Pigati, L.; Yaddanapudi, S.C.; Iyengar, R.; Kim, D.J.; Hearn, S.A.; Danforth, D.; Hastings, M.L.; Duelli, D.M. Selective release of microRNA species from normal and malignant mammary epithelial cells. *PLoS One* **2010**, *5*, e13515.
98. Jaiswal, R.; Luk, F.; Gong, J.; Mathys, J.M.; Grau, G.E.; Bebawy, M. Microparticle conferred microRNA profiles—Implications in the transfer and dominance of cancer traits. *Mol. Cancer* **2012**, *11*, 37.
99. Xiao, D.; Ohlendorf, J.; Chen, Y.; Taylor, D.D.; Rai, S.N.; Waigel, S.; Zacharias, W.; Hao, H.; McMasters, K.M. Identifying mRNA, microRNA and protein profiles of melanoma exosomes. *PLoS One* **2012**, *7*, e46874.
100. Simpson, R.J.; Jensen, S.S.; Lim, J.W. Proteomic profiling of exosomes: Current perspectives. *Proteomics* **2008**, *8*, 4083–4099.
101. de Gassart, A.; Geminard, C.; Fevrier, B.; Raposo, G.; Vidal, M. Lipid raft-associated protein sorting in exosomes. *Blood* **2003**, *102*, 4336–4344.
102. Thery, C.; Amigorena, S.; Raposo, G.; Clayton, A. Isolation and characterization of exosomes from cell culture supernatants and biological fluids. *Curr. Protoc. Cell Biol.* **2006**, doi:10.1002/0471143030.cb0322s30.
103. Cantin, R.; Diou, J.; Belanger, D.; Tremblay, A.M.; Gilbert, C. Discrimination between exosomes and HIV-1: Purification of both vesicles from cell-free supernatants. *J. Immunol. Methods* **2008**, *338*, 21–30.
104. Lamparski, H.G.; Metha-Damani, A.; Yao, J.Y.; Patel, S.; Hsu, D.H.; Ruegg, C.; Le Pecq, J.B. Production and characterization of clinical grade exosomes derived from dendritic cells. *J. Immunol. Methods* **2002**, *270*, 211–226.
105. System Biosciences. Available online: http://www.systembio.com/exoquick (accessed on 15 April 2013).

106. Taylor, D.D.; Atay, S.; Metzinger, D.S.; Gercel-Taylor, C. Characterization of humoral responses of ovarian cancer patients: Antibody subclasses and antigenic components. *Gynecol. Oncol.* **2010**, *116*, 213–221.
107. Koga, K.; Matsumoto, K.; Akiyoshi, T.; Kubo, M.; Yamanaka, N.; Tasaki, A.; Nakashima, H.; Nakamura, M.; Kuroki, S.; Tanaka, M.; Katano, M. Purification, characterization and biological significance of tumor-derived exosomes. *Anticancer Res.* **2005**, *25*, 3703–3707.
108. Tauro, B.J.; Greening, D.W.; Mathias, R.A.; Ji, H.; Mathivanan, S.; Scott, A.M.; Simpson, R.J. Comparison of ultracentrifugation, density gradient separation, and immunoaffinity capture methods for isolating human colon cancer cell line LIM1863-derived exosomes. *Methods* **2012**, *56*, 293–304.
109. Taylor, D.D.; Zacharias, W.; Gercel-Taylor, C. Exosome isolation for proteomic analyses and RNA profiling. *Methods Mol. Biol.* **2011**, *728*, 235–246.
110. Simona, F.; Laura, S.; Simona, T.; Riccardo, A. Contribution of proteomics to understanding the role of tumor-derived exosomes in cancer progression: State of the art and new perspectives. *Proteomics* **2013**, *13*, 1581–1594.
111. Eldh, M.; Lotvall, J.; Malmhall, C.; Ekstrom, K. Importance of RNA isolation methods for analysis of exosomal RNA: Evaluation of different methods. *Mol. Immunol.* **2012**, *50*, 278–286.
112. Lasser, C.; Eldh, M.; Lotvall, J. Isolation and characterization of RNA-containing exosomes. *J. Vis. Exp.* **2012**, e3037.
113. Lasser, C. Identification and analysis of circulating exosomal microRNA in human body fluids. *Methods Mol. Biol.* **2013**, *1024*, 109–128.
114. Huang, X.; Yuan, T.; Tschannen, M.; Sun, Z.; Jacob, H.; Du, M.; Liang, M.; Dittmar, R.L.; Liu, Y.; Liang, M.; *et al.* Characterization of human plasma-derived exosomal RNAs by deep sequencing. *BMC Genomics* **2013**, *14*, 319.
115. Nanosight. Available online: http://www.nanosight.com (accessed on 15 April 2013).
116. Zheng, Y.; Campbell, E.C.; Lucocq, J.; Riches, A.; Powis, S.J. Monitoring the Rab27 associated exosome pathway using nanoparticle tracking analysis. *Exp. Cell Res.* **2012**, doi:10.1016/j.yexcr.2012.10.006.
117. Hanahan, D.; Coussens, L.M. Accessories to the crime: Functions of cells recruited to the tumor microenvironment. *Cancer Cell* **2012**, *21*, 309–322.
118. Tian, T.; Wang, Y.; Wang, H.; Zhu, Z.; Xiao, Z. Visualizing of the cellular uptake and intracellular trafficking of exosomes by live-cell microscopy. *J. Cell. Biochem.* **2010**, *111*, 488–496.
119. Ristorcelli, E.; Beraud, E.; Verrando, P.; Villard, C.; Lafitte, D.; Sbarra, V.; Lombardo, D.; Verine, A. Human tumor nanoparticles induce apoptosis of pancreatic cancer cells. *FASEB J.* **2008**, *22*, 3358–3369.
120. Yang, Y.; Xiu, F.; Cai, Z.; Wang, J.; Wang, Q.; Fu, Y.; Cao, X. Increased induction of antitumor response by exosomes derived from interleukin-2 gene-modified tumor cells. *J. Cancer Res. Clin. Oncol.* **2007**, *133*, 389–399.
121. Dai, S.; Wei, D.; Wu, Z.; Zhou, X.; Wei, X.; Huang, H.; Li, G. Phase I clinical trial of autologous ascites-derived exosomes combined with GM-CSF for colorectal cancer. *Mol. Ther.* **2008**, *16*, 782–790.

122. Zhang, Y.; Wu, X.H.; Luo, C.L.; Zhang, J.M.; He, B.C.; Chen, G. Interleukin-12-anchored exosomes increase cytotoxicity of T lymphocytes by reversing the JAK/STAT pathway impaired by tumor-derived exosomes. *Int. J. Mol. Med.* **2010**, *25*, 695–700.
123. Zhang, Y.; Luo, C.L.; He, B.C.; Zhang, J.M.; Cheng, G.; Wu, X.H. Exosomes derived from IL-12-anchored renal cancer cells increase induction of specific antitumor response *in vitro*: A novel vaccine for renal cell carcinoma. *Int. J. Oncol.* **2010**, *36*, 133–140.
124. Chen, T.; Guo, J.; Yang, M.; Zhu, X.; Cao, X. Chemokine-containing exosomes are released from heat-stressed tumor cells via lipid raft-dependent pathway and act as efficient tumor vaccine. *J. Immunol.* **2011**, *186*, 2219–2228.
125. Peng, P.; Yan, Y.; Keng, S. Exosomes in the ascites of ovarian cancer patients: Origin and effects on anti-tumor immunity. *Oncol. Rep.* **2011**, *25*, 749–762.
126. Hakulinen, J.; Sankkila, L.; Sugiyama, N.; Lehti, K.; Keski-Oja, J. Secretion of active membrane type 1 matrix metalloproteinase (MMP-14) into extracellular space in microvesicular exosomes. *J. Cell. Biochem.* **2008**, *105*, 1211–1218.
127. McCready, J.; Sims, J.D.; Chan, D.; Jay, D.G. Secretion of extracellular hsp90alpha via exosomes increases cancer cell motility: A role for plasminogen activation. *BMC Cancer* **2010**, *10*, 294.
128. Janowska-Wieczorek, A.; Wysoczynski, M.; Kijowski, J.; Marquez-Curtis, L.; Machalinski, B.; Ratajczak, J.; Ratajczak, M.Z. Microvesicles derived from activated platelets induce metastasis and angiogenesis in lung cancer. *Int. J. Cancer* **2005**, *113*, 752–760.
129. Cho, J.A.; Park, H.; Lim, E.H.; Kim, K.H.; Choi, J.S.; Lee, J.H.; Shin, J.W.; Lee, K.W. Exosomes from ovarian cancer cells induce adipose tissue-derived mesenchymal stem cells to acquire the physical and functional characteristics of tumor-supporting myofibroblasts. *Gynecol. Oncol.* **2011**, *123*, 379–386.
130. Bobrie, A.; Krumeich, S.; Reyal, F.; Recchi, C.; Moita, L.F.; Seabra, M.C.; Ostrowski, M.; Thery, C. Rab27a supports exosome-dependent and -independent mechanisms that modify the tumor microenvironment and can promote tumor progression. *Cancer Res.* **2012**, *72*, 4920–4930.
131. Park, J.E.; Tan, H.S.; Datta, A.; Lai, R.C.; Zhang, H.; Meng, W.; Lim, S.K.; Sze, S.K. Hypoxic tumor cell modulates its microenvironment to enhance angiogenic and metastatic potential by secretion of proteins and exosomes. *Mol. Cell. Proteomics* **2010**, *9*, 1085–1099.
132. Kucharzewska, P.; Christianson, H.C.; Welch, J.E.; Svensson, K.J.; Fredlund, E.; Ringner, M.; Morgelin, M.; Bourseau-Guilmain, E.; Bengzon, J.; Belting, M. Exosomes reflect the hypoxic status of glioma cells and mediate hypoxia-dependent activation of vascular cells during tumor development. *Proc. Natl. Acad. Sci. USA* **2013**, doi:10.1073/pnas.1220998110.
133. Hong, B.S.; Cho, J.H.; Kim, H.; Choi, E.J.; Rho, S.; Kim, J.; Kim, J.H.; Choi, D.S.; Kim, Y.K.; Hwang, D.; *et al.* Colorectal cancer cell-derived microvesicles are enriched in cell cycle-related mRNAs that promote proliferation of endothelial cells. *BMC Genomics* **2009**, *10*, 556.
134. Ji, H.; Greening, D.W.; Barnes, T.W.; Lim, J.W.; Tauro, B.J.; Rai, A.; Xu, R.; Adda, C.; Mathivanan, S.; Zhao, W.; *et al.* Proteome profiling of exosomes derived from human primary and metastatic colorectal cells reveal differential expression of key metastatic factors and signal transduction components. *Proteomics* **2013**, *13*, 1672–1686.

135. Grange, C.; Tapparo, M.; Collino, F.; Vitillo, L.; Damasco, C.; Deregibus, M.C.; Tetta, C.; Bussolati, B.; Camussi, G. Microvesicles released from human renal cancer stem cells stimulate angiogenesis and formation of lung premetastatic niche. *Cancer Res.* **2011**, *71*, 5346–5356.
136. Zhu, W.; Huang, L.; Li, Y.; Zhang, X.; Gu, J.; Yan, Y.; Xu, X.; Wang, M.; Qian, H.; Xu, W. Exosomes derived from human bone marrow mesenchymal stem cells promote tumor growth *in vivo*. *Cancer Lett.* **2012**, *315*, 28–37.
137. Mineo, M.; Garfield, S.H.; Taverna, S.; Flugy, A.; De Leo, G.; Alessandro, R.; Kohn, E.C. Exosomes released by K562 chronic myeloid leukemia cells promote angiogenesis in a Src-dependent fashion. *Angiogenesis* **2012**, *15*, 33–45.
138. Peinado, H.; Aleckovic, M.; Lavotshkin, S.; Matei, I.; Costa-Silva, B.; Moreno-Bueno, G.; Hergueta-Redondo, M.; Williams, C.; Garcia-Santos, G.; Ghajar, C.; *et al.* Melanoma exosomes educate bone marrow progenitor cells toward a pro-metastatic phenotype through MET. *Nat. Med.* **2012**, *18*, 883–891.
139. Baer, C.; Squadrito, M.L.; Iruela-Arispe, M.L.; De Palma, M. Reciprocal interactions between endothelial cells and macrophages in angiogenic vascular niches. *Exp. Cell Res.* **2013**, doi:10.1016/j.yexcr.2013.03.026.
140. Cai, X.; Hagedorn, C.H.; Cullen, B.R. Human microRNAs are processed from capped, polyadenylated transcripts that can also function as mRNAs. *RNA* **2004**, *10*, 1957–1966.
141. Lee, Y.; Jeon, K.; Lee, J.T.; Kim, S.; Kim, V.N. MicroRNA maturation: Stepwise processing and subcellular localization. *EMBO J.* **2002**, *21*, 4663–4670.
142. Lee, Y.; Kim, M.; Han, J.; Yeom, K.H.; Lee, S.; Baek, S.H.; Kim, V.N. MicroRNA genes are transcribed by RNA polymerase II. *EMBO J.* **2004**, *23*, 4051–4060.
143. Zeng, Y.; Cullen, B.R. Structural requirements for pre-microRNA binding and nuclear export by Exportin 5. *Nucleic Acids Res.* **2004**, *32*, 4776–4785.
144. Bartel, D.P. MicroRNAs: Target recognition and regulatory functions. *Cell* **2009**, *136*, 215–233.
145. Griffiths-Jones, S. The microRNA registry. *Nucleic Acids Res.* **2004**, *32*, D109–D111.
146. Lim, L.P.; Lau, N.C.; Garrett-Engele, P.; Grimson, A.; Schelter, J.M.; Castle, J.; Bartel, D.P.; Linsley, P.S.; Johnson, J.M. Microarray analysis shows that some microRNAs downregulate large numbers of target mRNAs. *Nature* **2005**, *433*, 769–773.
147. Taft, R.J.; Pang, K.C.; Mercer, T.R.; Dinger, M.; Mattick, J.S. Non-coding RNAs: Regulators of disease. *J. Pathol.* **2010**, *220*, 126–139.
148. Melo, S.A.; Esteller, M. Dysregulation of microRNAs in cancer: Playing with fire. *FEBS Lett.* **2011**, *585*, 2087–2099.
149. Lu, J.; Getz, G.; Miska, E.A.; Alvarez-Saavedra, E.; Lamb, J.; Peck, D.; Sweet-Cordero, A.; Ebert, B.L.; Mak, R.H.; Ferrando, A.A.; *et al.* microRNA expression profiles classify human cancers. *Nature* **2005**, *435*, 834–838.
150. Guo, Y.; Chen, Z.; Zhang, L.; Zhou, F.; Shi, S.; Feng, X.; Li, B.; Meng, X.; Ma, X.; Luo, M.; *et al.* Distinctive microRNA profiles relating to patient survival in esophageal squamous cell carcinoma. *Cancer Res.* **2008**, *68*, 26–33.
151. Gibbings, D.J.; Ciaudo, C.; Erhardt, M.; Voinnet, O. Multivesicular bodies associate with components of miRNA effector complexes and modulate miRNA activity. *Nat. Cell Biol.* **2009**, *11*, 1143–1149.

152. Kosaka, N.; Iguchi, H.; Yoshioka, Y.; Takeshita, F.; Matsuki, Y.; Ochiya, T. Secretory mechanisms and intercellular transfer of microRNAs in living cells. *J. Biol. Chem.* **2010**, *285*, 17442–17452.
153. Yao, B.; La, L.B.; Chen, Y.C.; Chang, L.J.; Chan, E.K. Defining a new role of GW182 in maintaining miRNA stability. *EMBO Rep.* **2012**, *13*, 1102–1108.
154. Ohshima, K.; Inoue, K.; Fujiwara, A.; Hatakeyama, K.; Kanto, K.; Watanabe, Y.; Muramatsu, K.; Fukuda, Y.; Ogura, S.; Yamaguchi, K.; *et al.* Let-7 microRNA family is selectively secreted into the extracellular environment via exosomes in a metastatic gastric cancer cell line. *PLoS One* **2010**, *5*, e13247.
155. Cifuentes, D.; Xue, H.; Taylor, D.W.; Patnode, H.; Mishima, Y.; Cheloufi, S.; Ma, E.; Mane, S.; Hannon, G.J.; Lawson, N.D.; *et al.* A novel miRNA processing pathway independent of Dicer requires Argonaute2 catalytic activity. *Science* **2010**, *328*, 1694–1698.
156. Montecalvo, A.; Larregina, A.T.; Shufesky, W.J.; Stolz, D.B.; Sullivan, M.L.; Karlsson, J.M.; Baty, C.J.; Gibson, G.A.; Erdos, G.; Wang, Z.; *et al.* Mechanism of transfer of functional microRNAs between mouse dendritic cells via exosomes. *Blood* **2012**, *119*, 756–766.
157. Kogure, T.; Lin, W.L.; Yan, I.K.; Braconi, C.; Patel, T. Intercellular nanovesicle-mediated microRNA transfer: A mechanism of environmental modulation of hepatocellular cancer cell growth. *Hepatology* **2011**, *54*, 1237–1248.
158. Bryant, R.J.; Pawlowski, T.; Catto, J.W.; Marsden, G.; Vessella, R.L.; Rhees, B.; Kuslich, C.; Visakorpi, T.; Hamdy, F.C. Changes in circulating microRNA levels associated with prostate cancer. *Br. J. Cancer* **2012**, *106*, 768–774.
159. Takeshita, N.; Hoshino, I.; Mori, M.; Akutsu, Y.; Hanari, N.; Yoneyama, Y.; Ikeda, N.; Isozaki, Y.; Maruyama, T.; Akanuma, N.; *et al.* Serum microRNA expression profile: miR-1246 as a novel diagnostic and prognostic biomarker for oesophageal squamous cell carcinoma. *Br. J. Cancer* **2013**, *108*, 644–652.
160. Russo, F.; di Bella, S.; Nigita, G.; Macca, V.; Lagana, A.; Giugno, R.; Pulvirenti, A.; Ferro, A. miRandola: Extracellular circulating microRNAs database. *PLoS One* **2012**, *7*, e47786.
161. Umezu, T.; Ohyashiki, K.; Kuroda, M.; Ohyashiki, J.H. Leukemia cell to endothelial cell communication via exosomal miRNAs. *Oncogene* **2012**, *32*, 2747–2755.
162. Yang, M.; Chen, J.; Su, F.; Yu, B.; Su, F.; Lin, L.; Liu, Y.; Huang, J.D.; Song, E. Microvesicles secreted by macrophages shuttle invasion-potentiating microRNAs into breast cancer cells. *Mol. Cancer* **2011**, *10*, 117.
163. Zhou, R.; O'Hara, S.P.; Chen, X.M. MicroRNA regulation of innate immune responses in epithelial cells. *Cell. Mol. Immunol.* **2011**, *8*, 371–379.
164. Fabbri, M.; Paone, A.; Calore, F.; Galli, R.; Gaudio, E.; Santhanam, R.; Lovat, F.; Fadda, P.; Mao, C.; Nuovo, G.J.; *et al.* MicroRNAs bind to Toll-like receptors to induce prometastatic inflammatory response. *Proc. Natl. Acad. Sci. USA* **2012**, *109*, E2110–E2116.
165. Palma, J.; Yaddanapudi, S.C.; Pigati, L.; Havens, M.A.; Jeong, S.; Weiner, G.A.; Weimer, K.M.; Stern, B.; Hastings, M.L.; Duelli, D.M. MicroRNAs are exported from malignant cells in customized particles. *Nucleic Acids Res.* **2012**, *40*, 9125–9138.
166. Rana, S.; Malinowska, K.; Zoller, M. Exosomal tumor microRNA modulates premetastatic organ cells. *Neoplasia* **2013**, *15*, 281–295.
167. Quesenberry, P.J.; Aliotta, J.M. Cellular phenotype switching and microvesicles. *Adv. Drug Deliv. Rev.* **2010**, *62*, 1141–1148.

168. Weigelt, B.; Bissell, M.J. Unraveling the microenvironmental influences on the normal mammary gland and breast cancer. *Semin. Cancer Biol.* **2008**, *18*, 311–321.
169. Section on Breastfeeding. Breastfeeding and the use of human milk. *Pediatrics* **2012**, *129*, e827–e841.
170. Zhou, Q.; Li, M.; Wang, X.; Li, Q.; Wang, T.; Zhu, Q.; Zhou, X.; Wang, X.; Gao, X.; Li, X. Immune-related microRNAs are abundant in breast milk exosomes. *Int. J. Biol. Sci.* **2012**, *8*, 118–123.
171. Society, A.C. Breast Cancer Facts & Figures 2011–2012. Available online: http://www.cancer.org/research/cancerfactsfigures/breastcancerfactsfigures/breast-cancer-facts-and-figures-2011-2012 (accessed on 15 April 2013).
172. Locker, A.; Horrocks, C.; Gilmour, A.; Ellis, I.; Dowle, C.; Elston, C.; Blamey, R. Flow cytometric and histological analysis of ductal carcinoma in situ of the breast. *Br. J. Surg.* **1990**, *77*, 564–567.
173. Ottesen, G.; Christensen, I.; Larsen, J.; Christiansen, J.; Hansen, B.; Andersen, J. DNA analysis of in situ ductal carcinoma of the breast via flow cytometry. *Cytometry* **1995**, *22*, 168–176.
174. Leal, C.; Schmitt, F.; Bento, M.; Maia, N.; Lopes, C. Ductal carcinoma in situ of the breast. Histologic categorization and its relationship to ploidy and immunohistochemical expression of hormone receptors, p53, and c-erbB-2 protein. *Cancer* **1995**, *75*, 2123–2131.
175. Maguire, H.; Hellman, M.; Greene, M.; Yeh, I. Expression of c-erbB-2 in in situ and in adjacent invasive ductal adenocarcinomas of the female breast. *Pathobiology* **1992**, *60*, 117–121.
176. Vos, C.; Ter Haar, N.; Peterse, J.; Cornelisse, C.; van de Vijver, M. Cyclin D1 gene amplification and overexpression are present in ductal carcinoma in situ of the breast. *J. Pathol.* **1999**, *187*, 279–284.
177. Radford, D.; Fair, K.; Phillips, N.; Ritter, J.; Steinbrueck, T.; Holt, M.; Donis-Keller, H. Allelotyping of ductal carcinoma in situ of the breast: Deletion of loci on 8p, 13q, 16q, 17p and 17q. *Cancer Res.* **1995**, *55*, 3399–3405.
178. Radford, D.; Phillips, N.; Fair, K.; Ritter, J.; Holt, M.; Donis-Keller, H. Allelic loss and the progression of breast cancer. *Cancer Res.* **1995**, *55*, 5180–5183.
179. Ma, X.J.; Dahiya, S.; Richardson, E.; Erlander, M.; Sgroi, D.C. Gene expression profiling of the tumor microenvironment during breast cancer progression. *Breast Cancer Res.* **2009**, *11*, R7.
180. Liu, C.; Yu, S.; Zinn, K.; Wang, J.; Zhang, L.; Jia, Y.; Kappes, J.C.; Barnes, S.; Kimberly, R.P.; Grizzle, W.E.; *et al.* Murine mammary carcinoma exosomes promote tumor growth by suppression of NK cell function. *J. Immunol.* **2006**, *176*, 1375–1385.
181. Higginbotham, J.N.; Demory Beckler, M.; Gephart, J.D.; Franklin, J.L.; Bogatcheva, G.; Kremers, G.J.; Piston, D.W.; Ayers, G.D.; McConnell, R.E.; Tyska, M.J.; *et al.* Amphiregulin exosomes increase cancer cell invasion. *Curr. Biol.* **2011**, *21*, 779–786.
182. O'Brien, K.; Rani, S.; Corcoran, C.; Wallace, R.; Hughes, L.; Friel, A.M.; McDonnell, S.; Crown, J.; Radomski, M.W.; O'Driscoll, L. Exosomes from triple-negative breast cancer cells can transfer phenotypic traits representing their cells of origin to secondary cells. *Eur. J. Cancer* **2013**, *49*, 1845–1859.
183. Luga, V.; Zhang, L.; Viloria-Petit, A.M.; Ogunjimi, A.A.; Inanlou, M.R.; Chiu, E.; Buchanan, M.; Hosein, A.N.; Basik, M.; Wrana, J.L. Exosomes mediate stromal mobilization of autocrine Wnt-PCP signaling in breast cancer cell migration. *Cell* **2012**, *151*, 1542–1556.

184. Suetsugu, A.; Honma, K.; Saji, S.; Moriwaki, H.; Ochiya, T.; Hoffman, R.M. Imaging exosome transfer from breast cancer cells to stroma at metastatic sites in orthotopic nude-mouse models. *Adv. Drug Deliv. Rev.* **2013**, *65*, 383–390.
185. Sotiriou, C.; Pusztai, L. Gene-expression signatures in breast cancer. *N. Engl. J. Med.* **2009**, *360*, 790–800.
186. Blenkiron, C.; Goldstein, L.D.; Thorne, N.P.; Spiteri, I.; Chin, S.F.; Dunning, M.J.; Barbosa-Morais, N.L.; Teschendorff, A.E.; Green, A.R.; Ellis, I.O.; *et al.* MicroRNA expression profiling of human breast cancer identifies new markers of tumor subtype. *Genome Biol.* **2007**, *8*, R214.
187. Perou, C.M.; Sorlie, T.; Eisen, M.B.; van de Rijn, M.; Jeffrey, S.S.; Rees, C.A.; Pollack, J.R.; Ross, D.T.; Johnsen, H.; Akslen, L.A.; *et al.* Molecular portraits of human breast tumours. *Nature* **2000**, *406*, 747–752.
188. Ciravolo, V.; Huber, V.; Ghedini, G.C.; Venturelli, E.; Bianchi, F.; Campiglio, M.; Morelli, D.; Villa, A.; Della Mina, P.; Menard, S.; *et al.* Potential role of HER2-overexpressing exosomes in countering trastuzumab-based therapy. *J. Cell. Physiol.* **2012**, *227*, 658–667.
189. Marleau, A.M.; Chen, C.S.; Joyce, J.A.; Tullis, R.H. Exosome removal as a therapeutic adjuvant in cancer. *J. Transl. Med.* **2012**, *10*, 134.
190. Safaei, R.; Larson, B.J.; Cheng, T.C.; Gibson, M.A.; Otani, S.; Naerdemann, W.; Howell, S.B. Abnormal lysosomal trafficking and enhanced exosomal export of cisplatin in drug-resistant human ovarian carcinoma cells. *Mol. Cancer Ther.* **2005**, *4*, 1595–1604.
191. Ohno, S.; Takanashi, M.; Sudo, K.; Ueda, S.; Ishikawa, A.; Matsuyama, N.; Fujita, K.; Mizutani, T.; Ohgi, T.; Ochiya, T.; *et al.* Systemically injected exosomes targeted to EGFR deliver antitumor microRNA to breast cancer cells. *Mol. Ther.* **2013**, *21*, 185–191.
192. Kooijmans, S.A.; Vader, P.; van Dommelen, S.M.; van Solinge, W.W.; Schiffelers, R.M. Exosome mimetics: A novel class of drug delivery systems. *Int. J. Nanomedicine* **2012**, *7*, 1525–1541.

PERMISSIONS

All chapters in this book were first published by MDPI; hereby published with permission under the Creative Commons Attribution License or equivalent. Every chapter published in this book has been scrutinized by our experts. Their significance has been extensively debated. The topics covered herein carry significant findings which will fuel the growth of the discipline. They may even be implemented as practical applications or may be referred to as a beginning point for another development.

The contributors of this book come from diverse backgrounds, making this book a truly international effort. This book will bring forth new frontiers with its revolutionizing research information and detailed analysis of the nascent developments around the world.

We would like to thank all the contributing authors for lending their expertise to make the book truly unique. They have played a crucial role in the development of this book. Without their invaluable contributions this book wouldn't have been possible. They have made vital efforts to compile up to date information on the varied aspects of this subject to make this book a valuable addition to the collection of many professionals and students.

This book was conceptualized with the vision of imparting up-to-date information and advanced data in this field. To ensure the same, a matchless editorial board was set up. Every individual on the board went through rigorous rounds of assessment to prove their worth. After which they invested a large part of their time researching and compiling the most relevant data for our readers.

The editorial board has been involved in producing this book since its inception. They have spent rigorous hours researching and exploring the diverse topics which have resulted in the successful publishing of this book. They have passed on their knowledge of decades through this book. To expedite this challenging task, the publisher supported the team at every step. A small team of assistant editors was also appointed to further simplify the editing procedure and attain best results for the readers.

Apart from the editorial board, the designing team has also invested a significant amount of their time in understanding the subject and creating the most relevant covers. They scrutinized every image to scout for the most suitable representation of the subject and create an appropriate cover for the book.

The publishing team has been an ardent support to the editorial, designing and production team. Their endless efforts to recruit the best for this project, has resulted in the accomplishment of this book. They are a veteran in the field of academics and their pool of knowledge is as vast as their experience in printing. Their expertise and guidance has proved useful at every step. Their uncompromising quality standards have made this book an exceptional effort. Their encouragement from time to time has been an inspiration for everyone.

The publisher and the editorial board hope that this book will prove to be a valuable piece of knowledge for researchers, students, practitioners and scholars across the globe.

LIST OF CONTRIBUTORS

Akihiro Matsui, Anh Hai Nguyen and Kentaro Nakaminami
Plant Genomic Network Research Team, RIKEN Center for Sustainable Resource Science, 1-7-22 Suehiro-cho, Tsurumi-ku, Yokohama, Kanagawa 230-0045, Japan

Motoaki Seki
Plant Genomic Network Research Team, RIKEN Center for Sustainable Resource Science, 1-7-22 Suehiro-cho, Tsurumi-ku, Yokohama, Kanagawa 230-0045, Japan
Kihara Institute for Biological Research, Yokohama City University, 641-12 Maioka-cho, Totsukaku, Yokohama, Kanagawa 244-0813, Japan
CREST, JST, 4-1-8 Honcho, Kawaguchi, Saitama 332-0012, Japan

Mikhail M. Pooggin
University of Basel, Department of Environmental Sciences, Botany, Schönbeinstrasse 6, Basel 4056, Switzerland

Sung-Min Kang, Ji-Woong Choi and Su-Hyung Hong
Department of Oral Microbiology, School of Dentistry, Kyungpook National University, Daegu 700-412, Korea

Heon-Jin Lee
Department of Oral Microbiology, School of Dentistry, Kyungpook National University, Daegu 700-412, Korea
Brain Science and Engineering Institute, Kyungpook National University, Daegu 700-412, Korea

Claudio Iaconetti, Clarice Gareri, Alberto Polimeni and Ciro Indolfi
Division of Cardiology, Magna Graecia University, URT Consiglio Nazionale delle Ricerche (CNR), Catanzaro 88100, Italy

Xiangsheng Li, Zhichao Zhang, Ming Yu, Liqi Li, Guangsheng Du, Weidong Xiao and Hua Yang
Department of General Surgery, Xinqiao Hospital, Third Military Medical University, Xinqiao Road, Chongqing 400037, China

Michael Hecker
Department of Neurology, Division of Neuroimmunology, University of Rostock, Gehlsheimer Str. 20, 18147 Rostock, Germany
Steinbeis Transfer Center for Proteome Analysis, Schillingallee 68, 18057 Rostock, Germany

Madhan Thamilarasan, Ina Schröder and Uwe Klaus Zettl
Department of Neurology, Division of Neuroimmunology, University of Rostock, Gehlsheimer Str. 20, 18147 Rostock, Germany

Dirk Koczan, Kristin Flechtner and Hans-Jürgen Thiesen
Institute of Immunology, University of Rostock, Schillingallee 70, 18055 Rostock, Germany

Sherry Freiesleben and Georg Füllen
Institute for Biostatistics and Informatics in Medicine and Ageing Research, University of Rostock, Ernst-Heydemann-Str. 8, 18057 Rostock, Germany

Jing Li and Changning Liu
Bioinformatics Research Group, Key Laboratory of Intelligent Information Processing, Advanced Computer Research Center, Institute of Computing Technology, Chinese Academy of Sciences, Beijing 100190, China

Zhenyu Xuan
Department of Molecular and Cell Biology, Center for Systems Biology, University of Texas at Dallas, 800 W Campbell Road, Richardson, TX 75080, USA

Guangchao Sui
Department of Cancer Biology and Comprehensive Cancer Center, Wake Forest University School of Medicine, Winston-Salem, NC 27157, USA

Guorui Deng
Department of Cancer Biology and Comprehensive Cancer Center, Wake Forest University School of Medicine, Winston-Salem, NC 27157, USA
Hypertension and Vascular Research Center, Wake Forest University School of Medicine, Winston-Salem, NC 27157, USA

Graziella Curtale
Department of Medical Biotechnologies and Translational Medicine, University of Milan, Rozzano 20089, Italy
Humanitas Clinical and Research Center, Rozzano 20089, Italy

Franca Citarella
Department of Cellular Biotechnology and Hematology, Sapienza University of Rome, Rome 00185, Italy

Hao Zheng
School of Electrical and Computer Engineering, Georgia Institute of Technology, 225 North Avenue NW, Atlanta, GA 30301, USA
Department of Biomedical Science, Mercer University School of Medicine, Savannah Campus, 4700 Waters Ave, Savannah, GA 31404, USA

Shi-Wen Jiang
Department of Biomedical Science, Mercer University School of Medicine, Savannah Campus, 4700 Waters Ave, Savannah, GA 31404, USA

Rongguo Fu
Department of Nephrology, Second Affiliated Hospital, School of Medicine, Xi'an Jiaotong University, Xi'an 710004, Shannxi, China

Jin-Tao Wang
Department of Epidemiology, School of Public Health, Shanxi Medical University, 56 Xin Jian Nan Road, Taiyuan 030001, Shanxi, China

Qinyou Liu
Animal Reproduction Institute of Guangxi University, 100 Daxue Road, Nanning 530005, Guangxi, China

Haibin Chen
Department of Histology and Embryology, Shantou University Medical College, 22 Xinling Road, Shantou 515041, Guangdong, China

Bethany N. Hannafon and Wei-Qun Ding
Department of Pathology, University of Oklahoma Health Sciences Center, Oklahoma City, OK 73104, USA

Index

A

Abiotic Stress Response, 1, 4, 9

Adaptive Immunity, 170-173, 176, 191

Apoptosis, 54, 65-66, 68, 79, 83-85, 90, 92, 105, 107, 114, 118, 126, 134, 136, 143-144, 146, 151, 158, 160, 162, 177, 180, 183, 194-196, 210-211, 215, 217, 233

Argonaute, 3, 11, 17, 34, 41-42, 50, 141, 156, 172-173, 183, 194, 207

Artificial Target, 41-45, 47-48

B

Binding Sites, 43, 47, 67, 87, 108, 118, 125, 145, 152-154, 156, 201-203, 205-208

Biogenesis, 2-3, 9, 11-12, 15-19, 24, 26-28, 30, 32, 35, 37, 41-42, 49, 52, 108, 141-142, 145, 156, 158, 171, 173, 182, 184, 205, 210, 212, 218, 221, 229

Biomarker, 97-98, 115-116, 150, 199, 236

Breast Cancer, 93-94, 124-125, 132-134, 136, 138, 144, 149-152, 159-160, 167, 169, 209, 211, 214, 216, 218-219, 222-228, 231-232, 236-238

C

Carcinogenesis, 82-83, 90, 93, 132-133, 161-163

Cell Line, 42, 47, 83-84, 90-91, 128, 224-225, 227, 233, 236

Cell Proliferation, 56, 59-60, 62, 73, 75-76, 85, 91, 93-94, 127-128, 135-136, 144, 153, 158, 160-161, 163, 166, 175-176, 181, 200-201

Chemoresistance, 82-83, 86-87, 90-91, 94-95, 228

Chromatin, 1-2, 6-8, 12-18, 24, 53-54, 69-70, 72-73, 120-123, 125, 129-130, 132, 134, 138-140, 147, 149-150, 152, 155, 164-165, 171, 184-185, 187, 191

Coding Sequence, 47, 202

Complex Human Disease, 120

Cytoplasm, 20, 25, 29-30, 32-33, 46, 121, 141-142, 154-155, 172-173, 185, 210, 217, 222

Cytosine Methylation, 7, 14-17, 19, 22-26, 28, 31-33, 37

D

Disease Activity, 97-99, 114, 182

E

Ectopic Expression, 59, 63, 90-91, 128, 144-145, 178, 180

Endocytosis, 211, 221, 229

Endothelial Cell, 56-57, 60, 62, 74, 125, 162, 199, 222, 236

Epigenetic Silencing, 3, 6, 13, 34, 61, 149, 152-154

Exosomes, 188-190, 199, 209-238

F

Fibroblast, 47, 68, 81, 90, 94, 218, 227

G

Gastric Cancer, 42, 49, 82, 84, 93-95, 134, 160, 218, 224, 236

Gastric Mucosa, 84, 91

Geminivirus, 14, 20, 22-23, 25-28, 32, 36-38

Gene Expression, 1-2, 4, 11, 14-15, 19, 39, 52, 54, 58, 63, 67, 70, 73, 77-78, 80, 83, 139-140, 142, 146-149, 151-152, 154, 156, 165, 179, 182-185, 190-191, 196, 201, 206, 209, 212, 217-218, 221, 223, 228, 237

Gene Repression, 118, 131, 150, 152, 154, 167, 185

Genome, 1-2, 5, 7-9, 11-12, 15-17, 19, 24-33, 35, 37, 39, 51-53, 71-72, 77, 98-99, 110, 113, 115-116, 118, 120, 122-123, 125, 128-133, 140-141, 143-144, 148, 157, 164, 167, 171-172, 181-182, 184-185, 187, 190-192, 198, 204, 206-207, 238

H

Heart Pathophysiology, 51, 65, 69

Hepatocellular Carcinoma, 90, 94, 128, 134, 138, 149, 151, 159-160, 164-165, 218, 224

Histone Modification, 39, 73, 128, 138, 149, 163, 165, 201

Homeostasis, 4, 8, 10, 13, 75, 147, 170-172, 176-177, 179, 193, 222

Hybridization, 23, 28, 30, 48, 105, 188, 202-204

I

Immune Response, 74, 98, 116-117, 126, 155, 170-172, 175, 178, 180-182, 185, 189, 192, 197, 210, 215, 222, 226

Immunoprecipitation, 46, 48, 53, 150

Innate Immunity, 15, 32, 34, 40, 178

Interferon-beta, 96-97, 115-116

Isoform, 63, 78, 135, 138, 143, 152-154, 169

L

Ligands, 155, 215, 221-223, 225, 227

Lipid, 188, 199, 212-214, 221, 229-232, 234

Luciferase, 43, 47, 82-83, 87-88, 90, 92, 108, 224

Lung Adenocarcinoma, 127-128, 135, 150, 166, 219

Lung Cancer, 93, 124, 127, 135, 145, 151, 159, 163, 165-166, 168, 216, 222, 225, 232, 234

Lymph Node, 150, 159, 223, 225

Lymphocytes, 117-118, 126, 151, 171, 174-178, 180, 182-183, 185, 187-190, 192, 194, 210, 234

M

Malignant Transformation, 142, 161, 222-223, 227

Matrix Metalloproteinase, 68, 160, 234

Metastasis, 83-84, 86, 93-94, 127-128, 130, 134-135, 137, 146, 149-151, 154-155, 159-160, 165-168, 213, 215-216, 222-223, 225, 227, 234

Methylation, 2, 5, 7, 11, 14-19, 22-26, 28, 30-38, 53, 69-70, 128, 130, 138, 145-146, 149-151, 154, 166, 187, 196, 201

Microvesicles, 188-189, 199, 209-210, 216-217, 220, 222-224, 234-236

Multiple Sclerosis, 96-97, 104, 114-119

Myocardial Infarction, 52, 70-71, 80-81, 124-125, 135

N

Nutrient Deficiency, 3-4

O

Oligonucleotides, 47, 59, 61

Oncogene, 90, 134-135, 140, 143, 145, 147, 152-153, 157, 159-161, 163-165, 168, 178, 226, 236

Ovarian Cancer, 212, 214-216, 219-220, 228-229, 233-234

P

Pancreatic Cancer, 94, 130, 150, 165, 205, 215, 233

Pararetrovirus, 14, 29, 31-32, 39

Peripheral Blood, 96-98, 106, 109, 113, 115-116, 126, 178, 195, 199, 226

Phenotype, 55, 57-60, 62, 67, 75-77, 150, 156, 166, 175-176, 178-180, 182-183, 216-217, 235-236

Plant Gene, 39

Plant Virus, 14

Protein-coding Gene, 6, 126-127, 185

Pseudogene, 55, 73, 121-122, 131, 152-153, 167-168

R

Recipient Cell, 209, 212, 215, 218, 221-223, 227

S

Silencing Pathway, 26, 196

Suppressor Protein, 14, 55, 128

T

Therapeutic Strategies, 51, 57, 70

Transfection, 41, 43-45, 47-48, 70, 85-87, 89, 91-92, 108

Transgenes, 6, 15, 25-26, 31, 38

Tumor Cell, 47, 128, 149, 155, 213, 219-221, 223, 228, 231, 234

Tumor Microenvironment, 144, 209, 212, 215-217, 221, 227, 233-234, 237

Tumor Suppressor, 55, 81, 83, 90, 122, 127-128, 134, 137-138, 140, 143-145, 147, 152, 161-162, 178, 180, 183, 211, 224

U

Ubiquitination, 2, 147, 175

V

Vascular Development, 51, 55, 60-62, 75, 77

Vascular Disease, 51-52, 56, 58, 61-62

Printed in the USA
CPSIA information can be obtained
at www.ICGtesting.com
LVHW051102290923
759464LV00007B/196